AF361023

Techniques for Food Authentication: Trends and Emerging Approaches

Techniques for Food Authentication: Trends and Emerging Approaches

Editors

Margit Cichna-Markl
Isabel Mafra

MDPI • Basel • Beijing • Wuhan • Barcelona • Belgrade • Manchester • Tokyo • Cluj • Tianjin

Editors
Margit Cichna-Markl
University of Vienna
Austria

Isabel Mafra
Universidade do Porto
Portugal

Editorial Office
MDPI
St. Alban-Anlage 66
4052 Basel, Switzerland

This is a reprint of articles from the Special Issue published online in the open access journal *Foods* (ISSN 2304-8158) (available at: https://www.mdpi.com/journal/foods/special_issues/ Authentication_Techniques).

For citation purposes, cite each article independently as indicated on the article page online and as indicated below:

LastName, A.A.; LastName, B.B.; LastName, C.C. Article Title. *Journal Name* **Year**, *Volume Number*, Page Range.

ISBN 978-3-0365-7654-1 (Hbk)
ISBN 978-3-0365-7655-8 (PDF)

Contents

Contents

About the Editors

Margit Cichna-Markl

Margit Cichna-Markl is an analytical chemist at the Department of Analytical Chemistry, Faculty of Chemistry, University of Vienna, Austria. She is the head of the research group "Immunoanalytical and Molecular Biological Techniques". She has long-term experience in the development, optimization, and validation of a broad range of analytical methods, including chromatographic methods and enzyme-linked immunosorbent assays (ELISAs). In the last decade, she focused on nucleic acid analysis, aiming to detect food allergens or to differentiate between closely related species and/or cultivars for food authentication. She has strong expertise in primer design, real-time polymerase chain reaction (PCR), multiplexing, high-resolution melting analysis, and pyrosequencing. In addition, Margit Cichna-Markl has long-term experience in teaching, giving a variety of lectures, seminars, and practical courses on, e.g., "Food Analysis", "Nucleic Acid Analysis", and "Immunoanalytical Techniques".

Isabel Mafra

Isabel Mafra has a degree in Food Engineering (Portuguese Catholic University), a MSc in Biological Engineering (University of Minho) and a PhD in Chemistry (University of Aveiro). She is currently the Principal Researcher at the associated laboratory REQUIMTE-LAQV in the Faculty of Pharmacy of the University of Porto, where she has been developing research in food quality and safety, as a group leader in molecular biology applied to food authentication, analysis/characterization of food allergens and detection of genetically modified organisms. She has strong expertise in the development of DNA-based methods, including DNA extraction from complex food matrices, PCR, real-time PCR, HRM analysis, digital PCR, sequencing, among others. She has also expertise in the development of chromatographic (GC, HPLC) and spectroscopic methods (FTIR). She has published over 120 scientific articles in indexed international journals (H-index of 35) and participated in more than 30 scientific projects.

Editorial

Techniques for Food Authentication: Trends and Emerging Approaches

Margit Cichna-Markl [1] and Isabel Mafra [2,*]

1   Department of Analytical Chemistry, Faculty of Chemistry, University of Vienna, Währinger Straße 38, 1090 Vienna, Austria
2   REQUIMTE-LAQV, Faculdade de Farmácia, Universidade do Porto, Rua de Jorge Viterbo Ferreira, 228, 4050-313 Porto, Portugal
*   Correspondence: isabel.mafra@ff.up.pt

Citation: Cichna-Markl, M.; Mafra, I. Techniques for Food Authentication: Trends and Emerging Approaches. *Foods* **2023**, *12*, 1134. https://doi.org/10.3390/foods12061134

Received: 27 February 2023
Accepted: 3 March 2023
Published: 8 March 2023

Food producers and retailers are obliged to provide correct food information to consumers; however, despite national and international legislation, food labels frequently contain false or misleading statements regarding food composition, quality, geographic origin, and/or processing. Food authentication is very challenging, requiring highly selective, sensitive, accurate, reproducible, and robust analytical methods. This Special Issue of *Foods*, comprising ten research and two review articles, highlights recent advances in food authentication and clearly demonstrates that no single method is suitable for covering all aspects of food authenticity.

Undoubtedly, DNA-based methods that target nuclear or mitochondrial (mt) markers have played a key role in the identification and differentiation of species and/or cultivars in food. Real-time PCR continues to be the technique of choice for the authentication of diversified food commodities, owing to its high specificity, sensitivity, and reproducibility. This is also the case for species authentication in meat products, for which real-time PCR has been one of the most widely applied DNA-based techniques, mainly targeting mtDNA [1]; however, the quantification of meat species, or any other food, using real-time PCR is challenged by the accurate preparation of reference mixtures as calibrants for method development. The choice of DNA marker is also challenging, especially when the purpose is quantitative analysis. Despite the advantages of mtDNA regarding sensitivity and specificity, its variable copy number is a drawback for quantitative approaches. Accordingly, a TaqMan real-time PCR assay targeting the lactoferrin gene of roe deer was developed for its quantitative determination in meat products [2]. The assay was validated in-house by determining the roe deer content in model meat mixtures and a model sausage, after which it was applied to the analysis of commercial meat products [2]. Nevertheless, method standardization requires assessment through interlaboratory trials [3]. Therefore, the real-time PCR assay for roe deer was tested in an interlaboratory ring trial, including 14 laboratories from Austria, Germany, and Switzerland. The assay proved its applicability to detect and quantify roe deer in raw meat samples to detect food adulteration, though further trials are still needed to validate its application to thermally treated model foods [4].

The application of real-time PCR was also demonstrated in plant species authentication in a particularly challenging matrix—vegetable oil. For the first time, new qualitative and quantitative PCR assays were proposed to authenticate argan oil [5]. Argan oil is a premium product, commercialized worldwide as cosmetic- and food-grade, which is potentially adulterated with other vegetable oils. To address this problem, two real-time PCR calibration models were developed by using the normalized ΔCq method to estimate potential adulterations of argan oil with olive or soybean oils, after which it was validated in-house with blind mixtures [5].

DNA barcoding targeting the cytochrome c oxidase subunit I (COI) gene, as a relatively conserved region with sufficient variation among species, has been widely applied in the species authentication of seafood [6]. COI barcoding was employed to analyze sushi samples collected by means of the citizen science approach, involving people from eighteen different Italian cities (Northern, Central, and Southern Italy). Data indicated a substantial rate of species substitution—between 31.8% in Northern Italy and 40% in Central Italy. *Thunnus thynnus* was the species most frequently replaced, followed by flying fish roe substituted with the eggs of *Mallotus villosus* [7]. DNA metabarcoding, a combination of DNA barcoding with next-generation sequencing (NGS), is an emerging approach that allows for the detection of multiple species in complex and processed foods, overcoming the drawbacks of Sanger sequencing [6]. Bivalve species belonging to the Mytilidae (mussels), Pectinidae (scallops), and Ostreidae (oysters) families were successfully identified in foodstuffs via DNA metabarcoding using fragments as small as 150 bp of mitochondrial 16S rDNA [8]. The feasibility of DNA metabarcoding using 16S rDNA was also demonstrated in an analysis of mammalian and poultry species in food and pet food [9].

The varietal authentication of foods is another challenging task that has been successfully overcome by using DNA markers, namely microsatellites or simple sequence repeats (SSR). This was demonstrated in the case of olive species. Microsatellite markers have been employed in cultivar identification, characterization of autochthonous olives (ancient olive trees and oleasters), management of olive germplasm banks, phylogenetics, diversity analysis, and mapping, as reviewed by Yadav et al. [10].

In recent years, several studies have demonstrated the feasibility of using spectroscopy for food authentication purposes, taking advantage of its non-destructive character, simple sample preparation, and possibility of being operated by non-expert technicians. Nevertheless, the effective application of spectroscopic approaches in food authentication relies on the construction of suitable spectral databases and multivariate analysis. The combined techniques of Raman, near-infrared (NIR), and fluorescence spectroscopy were applied to the analysis of chia oils adulterated with sunflower oil [11]. Fourier transform mid-infrared spectroscopy with attenuated total reflection (ATR-FTMIR), coupled to multivariate analysis, was applied to discriminate doughs and 3D-printed baked snacks, enriched with edible insect (*Alphitobius diaperinus* and *Locusta migratoria*) powders [12].

Chromatographic techniques, particularly those hyphenated with mass spectrometry (MS), have provided some of the most powerful tools for the detection and identification of chemical markers for food authentication. Aiming at authenticating hay milk, a traditional dairy product recently launched on the market and protected as "traditional specialty guaranteed" (TSG), two chromatographic techniques were proposed. Gas chromatography with MS detection was used as a targeted approach to detect the cyclopropane fatty acid (dihydrosterculic acid, DHSA), a marker of the bacterial strains found in silage since hay milk should be obtained from a cow's feed ration free from silage. The detection of DHSA could be related to the presence of maize silage in feed, though it was ambiguous in the case of grass silage [13]. High-performance liquid chromatography (HPLC) coupled to high-resolution mass spectrometry (HRMS) was then used as an untargeted approach to characterize the lipidic profile, resulting in the identification of 14 triacylglycerol biomarkers in milk. The biomarker profiles, combined with a multivariate analysis, allowed to predict the use of maize and grass silage in a cow's diet with 100% recognition [13]. Comprehensive two-dimensional gas chromatography (GC×GC) is an advanced approach with high potential to characterize complex volatile fractions of foods, including enantiomeric recognition for authentication purposes, owing to its improved separation capacity, increased number of identified compounds, structured chromatograms, and significant signal enhancement [14]. Vyviurska et al. [15] exploited an enantioselective GC×GC analysis to assess botrytized wines in comparison to the corresponding varietal grape wines, selected essences, and varietal wines fermented with grape skins. After a hierarchic cluster analysis, the data showed that the varietal wines were successfully separated from the other types, and a correlation between the botrytized wines and the varietal wines fermented with grape skins could be observed [15].

The multi-element composition of an animal's tissues can reflect, to some extent, its diet, while in the case of plants it reflects the soil composition of the location where they grow. Therefore, direct information about the geographical origin of foods can be provided by the bioavailable nutrients underlying the soils [16]. Conversely, elements found in the tissues of aquatic animals, such as fish, are recognized as being derived from the elemental composition of the overall surrounding environment, the aquatic habitat to the production premises, which is particularly useful when intending to identify the country of origin of wild specimens [17]. The application of element profiling approaches to fish and seafood products has been gaining force due to the advances in the optimization of existing instrumentations for multi-elemental analysis, but also due to improved algorithms for statistical analysis. From the review of Varrà et al. [17], the discrimination of geographical origin has been the most frequently reported authenticity topic, while other aspects, such as farming systems, have been overlooked. From the available methodologies, inductively coupled plasma mass spectrometry (ICP-MS) for elemental speciation and ICP-MS/MS for interference-free determination as well as isotope ratio measurement are anticipated as turning points for the high-throughput analytical characterization of complex matrices, such as food [17]. Accordingly, the multi-elemental profiles of 237 walnut samples from 10 countries and 3 years of harvest were analyzed via ICP-MS, and the data were evaluated with chemometrics, including machine learning methods. The results showed that walnut cultivar and harvest year had no observable influence on origin differentiation and highlighted the high potential of element profiling for the origin authentication of walnuts [18].

Acknowledgments: The authors acknowledge the support of the FCT (Fundação para a Ciência e Tecnologia) through the strategic funding of UIDB/50006/2020 | UIDP/50006/2020 from the FCT/MCTES and the European Union (EU) through the European Regional Development Fund with the project Healthy&ValorFood (FEDER funds through NORTE-01-0145-FEDER-000052) and the SYSTEMIC project under ERA-NET ERA-HDHL (n° 696295). I. Mafra thanks the FCT for funding through the Individual Call to Scientific Employment Stimulus (2021.03670.CEECIND/CP1662/CT0011).

Conflicts of Interest: The authors declare no conflict of interest.

References

1. Amaral, J.; Meira, L.; Oliveira, M.B.P.P.; Mafra, I. 14-Advances in Authenticity Testing for Meat Speciation. In *Advances in Food Authenticity Testing*; Downey, G., Ed.; Woodhead Publishing: Duxford, UK, 2016; pp. 369–414. [CrossRef]
2. Druml, B.; Mayer, W.; Cichna-Markl, M.; Hochegger, R. Development and validation of a TaqMan real-time PCR assay for the identification and quantification of roe deer (*Capreolus capreolus*) in food to detect food adulteration. *Food Chem.* **2015**, *178*, 319–326. [CrossRef] [PubMed]
3. Taverniers, I.; De Loose, M.; Van Bockstaele, E. Trends in quality in the analytical laboratory. II. Analytical method validation and quality assurance. *Trends Anal. Chem.* **2004**, *23*, 535–552. [CrossRef]
4. Druml, B.; Uhlig, S.; Simon, K.; Frost, K.; Hettwer, K.; Cichna-Markl, M.; Hochegger, R. Real-time PCR assay for the detection and quantification of roe deer to detect food adulteration—Interlaboratory validation involving laboratories in Austria, Germany, and Switzerland. *Foods* **2021**, *10*, 2645. [CrossRef] [PubMed]
5. Amaral, J.S.; Raja, F.Z.; Costa, J.; Grazina, L.; Villa, C.; Charrouf, Z.; Mafra, I. Authentication of argan (*Argania spinosa* L.) oil using novel DNA-based approaches: Detection of olive and soybean oils as potential adulterants. *Foods* **2022**, *11*, 2498. [CrossRef] [PubMed]
6. Fernandes, T.J.R.; Amaral, J.S.; Mafra, I. DNA barcode markers applied to seafood authentication: An updated review. *Crit. Rev. Food Sci. Nutr.* **2021**, *61*, 3904–3935. [CrossRef] [PubMed]
7. Pappalardo, A.M.; Raffa, A.; Calogero, G.S.; Ferrito, V. geographic pattern of sushi product misdescription in Italy—A crosstalk between Citizen Science and DNA barcoding. *Foods* **2021**, *10*, 756. [CrossRef] [PubMed]
8. Gense, K.; Peterseil, V.; Licina, A.; Wagner, M.; Cichna-Markl, M.; Dobrovolny, S.; Hochegger, R. Development of a DNA metabarcoding method for the identification of bivalve species in seafood products. *Foods* **2021**, *10*, 2618. [CrossRef] [PubMed]
9. Preckel, L.; Brünen-Nieweler, C.; Denay, G.; Petersen, H.; Cichna-Markl, M.; Dobrovolny, S.; Hochegger, R. Identification of mammalian and poultry species in food and pet food samples using 16S rDNA metabarcoding. *Foods* **2021**, *10*, 2875. [CrossRef] [PubMed]
10. Yadav, S.; Carvalho, J.; Trujillo, I.; Prado, M. Microsatellite markers in olives (*Olea europaea* L.): Utility in the cataloging of germplasm, food authenticity and traceability studies. *Foods* **2021**, *10*, 1907. [CrossRef] [PubMed]

11. Mburu, M.; Komu, C.; Paquet-Durand, O.; Hitzmann, B.; Zettel, V. Chia oil adulteration detection based on spectroscopic measurements. *Foods* **2021**, *10*, 1798. [CrossRef] [PubMed]

12. García-Gutiérrez, N.; Mellado-Carretero, J.; Bengoa, C.; Salvador, A.; Sanz, T.; Wang, J.; Ferrando, M.; Güell, C.; Lamo-Castellví, S.D. ATR-FTIR Spectroscopy combined with multivariate analysis successfully discriminates raw doughs and baked 3D-Printed snacks enriched with edible insect powder. *Foods* **2021**, *10*, 1806. [CrossRef] [PubMed]

13. Imperiale, S.; Kaneppele, E.; Morozova, K.; Fava, F.; Martini-Lösch, D.; Robatscher, P.; Peratoner, G.; Venir, E.; Eisenstecken, D.; Scampicchio, M. Authenticity of hay milk vs. milk from maize or grass silage by lipid analysis. *Foods* **2021**, *10*, 2926. [CrossRef] [PubMed]

14. Cordero, C.; Schmarr, H.G.; Reichenbach, S.E.; Bicchi, C. Current Developments in analyzing food volatiles by multidimensional gas chromatographic techniques. *J. Agric. Food Chem.* **2018**, *66*, 2226–2236. [CrossRef]

15. Vyviurska, O.; Koljančić, N.; Thai, H.A.; Gorovenko, R.; Špánik, I. Classification of botrytized wines based on producing technology using flow-modulated comprehensive two-dimensional gas chromatography. *Foods* **2021**, *10*, 876. [CrossRef]

16. Drivelos, S.A.; Georgiou, C.A. Multi-element and multi-isotope-ratio analysis to determine the geographical origin of foods in the European Union. *TrAC Trends Anal. Chem.* **2012**, *40*, 38–51. [CrossRef]

17. Varrà, M.O.; Ghidini, S.; Husáková, L.; Ianieri, A.; Zanardi, E. Advances in troubleshooting fish and seafood authentication by inorganic elemental composition. *Foods* **2021**, *10*, 270. [CrossRef] [PubMed]

18. Segelke, T.; von Wuthenau, K.; Kuschnereit, A.; Müller, M.-S.; Fischer, M. Origin determination of walnuts (*Juglans regia* L.) on a worldwide and regional level by inductively coupled plasma mass spectrometry and chemometrics. *Foods* **2020**, *9*, 1708. [CrossRef] [PubMed]

Review

Advances in Troubleshooting Fish and Seafood Authentication by Inorganic Elemental Composition

Maria Olga Varrà [1], **Sergio Ghidini** [1], **Lenka Husáková** [2], **Adriana Ianieri** [1] and **Emanuela Zanardi** [1,*]

[1] Department of Food and Drug, University of Parma, Strada del Taglio 10, 43126 Parma, Italy; mariaolga.varra@studenti.unipr.it (M.O.V.); sergio.ghidini@unipr.it (S.G.); adriana.ianieri@unipr.it (A.I.)

[2] Department of Analytical Chemistry, Faculty of Chemical Technology, University of Pardubice, Studentska 573 HB/D, CZ-532 10 Pardubice, Czech Republic; lenka.husakova@upce.cz

[*] Correspondence: emanuela.zanardi@unipr.it; Tel.: +39-052-190-2760

Abstract: The demand for fish and seafood is growing worldwide. Meanwhile, problems related to the integrity and safety of the fishery sector are increasing, leading legislators, producers, and consumers to search for ways to effectively protect themselves from fraud and health hazards related to fish consumption. What is urgently required now is the availability of reliable, truthful, and reproducible methods assuring the correspondence between the real nature of the product and label declarations accompanying the same product during its market life. The evaluation of the inorganic composition of fish and seafood appears to be one of the most promising strategies to be exploited in the near future to assist routine and official monitoring operations along the supply chain. The present review article focuses on exploring the latest scientific achievements of using the multi-elemental composition of fish and seafood as an imprint of their authenticity and traceability, especially with regards to the geographical origin. The scientific literature of the last 10 years focusing on the analytical determination and statistical elaboration of elemental data (alone or in combination with methodologies targeting other compounds) to verify the identity of fishery products is summarized and discussed.

Keywords: fraud; authentication; traceability; fish; geographical origin; multi-elemental profile; stable isotopes; ICP-MS; ICP-OES; chemometrics

Citation: Varrà, M.O.; Ghidini, S.; Husáková, L.; Ianieri, A.; Zanardi, E. Advances in Troubleshooting Fish and Seafood Authentication by Inorganic Elemental Composition. *Foods* **2021**, *10*, 270. https://doi.org/10.3390/foods10020270

Academic Editor: Margit Cichna-Markl

Received: 28 December 2020
Accepted: 25 January 2021
Published: 29 January 2021

Publisher's Note: MDPI stays neutral with regard to jurisdictional claims in published maps and institutional affiliations.

1. Introduction

The verification of food authenticity and integrity is a complex topic which has become a matter of public interest in recent years. This issue involves many different aspects, from the identification of mislabeling and misrepresentation to adulteration and contamination of the product.

Today, the traceability of fish and seafood and detection of intentional and unintentional fraud is a challenging task, as the supply chain of fishery products is among the most diversified and globalized. As a matter of fact, fish is currently among the most frequently misdescribed foodstuffs worldwide, to a point that almost 20% of fish in the sail and restaurant sectors of 55 countries has been recently found to be misdescribed [1]. Specifically, the major economic losses affecting the sector derive from the substitution of highly valuable fish and seafood species for morphologically similar but lower-quality ones and from the increasingly common falsification in relation to the geographical origin. Albeit these fraudulent practices seem to have a negative impact only from an economical point of view, some health implications may arise, for example, from the replacement of certain fish species with cheaper but potentially poisonous ones [2], or from the sale of illegally caught fish originating from polluted areas [3].

In order to prevent fraud, protect producers and consumers, and promote high-quality fish products, the reinforcement of the international food monitoring program is not sufficient. Indeed, control measures are required to be undertaken in synergy with the implementation of proper vulnerability assessment systems and the development of rapid analytical tools, so as to confidently verify whether a product is genuine or counterfeit and to guarantee the integrity of the whole production chain.

Many different biological and chemical methods have been developed over the years to ascertain the authentic nature of a wide range of foodstuffs. These methods focus on the evaluation of the organic (DNA, proteins, lipids, sugars, and/or metabolites) and inorganic (elements, isotope ratios) fractions of food and exploit the principles of chromatography, mass spectrometry, and spectroscopy to identify, in a targeted way, few or multiple compounds acting as secondary markers of authenticity [4]. In this sense, the determination of the inorganic multi-elemental signatures (in terms of major, trace, and ultra-trace elements), accompanied by multivariate statistics, is increasingly applied to authenticate different foods of animal origin such as honey [5], pork meat [6], and cheese [7], especially in relation to the geographical origin and the method of production. In this context, the evaluation of the elemental content of fish and seafood is particularly advantageous, since it may allow for the simultaneously monitoring of mislabeling and of the maximum acceptable regulation limits for certain toxic elements established at the European level [8], thus pursuing both integrity and safety objectives.

Elements found in fish tissue are scientifically recognized to be a reflection of the elemental composition of the overall surrounding environment, from the aquatic habitat to the production premises, and this is particularly advantageous when the country of origin of wild specimens is thought to be identified. On the contrary, the elemental content of farmed specimens is inevitably affected by feeding stuffs, both from a qualitative and a quantitative point of view. Thus, using the elemental profile of the tissues of farmed specimens to trace back to the country of origin may be problematic by virtue of the fact that the same feed can be traded internationally and given to fish cultured in different parts of the world.

When working with the element composition of fish for authentication purposes, it should be taken into consideration that the presence of elements in the aquatic environment explored by the fish during life is not only dependent on the specific geochemical characteristics of the habitat, but it may be significantly influenced by other environmental factors either of natural (such as climate, water temperature, salinity, age, and sexual maturity of the animal) or anthropic (i.e., the exogenous pollution) origin [9]. In addition, after catch, fishery products are more frequently handled and enlivened compared to other foodstuffs. Therefore, the likelihood of unwanted and misleading elements being incorporated as contaminants throughout the whole production cycle increases significantly.

For these reasons, the overall elemental signature needs to be strictly evaluated before being used as a tool to address authenticity problems of fish and seafood. In this regard, chemometrics and machine learning have now become an essential support for increasing the strength and reliability of high-throughput analytical techniques. As a matter of fact, the advanced statistical elaboration of elemental data has already been proven to be a straightforward and effective means to study elements' behavior; identify common but hidden compositional characteristics among similar food samples; separate complementary, opposite, or redundant information enclosed into elemental data; define classification rules; and simplify the overall methodology by extracting the effectively significant elemental markers for classification [10].

The present review article was aimed at discussing the applications and advances in data mining of the multi-elemental profile of fish, mollusks, echinoderms, and crustaceans from the last 10 years as a strategy to verify whether mandatory labelling information matches the identity of these products. The survey took into consideration the elemental measurements performed only on edible tissues of fishery products, which, albeit being more rapidly subjected to variations induced by environment compared to hard structures such as otoliths, statoliths, skeleton, and scales, are retained in the final traded products and hence potentially monitorable in every phase of the production chain. As evidenced below, only raw products were discussed, since, as far as we know, no considerable breakthroughs in tracing and authenticating transformed (e.g., salted, smoked, marinated) fish and seafood products have been achieved.

2. Analytical and Chemometric Methodologies for Element and Stable-Isotope Analysis of Fish and Seafood

Various analytical techniques have been used for the determination of the elemental content of fish and seafood throughout the last years [9,11–13]. Among these, atomic spectroscopic methods, such as atomic absorption spectrometry (AAS) [14–20] with flame [16,19,20] or electrothermal atomization [16–18], atomic fluorescence spectrometry (AFS) [21] inductively coupled optical emission spectrometry (ICP-OES) [16,22], inductively coupled plasma mass spectrometry (ICP-MS) [23,24], and X-ray fluorescence (XRF) [19,25], have been the most frequently employed. On the contrary, electroanalytical [26] or neutron-activation-based techniques, such as neutron activation analysis (NAA), have been used to a lesser extent [27–29].

These techniques offer specific advantages and, at the same time, present some limitations which make their application preferable in some cases but not in others. The main characteristics and performances of the analytical methods that can be used for comparing the multi-element or stable-isotope composition of fish and seafood samples are examined in the text below, while a comprehensive and detailed overview of the main benefits and drawbacks is outlined in the Supplementary File (Table S1).

In the case of major and some minor elements, AAS and OES with flames (flame atomic absorption spectroscopy, FAAS, and flame optical emission spectroscopy, FOES), are still valuable and well-established techniques that are routinely and customarily applied in the area of fish and seafood analysis due to their robustness in relation to interferences and sample introduction problems, selectivity, straightforward application, and lower cost [16,19,20,30]. On the other hand, these methods still present limitations related to sensitivity. Thus, electrothermal AAS (ET-AAS) [16,17,28], hydride generation AAS (HG-AAS) [31], and cold vapor AAS (CV-AAS) [15,28] or direct thermal decomposition AAS [19,32] are employed in the lower concentration range. However, the main disadvantage is that AAS is primarily limited to the determination of metallic elements and is a single-element technique with a linear range typically less than two orders of magnitude. Despite being used in only in a very limited number of cases [18], high-resolution continuum source AAS (HR-CS-AAS) is overcoming some of the limitations of AAS, as it allows for the simultaneous evaluation of several absorption lines in the selected spectral range, accurate background correction, and the determination of nonmetals.

ICP-OES is by far the most commonly applied technique for the analysis of food samples [16,18,22,33] because it offers simultaneous multi-element measurement, capabilities for sensitive determination of refractory elements, quantification of nonmetals, and high analytical throughput. Microwave induced plasma optical emission spectrometry (MIP-OES) using the magnetically excited microwave plasma source has also been recently applied to fish and seafood [34,35], mainly because it is characterized by detection limits down to sub-ppb levels, significant cost reduction, and simpler spectra than ICP-OES. However, at present, both ICP-OES and MIP-OES fail to meet the needs required in routine applications when the determination of elements at trace or ultra-trace concentrations is in demand.

ICP-MS is better suited to meet this task and is currently a frontline technology, rapidly replacing other methods in many fields of food science. Unsurpassed advantages, such as high sensitivity, selectivity, wide dynamic concentration range up to 11 orders of magnitude, high sample throughput, and multi-analyte capabilities, make this method an ideal candidate for food authentication studies, since it might facilitate the discrimination and classification of samples [23,36,37].

More detailed technical aspects of the abovementioned methodologies can be retrieved from literature [37,38].

The analysis of biologic matrices such as foodstuffs by atomic and mass spectrometry methods, especially at trace and ultra-trace levels, is often a difficult and challenging task.

As a matter of fact, a quite complex biological matrix poses problems related not only to sample heterogeneity, the selection of proper sample treatment, and decomposition, but also matrix interferences.

In the ICP-MS analysis of fish and seafood samples, both spectral and non-spectral interferences are expected to be encountered [18,23,39]. Whereas non-spectral effects can be easily overcome using a proper calibration strategy, including the use of an internal standard [18,23], standard additions, and/or the isotope dilution, spectral effects due to the overlaps by different polyatomic ions (formed from the combination of species derived from the matrix elements, plasma gas and sample solvents) are more serious and difficult to handle [40].

High-resolution mass spectrometers with a sector field mass analyzer could be the ideal solution to bypass most of these problems [40]. However, owing to their high price, these instruments are not easily accessible for most laboratories.

Time-of-flight (TOF)-ICP-MS instruments have several advantages, such as fast simultaneous multi-elemental analysis, improved precision of measurements of the isotope ratios, very low volume of the sample needed for the analysis, and tolerance to higher salinity of samples. However, they do not have the adequate resolution to eliminate the spectral interferences typically encountered when analyzing biological samples. As a result, mathematical corrections must be employed, but this approach is less effective when performing trace analysis [41]. The absence for effective solutions related to the control of problematic spectral effects, which were not accessible to users until recently, has limited the widespread diffusion of this technique in routine practice. Nevertheless, there is an increasing trend in resorting to the use of collision cell technology for interference management during sample analysis in the current TOF-ICP-MS instrumentation [42]. At present, the quadrupole-based ICP-MS equipped with a collision/reaction cell (CRC) for the elimination of spectral interferences is the most popular ICP-MS instrumentation on the market. In the reaction cell mode, interfering ions are removed by the transformation into different species or uncharged atoms or molecules through specific chemical reactions with a supplementary reaction gas (H_2, NH_3, O_2, N_2O, or CH_4) [40]. Although this approach is more efficient for the removal of known spectral interferences, it may lead to a formation of new unwanted interfering polyatomic ions. The collision cell mode is instead more suitable for the multi-elemental analysis of unknown samples. For this purpose, He is widely used as a collision gas to slow down polyatomic interfering ions to a larger extent than the atomic analyte ions, such that the former could be selectively discriminated against on the basis of their lower kinetic energy.

With the introduction of an ICP-tandem mass spectrometer (MS/MS, often referred to as triple quadrupole ICP-MS or ICP-QQQ), the CRC technology in quadrupole-based ICP-MS has greatly improved [43]. This instrumentation, equipped with CRC located between two quadrupole mass filters, provides an elegant approach via a precursor ion and/or product ion scanning to solve even the most challenging cases of spectral overlap and interference. Moreover, it can determine a wider range of analytes at much lower concentrations with greater reliability and higher confidence [43].

In addition to total element determinations, the current ICP-MS instrumentation is suited also to isotope ratio analysis, even if the isotope ratio precision is strictly dependent on the type and the design of the instrument used. Considering that the simultaneous measurement of multiple isotopes provides a better precision in isotope ratio measurement, the use of TOF-ICP-MS or multi-collector mass spectrometer with a plasma source for ionization (MC-ICP-MS) is considerably more advantageous than the use of a single quadrupole ICP-MS for isotope analysis. However, the commonly used mass spectrometers typically do not provide the sensitivity and precision required for the determination of light isotopes ratios. In addition, they are susceptible to isotopic fractionation (mass bias). Therefore, isotope ratio mass spectrometry (IRMS) [44,45], nuclear magnetic resonance (NMR) [46,47], and thermal ionization mass spectrometry (TIMS) [12] are more suitable for this purpose.

Atomic fluorescence spectrometry (AFS) may represent an alternative to the other atomic and mass spectrometric techniques, as it provides low detection limits, wide linear calibration range, simplicity, and lower acquisition and running costs. These analytical features make AFS superior to AAS and equal to ICP-MS or ICP-OES [22,48], especially in speciation studies, as long as single element speciation studies are considered [36].

Recently, there has been an increase in the application of nondestructive multielement methods for analysis of seafood samples [25,49]. Methods based on X-ray spectrometry such as X-ray fluorescence (XRF) [19,25,49], energy dispersion-XRF (ED-XRF) [19], proton induced X-ray emission (PIXE), total reflection X-ray fluorescence spectrometry (TXRF), and synchrotron X-ray fluorescence (SXRF), as well as methods based on X-ray microanalysis, offer several benefits [50]. Among these, the selective detection and sensitivity (about mg kg^{-1} and below) for most of the elements [12,49], minimal sample preparation, high sample throughputs, and accuracy in quantification are worth mentioning [50]. In addition, field portable-XRF analyzers are becoming increasingly popular for a wide variety of elemental analysis applications [50].

Laser-based techniques also play an important role for the direct analysis of solid samples and, in the last years, they have become increasingly present in the food industry. Laser-induced breakdown spectrometry (LIBS) is considered a promising micro-destructive food analysis tool for rapid qualitative and quantitative chemical analysis [50,51]. However, the direct analysis of samples with complex organic matrices such as fresh food products is not easy [50]. As a matter of fact, it is often not possible to analyze the sample without any preparation, since the results might be misleadingly affected by any inhomogeneity of the material. On the other hand, the sample preparation for LIBS analysis is minimal when compared to reference methods such as AAS or ICP-MS. The major limitation of LIBS for practical applications results from its reduced sensitivity for minor mineral elements and heavy metals, with very low concentrations in a complex organic matrix.

The connection of laser ablation (LA) with ICP-MS [52–54] represents a quite versatile analytical tool, offering the fastest analytical speed compared to all the other techniques, favorable limits of detection (approaching ppb levels), capability for performing bulk analysis, depth profiling, and elemental/isotope mapping [12]. Nevertheless, LA-ICP-MS still lacks sufficiently matrix-matched reference materials for each considered matrix type, and the analysis accuracy is restricted by several factors, such as sensitivity drift, elemental/isotopic fractionation, and matrix effects [50,55].

Electrothermal vaporization (ETV) is also an efficient and powerful approach for a bulk analysis where solid samples can be directly turned into aerosols [50,55]. This strategy significantly boosts ICP-MS quantitative applications in desired field [56].

2.1. Sample Digestion Procedures for Elemental Analysis

The market of most of the abovementioned analytical apparatus, such as AAS, AFS, and those which make use of a plasma source for ionization, offers mainly instrumentation dedicated to the analysis of liquid samples. Consequently, digestion procedures for solid samples are necessarily required. Furthermore, sample preparation is a crucial issue for food products due to their inhomogeneity and matrix complexity.

Nowadays, the most used and useful digestion technique for a wide range of analytes and sample matrices is the high-pressure digestion using a closed-vessel microwave system [15–18,20,24,34,36,41,49,57]. This technique increases the sample throughput, minimizes analyte losses during the decomposition, reduces both contamination risk (especially for trace analytes) and consumption of reagents, and is more effective, resulting in low residual carbon content of digested samples [57]. In addition to high-pressure closed-vessel microwave digestion, digestion involving opened vessels or classical dry-ashing digestion is generally performed. In wet-acid digestion, HNO_3 alone [15,16,20,24] or combined with H_2O_2 [17,18,28,34,36,41,49] and, occasionally, HCl [35] or $HClO_4$ [22,23,33], is the most commonly used reagent. However, several novel approaches or adaptations to established procedures for sample preparation have been recently introduced. In particular, a growing interest toward the use of diluted and nonhazardous analytical reagents is now emerging, in accordance with green chemistry and the need to reduce the negative impact of chemical analyses on the environment [24,57]. From this standpoint, ultrasound-assisted extraction and microwave-assisted extraction [31,36,57] seem to be very promising approaches for sample preparation in the near future, allowing for the optimization of working times and consumption of analytical reagents.

2.2. Multivariate Data Analysis and Machine Learning

The growing interest in high-throughput element-based methods to characterize foodstuffs may be partly justified by the efforts in the field of multivariate data analysis and machine learning, which have significantly simplified data handling and improved the identification of food fraud. Multivariate qualitative methods are well established in the field of analytical chemistry oriented toward the authenticity and adulteration verification of foodstuffs, and the development of new algorithms for classification is continuously increasing [10]. Despite this, analysis of the literature revealed that the statistical analysis of the multi-elemental profile of fish has been mostly limited to the classical use of principal component analysis (PCA) and cluster analysis (CA) as exploratory (unsupervised) tools. As for sample classification purposes, hard modelling of data based on linear discriminant analysis (LDA) and canonical discriminant analysis (CDA) has been more frequently employed (see Table 1). This is probably due to the fact that the theoretical background of these data elaboration techniques is more consolidated among the scientific community compared to other more modern hard-modelling discriminant techniques such as partial least square discriminant analysis (PLS-DA) and soft-modelling techniques such as soft independent modelling of class-analogy (SIMCA) [58]. In addition, the applied methodologies appear to lack of proper validation protocols to be followed, which are necessary for the development of reliable and transferable multivariate-based models for foods classification [59].

Various techniques, including K-nearest neighbors (KNN), K-mean clustering, and artificial neural network (ANN), are crucial for future successful development of prediction models to food authentication solutions.

Further details on chemometrics and machine learning techniques applied to food science can be found in the literature [10,60].

Table 1. Overview of the literature dealing with multi-elemental profile for fish and seafood authenticity verification.

Product	Classification Objective	Input Data	Technique for Elemental Analysis	Elements	Data Analysis	Validation	Reference
				Fish			
Salmon	Production method	Elemental profile Stable isotope ratio	ICP-OES	As, Ba, Be, Ca, Cd, Co, Cr, Cu, Fe, K, Mg, Mn, Na, Ni, P, Pb, Sr, Ti, Zn	PCA, CDA, LDA, QDA, ANNs, PNNs, NNB	Cross-validation External validation	[61]
Catfish	Geographical origin	Elemental profile	ICP-OES	Al, Ca, Cr, Cu, Fe, K, Mg, Na, P, S, Zn	PCA, CDA, k-NN	Cross-validation	[33]
Croacker	Geographical origin Seasonality	Elemental profile Stable isotope ratio Proximate composition	EDXRF	As, Br, Ca, Cd, Cl, Cu, Fe, Hg, K, Pb, Rb S, Se, Zn,	PCA	–	[19]
European seabass	Geographical origin Production method	Elemental profile Stable isotope ratio Biometric measures Fatty acids	ICP-OES	As, Ca, Cd, Co, Cr, Cu, Fe, Hg, K, Mg, Mn, Na, Ni, P, Pb, S, Se, Zn	PCA	Cross-validation	[48]
Asian seabass	Geographical origin Production method	Elemental profile Stable isotope ratio	XRF	Al, As, At, Bi, Br, Ca, Cd, Cl, Cr, Cu, Fe, Hf, K, Mg, Mn, Nd, Ni, P, Pb, Rb, S, Sb, Se, Si, Sn, Sr, Ti, U, Y, Zn, Zr	PCA, LDA, RF	Cross-validation External validation	[49]
European seabass	Geographical origin Production method	Element profile Stable isotope ratio	ICP-MS	Er, Eu, Ho, La, Lu, Tb	PCA, OLPS-DA	Cross-validation External validation	[62]
				Echinoderms			
Sea cucumber	Geographical origin	Elemental profile	ICP-MS	Al, As, Cd, Co, Cr, Cu, Fe, Hg, Mn, Mo, Ni, Pb, Se, V, Zn	PCA, CA, LDA	Cross-validation	[23]
Sea cucumber	Geographical origin	Elemental profile	ICP-OES ICP-MS	Ag, Al, As, Ba, Bi, Ca, Cd, Ce, Co, Cr, Cu, Dy, Er, Eu, Fe, Gd, Ho, K, La, Li, Lu, Mg, Mn, Na, Nd, Ni, Pb, Pr, Sc, Se, Sm, Sn, Sr, Tb, Tm, V, Y, Yb, Zn	PCA, LDA	Cross-validation	[63]

Table 1. *Cont.*

Product	Classification Objective	Input Data	Technique for Elemental Analysis	Elements	Data Analysis	Validation	Reference
				Crustaceans			
Pacific white shrimp	Geographical origin	Elemental profile	ICP-OES	Al, As, Ba, Ca, Co, Cr, Cu, Fe, K, Mg, Mn, Mo, Na, Ni, P, S, Se, Ti, Zn, Zr	PCA, CDA, k-NN	Cross-validation	[64]
Shrimps	Geographical origin Production method Species	Elemental profile Stable isotope ratio	ICP-OES ICP-MS	As, Cd, P, Pb, S	PCA, CA, LDA, k-NN	Cross-validation	[65]
Prawns	Geographical origin	Elemental profile Stable isotope ratio	ICP-MS	Al, As, B, Cd, Co, Cr, Cu, Fe, Hg, K, Li, Mn, Mo, Ni, Se, Sr, Ti, V, Zn	PCA, CDA	Cross-validation	[66]
Pacific white shrimps	Geographical origin	Elemental profile	ICP-OES	Al, As, B, Ba, Ca, Cd, Co, Cr, Cu, Fe, K, Mg, Mn, Na, Ni, P, Pb, S, Se, Si, Ti, Zn, Zr	PCA, CDA, S-LDA	Cross-validation	[22]
Chinese mitten crab	Geographical origin	Elemental profile Stable isotope ratio	ICP-MS	Al, Ba, Ca, Cu, K, Mg, Mn, Na, Sr, Zn	LDA, SVM	Cross-validation External validation	[67]
Pacific white shrimps	Seawater vs. Freshwater	Elemental profile Stable isotope ratio	ICP-MS	Ag, Al, As, Ba, Cd, Ce, Co, Cr, Cs, Cu, Dy, Er, Eu, Fe, Ga, Gd, Ho, Li, Lu, Mn, Nd, Ni, Pb, Pr, Rb, Sm, Sr, Tb, Th, Tm, U, V, Y, Yb, Zn	PCA, CDA, S-LDA	Cross-validation	[68]
Black tiger prawn	Geographical origin Production method	Elemental profile Stable isotope ratio	XRF	Al, As, At, Bi, Br, Ca, Cd, Cl, Cr, Cu, Fe, Hf, K, Mg, Mn, Nd, Ni, P, Pb, Rb, S, Sb, Se, Si, Sn, Sr, Ti, U, Y, Zn, Zr	LDA, RF	Cross-validation External validation	[25]

Table 1. *Cont.*

Product	Classification Objective	Input Data	Technique for Elemental Analysis	Elements	Data Analysis	Validation	Reference
Mollusks							
Mussels	Geographical origin	Elemental profile	ICP-MS	Ag, As, Ba, Cd, Ce, Co, Cr, Cu, Dy, Er, Eu, Ga, Gd, Ho, La, Lu, Mn, Mo, Nb, Nd, Ni, Pb, Pr, Rb, Sb, Se, Sm, Sn, Sr, Ta, Te, Th, Tl, Tm, U, V, Y, Yb, Zn, Zr	LDA, SIMCA, ANNs	Cross-validation	[24]
Manila clams	Geographical origin	Elemental profile	ICP-MS	Al, As, Ba, Cd, Ce, Co, Cs, Cu, Fe, K, La, Mg, Mn, Na, Mo, Pb, Pd, Rb, Sb, Se, Sr, Sn, U, V, Zn	S-LDA	Cross-validation	[69]
Cuttlefish (ink)	Geographical origin	Elemental profile	ICP-MS	As, Ca, Cd, Co, Cr, Cu, Fe, Hg, K, Mg, Mn, Mo, Na, Ni, P, Pb, V, Zn	PCA	–	[70]

ANNs = artificial neural networks. CDA = canonical discriminant analysis. EDXRF = energy dispersive X-ray fluorescence spectroscopy. ICP-OES = inductively coupled plasma optical emission spectroscopy. ICP-MS = inductively coupled plasma mass spectrometry. k-NN = k-nearest neighbors. LDA = linear discriminant analysis. NNS = neural network bagging. PCA = principal component analysis. PNNs = probabilistic neural networks. QDA = quadratic discriminant analysis. RF = random forest. SIMCA = soft independent modelling of class analogy. S-LDA = stepwise-linear discriminant analysis. XRF = X-ray fluorescence spectroscopy.

3. Authentic Elemental Signature of Fish and Seafood

As discussed below, authentication and traceability studies have often been performed by coupling elemental analysis (major, trace, and/or ultra-trace elements) with other techniques targeting other compounds, with the objective to increase the specificity of discrimination and obtain better results.

The merging of data from the isotopic analysis and elemental analysis of light (H, C, N, O, and S) and heavy elements (Sr, Pb) has been the most frequently investigated analytical strategy to approach traceability problems of fish and seafood.

The rationale behind this research trend over the years lies in the strong correlation between any variation in isotope fractionation (ratio between isotopes of a specific elements) and the geological, pedological, and wheatear characteristics of a given geographical area [71]. Among these, the isotopic distribution of light elements such as O (δ^{18}O, ^{18}O/^{16}O), and H (δ^{1}H, ^{1}H/^{2}H) in fishery products is influenced by the original isotopic distribution of the same elements in the water basin from which the fish come from, which, in turn, is the reflection of the isotopic distribution in the rainfall of the specific area [72]. More, the isotope ratio of C (δ^{13}C, ^{13}C/^{12}C) in fish tissues may be related to the type of vegetation eaten by the fish during its life. In particular, the plants are characterized by a C3, C4, or Crassulacean Acid Metabolism (CAM) photosynthetic metabolism. Considering that each type of these plants typically grows at certain latitudes, the isotopic distribution of C may be, at first instance, indirectly exploited as a marker of origin. Since fractionation of C is expected to vary between the artificial feed used to rare aquaculture fish and the natural food of wild fish, its isotopic ratio may also be exploited to distinguish the production method of fish [72]. Indeed, isotopes of N are good indicators of the feeding regime of fish and of the position occupied by the fish in the food chain, thus being ideal markers of the production methods. Wild fish at higher trophic levels is in fact characterized by a greater enrichment in δ^{15}N (^{15}N/^{14}N), and δ^{15}N enrichment in artificial feeding given to farmed fish is expected to be significantly different compared to those present into the natural food eaten by wild fish [72].

In the present review, recent research in the field of multi-elemental analysis applied to edible tissues of fish and seafood was taken into consideration and reviewed. The scientific literature herein includes research articles pertinent to the topic of the present review and published between 2010 and 2020. Articles were retrieved from the Web of Science and Scopus databases (search terms: 'fish,' 'seafood,' 'authentication,' 'elemental analysis,' 'elemental profile,' 'elemental fingerprinting,' 'chemometrics').

For the sake of clarity, the next paragraphs are structured to enclose the same type of product. Therefore fish, mollusks (both bivalve and cephalopods), crustaceans, and echinoderms are discussed separately. The most frequently measured elemental markers of both geographical origin and method of production, retrieved from the reviewed scientific literature discussed below, are graphically shown in the radial bar chart reported in Figure 1. For a quick comparison, a summary overview of the methodological and technical aspects of the published works is given in Table 1. The concentrations of the elements measured in each work were deepened and provided in the Supplementary File (Table S2).

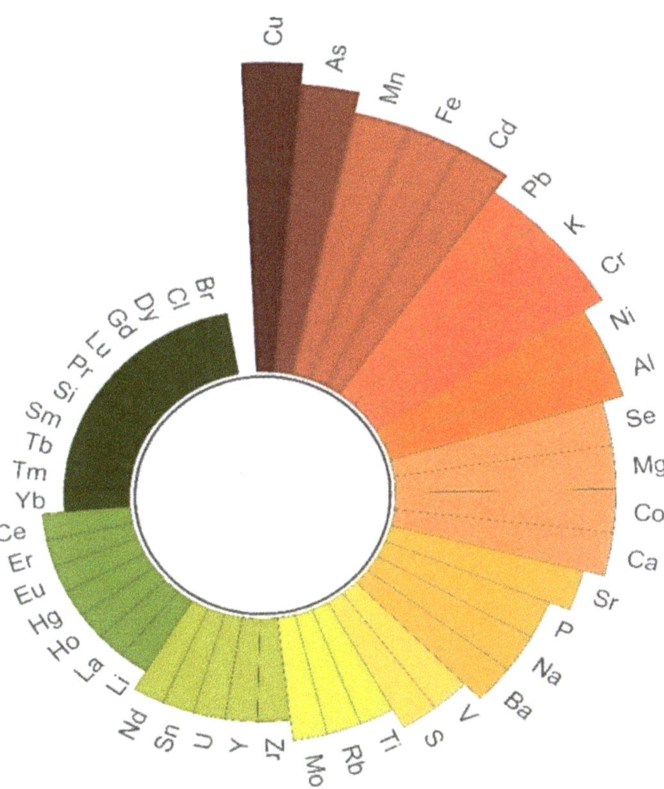

Figure 1. Radial bar chart showing the most widely used elemental markers in the works from the last 10 years dealing with authenticity and traceability of fish and seafood products. Data were elaborated from the scientific literature (published in 2010–2020) and collected from the Scopus and Web of Science search engines, using 'fish,' 'seafood,' 'authentication,' 'elemental analysis,' 'elemental profile,' 'elemental fingerprinting,' and/or 'chemometrics' as search terms.

3.1. Fish

The maximum guarantee of transparency about the method of production, intended as catching wild fish or raising aquaculture fish, is of extreme importance, given that the two products have a differing economic value. In addition, certain farmed fish such as salmonids are reported to be more prone to accumulate environmental toxic substances, especially of organic nature [73], thus questioning the overall wholesomeness of these products. Tracing the geographical origin of aquaculture products may be, in some ways, more complicated than tracing that of wild-caught products. In fact, despite the fact that the feeding habits and prey availability for wild fish are highly variable and cannot be controlled, it should be emphasized that feeds used in aquaculture practices (which significantly affect mineral and trace element contents of fish tissues) are not only extremely variable in terms of composition but are frequently used worldwide to raise fish of different geographical origin [74], thus masking any discriminant potential of the elemental profile.

Despite these hurdles, different species of both wild and farmed salmons corresponding to king salmon (*Oncorhynchus tshawytscha*), coho salmon (*Oncorhynchus kysutch*), and Atlantic salmon (*Salmo salar*) were analyzed for their major and trace elemental content and isotope ratio profile of carbon (δ^{13}C) and nitrogen (δ^{15}N) in order to develop a model suited for their classification [61]. As for the type of employed tracers, it was verified that using elements or isotope ratios has no bearing on the overall performances of salmon classification, but on the contrary, the outcomes are strongly influenced by the number of samples employed to train the classification model as well as by the chosen classification algorithm. On that note, using machine learning algorithms as artificial neural networks (ANNs) and neural network bagging (NNB) gave a 94% and 92% correct classification rate, respectively, when applied to elements only, and 94% and 87% when using stable isotope ratios only [61].

The possibility of using rare earth elemental profile and/or light stable isotope ratios to identify fish production methods was also recently investigated for European sea bass (*Dicentrarchus labrax*, L.) samples [62]. In this case, the concentrations of lanthanum, europium, holmium, erbium, lutetium, and terbium elaborated by PCA and orthogonal partial least square discriminant analysis (OPLS-DA) did not impact the differentiation of wild from farmed specimens in contrast to light isotope ratios of carbon (δ^{13}C) and nitrogen (δ^{15}N), which had a higher influence. However, the authors verified that holmium and lanthanum, due to their natural variability in the marine environment, had a significant influence on the discrimination of the same samples by geographic origin. As a matter of fact, almost 89% of unlabeled samples from three different fishing areas in the Mediterranean Sea (used to test the validity of the developed model) were correctly discriminated [62].

The truthfulness of the label description of European seabass (*Dicentrarchus labrax*, L.) was also analyzed in another study which took into consideration the outputs obtained through the measurement of several parameters, corresponding to the biometric indices, fatty acids profile, analysis of 18 elements, and stable isotope ratios of carbon and nitrogen [48]. The method of production, the intensity of farming system, and the geographical provenance of sea bass were better discriminated using the fatty acids composition, while the use of elements alone outperformed compared to the other analytical data. Only the concentrations of Ca were in found to be significantly affected by feeding system and geographical origin, but the differences in fish tissues were not sufficient to achieve satisfying discrimination results, which settled around 79% for production method and 57% for the origin. On the contrary, stable isotope ratio data performed well in discriminating the production method of samples due to the strong influence of the feeding inputs on these parameters, but they were not able to classify samples according to provenance [48].

The merging of the results of multi-elemental and stable isotope ratios analyses has been successful also for the discrimination by origin and production method of Asian sea bass (*Lates calcarifer*) [49] and, when adding proximate composition, for the discrimination by origin of croaker [19]. Unlike the previously reported studies, XFR was the chosen technique to determine the elemental content of fish samples, mainly offering advantages in terms of the speed of operation. In this case, although the origin discriminant models for sea bass created by applying LDA or RF to stable isotope data were more accurate than those computed using elemental data only, isotopic analysis was less performant when used alone to predict both the origin and the production method of unknown samples, thus suggesting that information provided by elements is essential to achieve satisfying discrimination accuracy for the identification of geographical provenances [49].

One single attempt to discriminate the origin of freshwater cultured fish was found in the literature. In this case, fillets of channel catfish (*Ictalurus punctatus*) and hybrid catfish (*Ictalurus furcatus*) from 3 geographic areas were subjected to ICP-OES to measure a total of 11 elements [33]. Although the authors did not find a direct influence of water and feed used to raise the catfish on the final elemental composition of the fillets, the products were separated by origin, with 100% accuracy whether canonical discriminant analysis or *K*-nearest-neighbor analysis were used. Despite this, it should be noted that provenances considered in this study are of geopolitical rather than of geochemical nature. Therefore, the validity of the discrimination is limited by the fact that aquaculture catfish can be raised elsewhere in waters with an equivalent elemental composition [33].

3.2. Echinoderms and Crustaceans

Mislabeling of echinoderms has been poorly treated by the scientific community, probably because the consumption of these products, however high, is mainly limited to Asian countries. Two applications regarding the authentication of sea cucumber (*Apostichopus japonicus*) through elemental profiles have aimed at classifying the samples according to three [23] and five [63] sampling areas in China. However, these applications used a different number of elements, with 15 elements in the first case and to 39 in the second one. In both works, a stepwise-LDA was used to concomitantly sort elements by their relative importance in discrimination and build classification models. Concentrations of Al, Mn, Fe, Co, Ni, Cu, As, Se, Cd, and Hg were found to be appropriate to differentiate 100% of sea cucumbers in relation to the three sampling areas [23], while concentrations of Li, Na, Al, K, Co, Cu, Cd, and Sc made it possible to achieve 88% accuracy in differentiating samples originating from the five areas [63]. So, despite the higher number of elements measured in the second study, measuring a higher number of elements is not always a straightforward matter to achieve better discrimination results. If redundant or noise elements are not strictly evaluated and removed by proper statistics, models built using many elements as variables are likely to outperform, especially with an increasing number of origins to be identified.

Reviewing the literature, crustaceans emerged as the most frequently analyzed category of seafood products intended to be authenticated by their elemental composition (see Table 1). More specifically, six out of seven works analyzing the multi-elemental profile of crustaceans and taken into consideration in the present review dealt with the authentication of the origin of shrimps or prawns [22,25,64–66,68]. Among these, only two works concurrently investigated the possibility of using the same profile to address other problems, such as the production method and the species identifications [25,65].

The use of the elemental profile alone was demonstrated to be an optimal strategy to accurately assess the traceability of Pacific white shrimps (*Litopenaeus vannamei*) from different sampling sites in the USA [64] to differentiate shrimps obtained from Vietnam, Thailand, and India, which represent the biggest producing countries in the world [22]. When used in combination with light stable isotope ratios of carbon and nitrogen, the elemental profile was able to discriminate shrimps according to different sampling areas in China [64]. In general, despite the combination of major, minor, and trace elements (especially K, Mg, Na, P, Ca, Ba, Cr, Pb, Se, Si Cd, Co, and Zr), the elemental profile was successful in solving the origin discrimination problems in all cases. When concentrations of REEs were determined and used as discriminant variables, it was found that these elements had a greater analytical significance in determining the provenance of shrimps compared to other variables [64].

To some extent, the superiority of the element composition over stable isotope ratios of C and N to assess the traceability of shrimps was also demonstrated when farmed and wild samples of seven different biological species, obtained from nine sampling zones, were investigated [65]. Stable isotope analysis alone yielded to 100%, 71%, and 58% of samples to be correctly classified using LDA by production method, origin, and biological species, respectively. However, with an increasing number of samples into the models, the origin discrimination accuracy decreased or did not significantly increase. On the contrary, As, Cd, Pb, P and S concentrations alone showed greater accuracy in classifying samples by origin (94%) and species (74%) and, when merged with stable isotopes ratios, the two techniques showed the maximum discrimination power [65]. Similarly, both the production method and the origin traceability of prawn (*Penaeus monodon*) were assessed, with 100% accuracy when the multi-elemental profile and stable isotopes ratios were used complementarily [25].

Advantages of coupling elements and light stable isotope ratio analyses outputs to verify the exact provenances of high-value crustaceans are even more evident when powerful classification machine learning techniques are applied. The contents of Na, Mg, Al, K, Ca, Mn, Cu, Zn, Sr, and Ba, plus δ^{13}C and δ^{15}N measured on limited sample material and elaborated by means of SVM, allowed for the tracing of Chinese mitten crabs (*Eriocheir sinensis*) according to eight different geographical origins around China, with 100% and 97% accuracy in cross-validation and external validation, respectively [67].

3.3. Mollusks

To date, mollusks have appeared to be the least frequently studied aquatic products in terms of the evaluation of authenticity and fraud verification. This is particularly remarkable considering that, according to the latest available data, the worldwide supply of cephalopods and other mollusks has reached values of 3,535,732 tons and of 17,500,801 tons per year, respectively [75].

Historically, the elemental profiles of bivalve or cephalopods mollusks were employed to assess their geographical authenticity but performing such analyses on nonedible hard parts of the animals (e.g., shells, statoliths, beaks) [76,77] does not guarantee the possibility to apply the same methods to ready-to-cook products (eviscerated, beheaded, shelled), which are rising in popularity in international markets.

An interesting study used ICP-MS in combination with LDA, SIMCA, and ANNs to quantify and elaborate a total of forty elements in order to authenticate Galician mussels (*Mytilus galloprovincialis*) under the European Protected Designation of Origin (PDO) and protect the products from similar but lower-quality mussels [24]. A strong relation between element composition of PDO mussels and the geomorphology and lithology of the specific production zone, as well as with external contamination sources, was found. Whereas the Se, Zn, Pb, Co, Mo, Ag, and Ba elemental signature was attributed to the metabolic activity of the animals, the Ga, Zr, Eu, Lu, Th, and U signature was specifically related to mineralogical sources of the area, and the V, Cd, and Sb signature was related to the anthropogenic pollutant activities characterizing the area [24]. Keeping the complementary information provided by all these elements, PDO from non-PDO products were 100% accurately classified by LDA and SIMCA. On the contrary, the use of ANNs was found to be more effective in discriminating the five different sampling zones from which the PDO mussels were obtained.

In another work, particular attention was paid toward any effect that the seasonality had on the elemental composition of bivalves [69]. Since season variations were misleadingly reflected on the Mg, Rb, Pd, Cd, Sn, Ba, La, and Ce distribution into the mollusks, the authors were able to authenticate samples of Manila clams (*Ruditapes philippinarum*) using a different pattern of elements composed by Mg, Rb, Pd, Cd, Sn, Ba, La, and Ce, which, in contrast, was found to be more strongly linked to the geographical origin of clams [69].

As far as we know, no works oriented toward the evaluation of cephalopods mislabeling by measuring element composition of edible tissues such as mantles and fins are available. Nevertheless, the inorganic composition of ink derived from cuttlefish (commonly used in the Mediterranean and Japanese gastronomy) showed some potential ability to enclose geographical-related information [70]. Although no classification analysis was performed, some elements, such as Cr, Ni, V, Cd, Pb, As, and Hg, were significantly different among ink samples of cuttlefish (*Sepia officinalis*) of different sampling sites in the central Mediterranean Sea, suggesting that the contribution of the environmental pollution should be further investigated in these kind of studies to verify whether it can reveal actionable insights.

4. Why Are Aquatic Animals Ideal Candidates for Multi-Elemental Analysis?

The evaluation of the organic composition of foodstuffs continues to be the first choice when the identification of individual markers or patterns of markers for authenticity and traceability of fishery products is the main research goal. Nevertheless, measuring of a high number of organic components without carefully considering their origin, significance, sources of variations, and the general framework within which they are evaluated frequently puts their outright specificity as markers of origin into question. Indeed, the concentration and distribution of certain classes of organic compounds, such as fatty acids, peptides, and enzymes, are concurrently affected by so many aspects and circumstances that it is often challenging to univocally relate them to the sole species and geographic and/or farming origin. Pre-catch conditions, such as seasonality, climatic conditions, fishing period, fish size, fish physiology and metabolism, and fishing gear, as well as human's post-catch manipulation and storage operations (storage temperature, packaging, lifetime of the product, and so forth), are just a few examples of factors affecting the organic composition of fish [9]. Similar considerations are valid also for inorganic constituents, but the correlation between the elemental compositions of fish tissues and the surrounding aquatic environment has been demonstrated to be more stable and consistent over time. Therefore, the probability that inorganic markers of fish origin are hidden by misleading factors may be considered lower compared to organic markers. Based on the concentrations found in the matrix, elements are normally categorized as major and minor (trace and ultra-trace) elements. A detailed definition has been reported only for trace elements, defined as those elements whose concentrations in the matrix are lower than $100 \, \text{mg kg}^{-1}$ [78], and which are mainly represented in food by B, Al, Fe, Mn, Ni, Cu, Zn, As, Se, Sr, and sometimes La and Ce. Consequently, major elements have mass fractions above $100 \, \text{mg kg}^{-1}$ (Na, Mg, P, K, Ca, Mg), while ultra-trace elements have mass fractions generally below $1 \, \text{mg kg}^{-1}$ [37] (Li, V, Cr, Co, Rb, Y, Zr, Mo, Ru, Pd, Cd, Sn, Ag, Cd, Sn, Sb, Cs, Ba, lanthanides, Hf, Re, Pt, Bi, Hg, Th, U, Hg). Rare earth elements (REEs), usually including Y, La, and lanthanides (Ce, Pr, Nd, Pm, Sm, Eu, Gd, Tb, Dy, Ho, Er, Tm, Yb, and Lu) [79], are emerging as very promising inorganic markers of fish authenticity, despite the fact that their quantification into foodstuffs is still limited to the very low abundance and consequent obstacles in quantification by modern instrumentations. Concentrations of REEs in both surface water and groundwater were found to vary significantly in relation to the geographical areas, with the Asiatic continent (and, in particular, China) showing the highest levels, followed by Europe, Africa, USA, and Australia [80]. These variations may be attributed both to the natural release of REEs from the parental soil (weathering of black shale is a common cause of increasing REEs composition in water) and to some anthropic activities (metallurgy, glass and ceramic industry, electronics) responsible for the REEs release into the aquatic environment and the consequent uptake by the aquatic fauna [81].

The overall major and minor elemental composition of fish is largely related to the elemental content of the eaten preys, vegetation, or fodder. In turn, the content of elements in animal and vegetable feeds is the result of the bioavailable elements which have been mobilized from the soil and which reflect the overall characteristics of the geographical area [12,82]. For example, as some alkaline metals (e.g., Rb and Cs) can be easily mobilized from the underlying soils, the probability that their incorporation into fish tissues is variable according to the geographical site is very high [83]. Other trace elements, such as B and As, naturally enter the aquatic environment from volcanic and geothermal activities [84,85]. Therefore, their concentrations in fish and seafood tissues may be exploited to discriminate animals from marine areas with specific geochemical characteristics. Moreover, concentrations of some major and trace elements, such as Li, Mg, Ca, Sr, Zn, Mn, and Cu, are strictly regulated by the salinity of the marine basin, and this characteristic makes them suitable to be potentially used for marine fish-tracing purposes [86]. In this setting, it is not unexpected that traceability studies concerning marine fish species are, to some extent, more standardizable, and thus more reliable compared to those dealing with freshwater species. Along with some concern deriving from the closeness to the anthropic environment, this aspect is attributable to the higher degree of dynamism of the marine systems compared to freshwater ones. Since this dynamism is biologically, chemically, and physically controlled, a more uniform element concentration from both a temporal and spatial point of view can therefore be found in the marine environment, especially in open ocean waters. Nevertheless, when performing authentication studies, consideration must be given to the fact that the distribution ratio of certain elements between fish tissues and seawater is altered by the metabolic activity of animals [74]. Specifically, the uptake of many essential elements, such as Na, K, Mg, and Ca, is metabolically regulated by the same fish, since it is necessary for regulating physiological functions. Hence, the potential for variation of these elements in relation to origin is masked by physiological 'noise' [87]. Consequently, they are hardly ever used in fish authentication studies.

Finally, a greater compositional heterogeneity is encountered in waters of coastal areas compared to deep seawaters, where the proximity of anthropic releasing sources leads some trace and ultra-trace elements to be variably introduced into the marine environment. Nickel, zinc, arsenic, lead, mercury, and cadmium are well known for their higher concentrations along shorelines [88], since they are derived from certain agricultural practices or industrial activities. On the other side, fish and seafood are not able to physiologically regulate the concentration of these nonessential (and often toxic) elements, which, consequently, are passively accumulated into the animal's tissues. If properly evaluated, anthropic elements can also be used for origin authentication purposes [89].

To conclude, although introduced through different sources, elements can be successfully employed as authenticity markers of fish and seafood if the same introduction sources are systematic, identifiable, manageable, and suggestive of the geographical origin or production process [74]. In this context, when dealing with authentication of transformed fish products, particular attention should be paid against the introduction of elements from the production chain. These obstacles may be often overcome by comparing the effective concentrations of elements in the final products with those found along as many stages as possible of the transformation process, so as to be able to verify whether distribution trends are retained along the production stages [74].

5. Final Remarks and Conclusions

The application of element profiling approaches to fish and seafood products has been gaining momentum, and the scientific community has been working on the optimization of both existing instrumentations for multi-elemental analysis and algorithms for statistical analysis. The greater thrust has come from advances in chemometrics and machine learning techniques, which now provide great support to the identification of maximum relevant chemical information from large datasets not otherwise accessible.

From the analysis of the literature presented in this review, it is clear that the discrimination of the geographical origin has been the most frequently discussed authenticity topic, while other aspects, such as the farming systems, have been overlooked. In addition, crustaceans have emerged as the most frequently investigated category of products, while less emphasis has been placed on fish, echinoderms, and mollusks, especially cephalopods, probably due to difficulties in drawing up an adequate sampling plan to build representative datasets. Regarding the statistical data treatment, PCA and LDA have been more widely used, while machine learning algorithms have been neglected, despite their great potential in discovering hidden discriminant patterns among data.

As for the selected methodologies, ICP-MS, followed by ICP-OES, has been the first choice, accounting for the vast majority of the published research. Especially in the last years, ICP-MS has been gaining popularity within the scientific community because it is less complicated, less expensive, and undoubtedly the fastest and most universal trace element technique commercially available today. This is mainly due to the advances in collision/reaction cell technology, which offers an effective way to reduce spectral effects from different polyatomic ions. Quadrupole mass spectrometers, in particular, are increasingly being used and, until recently, it seemed impossible that a single technique would fit perfectly to the needs of all the laboratories. For this reason, these instruments are expected to supersede most of the ICP-OES and AAS applications in the near future. In addition, it may be expected that various solid-sampling techniques, such as ETV-ICP-MS, LA-ICP-MS, and XRF, may succeed more in the field of food authentication, with the advantage of a reduced sample preparation.

Another peculiarity emerging from the published literature is the tendency to couple element profiles of fish and seafood with other analytical parameters, especially stable isotopes of carbon, hydrogen, and nitrogen, which has probably been motivated by the need to increase the accuracy of discrimination. Despite the benefits deriving from the fusion of complementary or synergistic information, it is worth highlighting that multi-elemental analysis may be sufficient to achieve equivalent results with an optimal cost-performance ratio.

Looking forward, the increased use of ICP-MS-hyphenated techniques for elemental speciation and ICP-MS/MS for interference-free determination and isotope ratio measurement would represent a turning point for the high-throughput analytical characterization of complex matrices such as food. Nevertheless, the reduction of the cost of the equipment for multi-elemental analysis would certainly be desirable to further encourage the spreading of multi-elemental analytical approaches in a different context from that of the specialized laboratories dealing with food surveillance. Before getting to this point, the validity and robustness of elemental markers to ascertain fish and seafood authenticity must be increased. Further work on these issues is therefore encouraged in order to integrate information relating to any possible variable influencing the inorganic profile of fishery products with the elemental information relating to the origin into adequately defined reference databases. At the same time, continuous technological improvements, as well as the shift toward a progressive miniaturization of the instruments, may be a major turning point, helping to concomitantly monitor health risks associated with the occurrence of toxic metals such as cadmium, lead, mercury, and arsenic, and to meet the demand for cost-effective and energy- and reagent-saving instruments.

Supplementary Materials: The following are available online at https://www.mdpi.com/2304-8158/10/2/270/s1, Table S1: A summary of the advantages and disadvantages of the main each analytical methodology examined in the present review, Table S2: Concentrations of the elements (means and standard deviations in brackets, concentrations expressed as mg kg^{-1}) in the reviewed studies.

Author Contributions: All authors contributed equally to this work. All authors have read and agreed to the published version of the manuscript.

Funding: This research received no external funding.

Conflicts of Interest: The authors declare no conflict of interest.

References

1. Warner, K.; Mustain, P.; Lowell, B.; Geren, S.; Talmage, S. *Deceptive Dishes: Seafood Swaps Found Worldwide*; Oceana Reports; Oceana: New York, NY, USA, 2016; Available online: https://usa.oceana.org/publications/reports/deceptive-dishes-seafood-swaps-found-worldwide (accessed on 16 September 2020).
2. Giusti, A.; Ricci, E.; Guarducci, M.; Gasperetti, L.; Davidovich, N.; Guidi, A.; Armani, A. Emerging risks in the European seafood chain: Molecular identification of toxic Lagocephalus spp. in fresh and processed products. *Food Control* **2019**, *91*, 311–320. [CrossRef]
3. Kusche, H.; Hanel, R. Consumers of mislabeled tropical fish exhibit increased risks of ciguatera intoxication: A report on substitution patterns in fish imported at Frankfurt Airport, Germany. *Food Control* **2021**, *121*, 107647. [CrossRef]
4. Ballin, N.Z.; Laursen, K.H. To target or not to target? Definitions and nomenclature for targeted versus non-targeted analytical food authentication. *Trends Food Sci. Technol.* **2019**, *86*, 537–543. [CrossRef]
5. Zhou, X.; Taylor, M.P.; Salouros, H.; Prasad, S. Authenticity and geographic origin of global honeys determined using carbon isotope ratios and trace elements. *Sci. Rep.* **2018**, *8*, 1–11. [CrossRef] [PubMed]
6. Kim, J.S.; Hwang, I.M.; Lee, G.H.; Park, Y.M.; Choi, J.Y.; Jamila, N.; Khan, N.; Kim, K.S. Geographical origin authentication of pork using multi-element and multivariate data analyses. *Meat Sci.* **2017**, *123*, 13–20. [CrossRef] [PubMed]
7. Magdas, D.A.; Feher, I.; Cristea, G.; Voica, C.; Tabaran, A.; Mihaiu, M.; Cordea, V.D.; Balteaunu, V.A.; Dan, D.S. Geographical origin and species differentiation of Transylvanian cheese. Comparative study of isotopic and elemental profiling vs. DNA results. *Food Chem.* **2019**, *277*, 307–313. [CrossRef] [PubMed]
8. Commission regulation (EC) No 1881/2006 of 19 December 2006 setting maximum levels for certain contaminants in foodstuffs. (20/12/2006). *Off. J. Eur. Union* **2006**, *L364*, 5–24.
9. Danezis, G.P.; Tsagkaris, A.S.; Camin, F.; Brusic, V.; Georgiou, C.A. Food authentication: Techniques, trends & emerging approaches. *TrAC Trends Anal. Chem.* **2016**, *85*, 123–132. [CrossRef]
10. Callao, M.P.; Ruisánchez, I. An overview of multivariate qualitative methods for food fraud detection. *Food Control* **2018**, *86*, 283–293. [CrossRef]

11. Danezis, G.P.; Tsagkaris, A.S.; Brusic, V.; Georgiou, C.A. Food authentication: State of the art and prospects. *Curr. Opin. Food Sci.* **2016**, *10*, 22–31. [CrossRef]

12. Drivelos, S.A.; Georgiou, C.A. Multi-element and multi-isotope-ratio analysis to determine the geographical origin of foods in the European Union. *TrAC Trends Anal. Chem.* **2012**, *40*, 38–51. [CrossRef]

13. Gopi, K.; Mazumder, D.; Sammut, J.; Saintilan, N. Determining the provenance and authenticity of seafood: A review of current methodologies. *Trends Food Sci. Technol.* **2019**, *91*, 294–304. [CrossRef]

14. Adebiyi, F.M.; Ore, O.T.; Ogunjimi, I.O. Evaluation of human health risk assessment of potential toxic metals in commonly consumed crayfish (Palaemon hastatus) in Nigeria. *Heliyon* **2020**, *6*, e03092. [CrossRef]

15. Jinadasa, B.K.K.K.; Chathurika, G.S.; Jayasinghe, G.D.T.M.; Jayaweera, C.D. Mercury and cadmium distribution in yellowfin tuna (*Thunnus albacares*) from two fishing grounds in the Indian Ocean near Sri Lanka. *Heliyon* **2019**, *5*, e01875. [CrossRef]

16. Stancheva, M.; Makedonski, L.; Peycheva, K. Determination of heavy metal concentrations of most consumed fish species from Bulgarian Black Sea coast. *Bulg. Chem. Comm.* **2014**, *46*, 195–203.

17. Santos, L.F.P.; Trigueiro, I.N.S.; Lemos, V.A.; da Nóbrega Furtunato, D.M.; Cardoso, R.D.C.V. Assessment of cadmium and lead in commercially important seafood from São Francisco do Conde, Bahia, Brazil. *Food Control* **2013**, *33*, 193–199. [CrossRef]

18. De Andrade, R.M.; De Gois, J.S.; Toaldo, I.M.; Batista, D.B.; Luna, A.S.; Borges, D.L. Direct determination of trace elements in meat samples via high-resolution graphite furnace atomic absorption spectrometry. *Food Anal. Methods* **2017**, *10*, 1209–1215. [CrossRef]

19. Chaguri, M.P.; Maulvault, A.L.; Nunes, M.L.; Santiago, D.A.; Denadai, J.C.; Fogaça, F.H.; Sant'Ana, L.S.; Ducatti, C.; Bandarra, N.; Carvalho, M.L.; et al. Different tools to trace geographic origin and seasonality of croaker (*Micropogonias furnieri*). *LWT Food Sci. Technol.* **2015**, *61*, 194–200. [CrossRef]

20. Skałecki, P.; Florek, M.; Kędzierska-Matysek, M.; Poleszak, E.; Domaradzki, P.; Kaliniak-Dziura, A. Mineral and trace element composition of the roe and muscle tissue of farmed rainbow trout (*Oncorhynchus mykiss*) with respect to nutrient requirements: Elements in rainbow trout products. *J. Trace Elem. Med. Biol.* **2020**, *62*, 126619. [CrossRef]

21. Luo, H.; Wang, X.; Dai, R.; Liu, Y.; Jiang, X.; Xiong, X.; Huang, K. Simultaneous determination of arsenic and cadmium by hydride generation atomic fluorescence spectrometry using magnetic zero-valent iron nanoparticles for separation and pre-concentration. *Microchem. J.* **2017**, *133*, 518–523. [CrossRef]

22. Li, L.; Boyd, C.E.; Racine, P.; McNevin, A.A.; Somridhivej, B.; Minh, H.N.; Tinh, H.Q.; Godumala, R. Assessment of elemental profiling for distinguishing geographic origin of aquacultured shrimp from India, Thailand and Vietnam. *Food Control* **2017**, *80*, 162–169. [CrossRef]

23. Liu, X.; Xue, C.; Wang, Y.; Li, Z.; Xue, Y.; Xu, J. The classification of sea cucumber (*Apostichopus japonicus*) according to region of origin using multi-element analysis and pattern recognition techniques. *Food Control* **2012**, *23*, 522–527. [CrossRef]

24. Costas-Rodríguez, M.; Lavilla, I.; Bendicho, C. Classification of cultivated mussels from Galicia (Northwest Spain) with European Protected Designation of Origin using trace element fingerprint and chemometric analysis. *Anal. Chim. Acta* **2010**, *664*, 121–128. [CrossRef] [PubMed]

25. Gopi, K.; Mazumder, D.; Sammut, J.; Saintilan, N.; Crawford, J.; Gadd, P. Combined use of stable isotope analysis and elemental profiling to determine provenance of black tiger prawns (*Penaeus monodon*). *Food Control* **2019**, *95*, 242–248. [CrossRef]

26. Melucci, D.; Casolari, S.; De Laurentiis, F.; Locatelli, C. Determination of toxic metals in marine bio-monitors by multivariate analysis of voltametric and spectroscopic data. Application to mussels, algae and fishes of the Northern Adriatic Sea. *Fresenius Environ. Bull.* **2017**, *26*, 3756–3763.

27. Moon, J.H.; Ni, B.F.; Theresia, R.M.; Salim, N.A.; Arporn, B.; Vu, C.D. Analysis of consumed fish species by neutron activation analysis in six Asian countries. *J. Radioanal. Nucl.* **2015**, *303*, 1447–1452. [CrossRef]

28. Fabiano, K.C.; Lima, A.P.; Vasconcellos, M.B.; Moreira, E.G. Contribution to food safety assurance of fish consumed at São Paulo city by means of trace element determination. *J. Radioanal. Nucl.* **2016**, *309*, 383–388. [CrossRef]

29. Rentería-Cano, M.E.; Sánchez-Velasco, L.; Shumilin, E.; Lavín, M.F.; Gómez-Gutiérrez, J. Major and trace elements in zooplankton from the Northern Gulf of California during summer. *Biol. Trace Elem. Res.* **2011**, *142*, 848–864. [CrossRef]

30. Brown, R.J.C.; Milton, M.J.T. Analytical techniques for trace element analysis: An overview. *TrAC Trends Anal. Chem.* **2005**, *24*, 266–274. [CrossRef]

31. Rasmussen, R.R.; Hedegaard, R.V.; Larsen, E.H.; Sloth, J.J. Development and validation of an SPE HG-AAS method for determination of inorganic arsenic in samples of marine origin. *Anal. Bioanal. Chem.* **2012**, *403*, 2825–2834. [CrossRef]

32. Panichev, N.A.; Panicheva, S.E. Determination of total mercury in fish and sea products by direct thermal decomposition atomic absorption spectrometry. *Food Chem.* **2015**, *166*, 432–441. [CrossRef] [PubMed]

33. Li, L.; Boyd, C.E.; Odom, J.; Dong, S. Identification of Ictalurid Catfish Fillets to Rearing Location Using Elemental Profiling. *J. World Aquac. Soc.* **2013**, *44*, 405–414. [CrossRef]

34. De Sá, I.P.; Higuera, J.M.; Costa, V.C.; Costa, J.A.S.; Da Silva, C.M.P.; Nogueira, A.R.A. Determination of trace elements in meat and fish samples by MIP OES using solid-phase extraction. *Food Anal. Methods* **2020**, *13*, 238–248. [CrossRef]

35. Ríos, S.E.G.; Peñuela, G.A.; Botero, C.M.R. Method validation for the determination of mercury, cadmium, lead, arsenic, copper, iron, and zinc in fish through microwave-induced plasma optical emission spectrometry (MIP OES). *Food Anal. Methods* **2017**, *10*, 3407–3414. [CrossRef]

36. Zmozinski, A.V.; Carneado, S.; Ibáñez-Palomino, C.; Sahuquillo, À.; López-Sánchez, J.F.; Da Silva, M.M. Method development for the simultaneous determination of methylmercury and inorganic mercury in seafood. *Food Control* **2014**, *46*, 351–359. [CrossRef]

37. Bulska, E.; Ruszczyńska, A. Analytical techniques for trace element determination. *Phys. Sci. Rev.* **2017**, 1–14. [CrossRef]

38. Yeung, V.; Miller, D.D.; Rutzke, M.A. *Atomic Absorption Spectroscopy, Atomic Emission Spectroscopy, and Inductively Coupled Plasma-Mass Spectrometry*, 1st ed.; Nielsen, S., Ed.; Springer: Cham, Switzerland, 2017; pp. 129–150.

39. Sneddon, J.; Vincent, M.D. ICP-OES and ICP-MS for the determination of metals: Application to oysters. *Anal. Lett.* **2008**, *41*, 1291–1303. [CrossRef]

40. Lum, T.S.; Leung, K.S.Y. Strategies to overcome spectral interference in ICP-MS detection. *J. Anal. At. Spectrom.* **2016**, *31*, 1078–1088. [CrossRef]

41. Husáková, L.; Urbanová, I.; Šrámková, J.; Černohorský, T.; Krejčová, A.; Bednaříková, M.; Frýdová, E.; Nedělková, I.; Pilařová, L. Analytical capabilities of inductively coupled plasma orthogonal acceleration time-of-flight mass spectrometry (ICP-oa-TOF-MS) for multi-element analysis of food and beverages. *Food Chem.* **2011**, *129*, 1287–1296. [CrossRef]

42. Burger, M.; Hendriks, L.; Kaeslin, J.; Gundlach-Graham, A.; Hattendorf, B.; Günther, D. Characterization of inductively coupled plasma time-of-flight mass spectrometry in combination with collision/reaction cell technology–insights from highly time-resolved measurements. *J. Anal. At. Spectrom.* **2019**, *34*, 135–146. [CrossRef]

43. Balcaen, L.; Bolea-Fernandez, E.; Resano, M.; Vanhaecke, F. Inductively coupled plasma–Tandem mass spectrometry (ICP-MS/MS): A powerful and universal tool for the interference-free determination of (ultra) trace elements—A tutorial review. *Anal. Chim. Acta* **2015**, *894*, 7–19. [CrossRef] [PubMed]

44. Zhao, X.; Liu, Y.; Li, Y.; Zhang, X.; Qi, H. Authentication of the sea cucumber (*Apostichopus japonicus*) using amino acids carbon stable isotope fingerprinting. *Food Control* **2018**, *91*, 128–137. [CrossRef]

45. Molkentin, J.; Lehmann, I.; Ostermeyer, U.; Rehbein, H. Traceability of organic fish–Authenticating the production origin of salmonids by chemical and isotopic analyses. *Food Control* **2015**, *53*, 55–66. [CrossRef]

46. Remaud, G.S.; Giraudeau, P.; Lesot, P.; Akoka, S. Isotope Ratio Monitoring by NMR: Part 1—Recent Advances. In *Modern Magnetic Resonance*; Webb, G., Ed.; Springer: Cham, Switzerland, 2018; pp. 1353–1378.

47. Aursand, M.; Mabon, F.; Martin, G.J. Characterization of farmed and wild salmon (*Salmo salar*) by a combined use of compositional and isotopic analyses. *J. Am. Oil Chem. Soc.* **2000**, *77*, 659–666. [CrossRef]

48. Farabegoli, F.; Pirini, M.; Rotolo, M.; Silvi, M.; Testi, S.; Ghidini, S.; Zanardi, E.; Remondini, D.; Bonaldo, A.; Parma, L.; et al. Toward the Authentication of European Sea Bass Origin through a Combination of Biometric Measurements and Multiple Analytical Techniques. *J. Agric. Food Chem.* **2018**, *66*, 6822–6831. [CrossRef]

49. Gopi, K.; Mazumder, D.; Sammut, J.; Saintilan, N.; Crawford, J.; Gadd, P. Isotopic and elemental profiling to trace the geographic origins of farmed and wild-caught Asian seabass (*Lates calcarifer*). *Aquaculture* **2019**, *502*, 56–62. [CrossRef]

50. Machado, R.C.; Andrade, D.F.; Babos, D.V.; Castro, J.P.; Costa, V.C.; Sperança, M.A.; Garcia, J.A.; Gamela, R.R.; Pereira-Filho, E.R. Solid sampling: Advantages and challenges for chemical element determination—A critical review. *J. Anal. At. Spectrom.* **2020**, *35*, 54–77. [CrossRef]

51. Markiewicz-Keszycka, M.; Cama-Moncunill, X.; Casado-Gavalda, M.P.; Dixit, Y.; Cama-Moncunill, R.; Cullen, P.J.; Sullivan, C. Laser-induced breakdown spectroscopy (LIBS) for food analysis: A review. *Trends Food Sci. Technol.* **2017**, *65*, 80–93. [CrossRef]

52. Pozebon, D.; Scheffler, G.L.; Dressler, V.L.; Nunes, M.A. Review of the applications of laser ablation inductively coupled plasma mass spectrometry (LA-ICP-MS) to the analysis of biological samples. *J. Anal. At. Spectrom.* **2014**, *29*, 2204–2228. [CrossRef]

53. Dunphy, B.J.; Millet, M.A.; Jeffs, A.G. Elemental signatures in the shells of early juvenile green-lipped mussels (*Perna canaliculus*) and their potential use for larval tracking. *Aquaculture* **2011**, *311*, 187–192. [CrossRef]

54. Flem, B.; Moen, V.; Finne, T.E.; Viljugrein, H.; Kristoffersen, A.B. Trace element composition of smolt scales from Atlantic salmon (*Salmo salar L.*), geographic variation between hatcheries. *Fish. Res.* **2017**, *190*, 183–196. [CrossRef]

55. Limbeck, A.; Bonta, M.; Nischkauer, W. Improvements in the direct analysis of advanced materials using ICP-based measurement techniques. *J. Anal. At. Spectrom.* **2017**, *32*, 212–232. [CrossRef]

56. Von der Au, M.; Karbach, H.; Bell, A.M.; Bauer, O.B.; Karst, U.; Meermann, B. Determination of metal uptake in single organisms, *Corophium volutator*, via complementary electrothermal vaporization/inductively coupled plasma mass spectrometry and laser ablation/inductively coupled plasma mass spectrometry. *Rapid Commun. Mass Spectrom.* **2021**, *35*, e8953. [CrossRef] [PubMed]

57. Bizzi, C.A.; Pedrotti, M.F.; Silva, J.S.; Barin, J.S.; Nóbrega, J.A.; Flores, E.M.M. Microwave-assisted digestion methods: Towards greener approaches for plasma-based analytical techniques. *J. Anal. At. Spectrom.* **2017**, *32*, 1448–1466. [CrossRef]

58. Marini, F. Classification Methods in Chemometrics. *Curr. Anal. Chem.* **2010**, *6*, 72–79. [CrossRef]

59. López, M.I.; Colomer, N.; Ruisánchez, I.; Callao, M.P. Validation of multivariate screening methodology. Case study: Detection of food fraud. *Anal. Chim. Acta* **2014**, *827*, 28–33. [CrossRef]

60. Jiménez-Carvelo, A.M.; González-Casado, A.; Bagur-González, M.G.; Cuadros-Rodríguez, L. Alternative data mining / machine learning methods for the analytical evaluation of food quality and authenticity—A review. *Food Res. Int.* **2019**, *122*, 25–39. [CrossRef]

61. Anderson, K.A.; Hobbie, K.A.; Smith, B.W. Chemical profiling with modeling differentiates wild and farm-raised salmon. *J. Agric. Food Chem.* **2010**, *58*, 11768–11774. [CrossRef]

62. Varrà, M.O.; Ghidini, S.; Zanardi, E.; Badiani, A.; Ianieri, A. Authentication of European sea bass according to production method and geographical origin by light stable isotope ratio and rare earth elements analyses combined with chemometrics. *Ital. J. Food Saf.* **2019**, *8*. [CrossRef]

63. Kang, X.; Zhao, Y.; Shang, D.; Zhai, Y.; Ning, J.; Sheng, X. Elemental analysis of sea cucumber from five major production sites in China: A chemometric approach. *Food Control* **2018**, *94*, 361–367. [CrossRef]

64. Li, L.; Boyd, C.E.; Odom, J. Identification of Pacific white shrimp (*Litopenaeus vannamei*) to rearing location using elemental profiling. *Food Control* **2014**, *45*, 70–75. [CrossRef]

65. Ortea, I.; Gallardo, J.M. Investigation of production method, geographical origin and species authentication in commercially relevant shrimps using stable isotope ratio and/or multi-element analyses combined with chemometrics: An exploratory analysis. *Food Chem.* **2015**, *170*, 145–153. [CrossRef] [PubMed]

66. Carter, J.F.; Tinggi, U.; Yang, X.; Fry, B. Stable isotope and trace metal compositions of Australian prawns as a guide to authenticity and wholesomeness. *Food Chem.* **2015**, *170*, 241–248. [CrossRef] [PubMed]

67. Luo, R.; Jiang, T.; Chen, X.; Zheng, C.; Liu, H.; Yang, J. Determination of geographic origin of Chinese mitten crab (*Eriocheir sinensis*) using integrated stable isotope and multi-element analyses. *Food Chem.* **2019**, *274*, 1–7. [CrossRef]

68. Li, L.; Han, C.; Dong, S.; Boyd, C.E. Use of elemental profiling and isotopic signatures to differentiate Pacific white shrimp (*Litopenaeus vannamei*) from freshwater and seawater culture areas. *Food Control* **2019**, *95*, 249–256. [CrossRef]

69. Zhao, H.; Zhang, S. Effects of sediment, seawater, and season on multi-element fingerprints of Manila clam (*Ruditapes philippinarum*) for authenticity identification. *Food Control* **2016**, *66*, 62–68. [CrossRef]

70. Bua, G.D.; Albergamo, A.; Annuario, G.; Zammuto, V.; Costa, R.; Dugo, G. High-Throughput ICP-MS and Chemometrics for Exploring the Major and Trace Element Profile of the Mediterranean Sepia Ink. *Food Anal. Methods* **2017**, *10*, 1181–1190. [CrossRef]

71. Kelly, S.; Brodie, C.; Hilkert, A. Isotopic-Spectroscopic Technique: Stable Isotope-Ratio Mass Spectrometry (IRMS). In *Modern Techniques for Food Authentication*, 2nd ed.; Academic Press: Cambridge, MA, USA, 2018; pp. 349–413. [CrossRef]

72. Kelly, S.; Heaton, K.; Hoogewerff, J. Tracing the geographical origin of food: The application of multi-element and multi-isotope analysis. *Trends Food Sci. Tech.* **2005**, *16*, 555–567. [CrossRef]

73. Hites, R.A.; Foran, J.A.; Carpenter, D.O.; Hamilton, M.C.; Knuth, B.A.; Schwager, S.J. Global Assessment of Organic Contaminants in Farmed Salmon. *Science* **2004**, *303*, 226–229. [CrossRef]

74. Aceto, M. The Use of ICP-MS in Food Traceability. In *Advances in Food Traceability Techniques and Technologies, Improving Quality Throughout the Food Chain*, 1st ed.; Espiñeira, M., Santaclara, F.J., Eds.; Woodhead Publishing: Cambridge, UK, 2016; pp. 137–164.

75. Food and Agriculture Organization (FAO). FAO Yearbook. Fishery and Aquaculture Statistics 2018. 2020. Available online: http://www.fao.org/fishery/static/Yearbook/YB2018_USBcard/booklet/web_CB1213T.pdf (accessed on 16 September 2020).

76. Iguchi, J.; Takashima, Y.; Namikoshi, A.; Yamashita, Y.; Yamashita, M. Origin identification method by multiple trace elemental analysis of short-neck clams produced in Japan, China, and the Republic of Korea. *Fish. Sci.* **2013**, *79*, 977–982. [CrossRef]

77. Northern, T.J.; Smith, A.M.; McKinnon, J.F.; Bolstad, K.S.R. Trace elements in beaks of greater hooked squid Onykia ingens: Opportunities for environmental tracing. *Molluscan Res.* **2019**, *39*, 29–34. [CrossRef]

78. International Union of Pure and Applied Chemistry (IUPAC). *Compendium of Chemical Terminology (the "Gold Book")*, 2nd ed.; Blackwell Scientific Publications: Oxford, UK, 1997; p. 1551. Available online: http://goldbook.iupac.org/ (accessed on 20 September 2020).

79. Tyler, G. Rare earth elements in soil and plant systems—A review. *Plant Soil* **2004**, *267*, 191–206. [CrossRef]

80. Adeel, M.; Lee, J.Y.; Zain, M.; Rizwan, M.; Nawab, A.; Ahmad, M.A.; Shafiq, M.; Yi, H.; Jilani, G.; Javed, R.; et al. Cryptic footprints of rare earth elements on natural resources and living organisms. *Environ. Int.* **2019**, *127*, 785–800. [CrossRef] [PubMed]

81. Migaszewski, Z.M.; Galuszka, A. The characteristics, occurrence, and geochemical behavior of rare earth elements in the environment: A review. *Crit. Rev. Environ. Sci. Technol.* **2015**, *45*, 429–471. [CrossRef]

82. Reinholds, I.; Bartkevics, V.; Silvis, I.C.J.; van Ruth, S.M.; Esslinger, S. Analytical techniques combined with chemometrics for authentication and determination of contaminants in condiments: A review. *J. Food Compos. Anal.* **2015**, *44*, 56–72. [CrossRef]

83. Kelly, B.C.; Ikonomou, M.G.; Higgs, D.A.; Oakes, J.; Dubetz, C. Mercury and other trace elements in farmed and wild salmon from British Columbia, Canada. *Environ. Toxicol. Chem.* **2008**, *27*, 1361–1370. [CrossRef]

84. Parks, J.L.; Edwards, M. Critical Reviews in Environmental Science and Technology Boron in the Environment. *Crit. Rev. Environ. Sci. Technol.* **2005**, *35*, 81–114. [CrossRef]

85. Zkeri, E.; Aloupi, M.; Gaganis, P. Seasonal and spatial variation of arsenic in groundwater in a rhyolithic volcanic area of Lesvos Island, Greece. *Environ. Monit. Assess.* **2018**, *190*. [CrossRef]

86. Hanson, P.J.; Zdanowicz, V.S. Elemental composition of otoliths from Atlantic croaker along an estuarine pollution gradient. *J. Fish Biol.* **1999**, *54*, 656–668. [CrossRef]

87. Sturrock, A.M.; Hunter, E.; Milton, J.A.; Johnson, R.C.; Waring, C.P.; Trueman, C.N. Quantifying physiological influences on otolith microchemistry. *Methods Ecol. Evol.* **2015**, *6*, 806–816. [CrossRef]

88. Rainbow, P.S. *Trace Metals in the Environment and Living Organisms: The British Isles as a Case Study*; Cambridge University Press: Cambridge, UK, 2018; pp. 1–472.

89. Cubadda, F.; Raggi, A.; Coni, E. Element fingerprinting of marine organisms by dynamic reaction cell inductively coupled plasma mass spectrometry. *Anal. Bioanal. Chem.* **2006**, *384*, 887–896. [CrossRef] [PubMed]

Review

Microsatellite Markers in Olives (*Olea europaea* L.): Utility in the Cataloging of Germplasm, Food Authenticity and Traceability Studies

Shambhavi Yadav [1,*], Joana Carvalho [2,3], Isabel Trujillo [4,*] and Marta Prado [2]

1   Genetics and Tree Improvement Division, Forest Research Institute, P.O. New Forest, Dehradun 248001, India
2   Food Quality and Safety Research Group, International Iberian Nanotechnology Laboratory (INL), 4715-330 Braga, Portugal; joana.rr.carvalho@gmail.com (J.C.); marta.prado@inl.int (M.P.)
3   Department of Analytical Chemistry, Nutrition and Food Science, Campus Vida, College of Pharmacy/School of Veterinary Sciences, University of Santiago de Compostela, E-15782 Santiago de Compostela, Spain
4   Excellence Unit of Maria de Maeztu, Department of Agronomy, Rabanales Campus, International Campus of Excellence on Agrofood (ceiA3), University of Córdoba, 14014 Córdoba, Spain
*   Correspondence: shambhaviy@icfre.org (S.Y.); ag2trnai@uco.es (I.T.)

Abstract: The olive fruit, a symbol of Mediterranean diets, is a rich source of antioxidants and oleic acid (55–83%). Olive genetic resources, including cultivated olives (cultivars), wild olives as well as related subspecies, are distributed widely across the Mediterranean region and other countries. Certain cultivars have a high commercial demand and economical value due to the differentiating organoleptic characteristics. This might result in economically motivated fraudulent practices and adulteration. Hence, tools to ensure the authenticity of constituent olive cultivars are crucial, and this can be achieved accurately through DNA-based methods. The present review outlines the applications of microsatellite markers, one of the most extensively used types of molecular markers in olive species, particularly referring to the use of these DNA-based markers in cataloging the vast olive germplasm, leading to identification and authentication of the cultivars. Emphasis has been given on the need to adopt a uniform platform where global molecular information pertaining to the details of available markers, cultivar-specific genotyping profiles (their synonyms or homonyms) and the comparative profiles of oil and reference leaf samples is accessible to researchers. The challenges of working with microsatellite markers and efforts underway, mainly advancements in genotyping methods which can be effectively incorporated in olive oil varietal testing, are also provided. Such efforts will pave the way for the development of more robust microsatellite marker-based olive agri-food authentication platforms.

Keywords: authentication; cultivar identification; *Olea europaea*; olive oil; simple sequence repeats; traceability; table olive

Citation: Yadav, S.; Carvalho, J.; Trujillo, I.; Prado, M. Microsatellite Markers in Olives (*Olea europaea* L.): Utility in the Cataloging of Germplasm, Food Authenticity and Traceability Studies. *Foods* **2021**, *10*, 1907. https://doi.org/10.3390/foods10081907

Academic Editors: Margit Cichna-Markl and Isabel Mafra

Received: 29 June 2021
Accepted: 28 July 2021
Published: 17 August 2021

Publisher's Note: MDPI stays neutral with regard to jurisdictional claims in published maps and institutional affiliations.

1. Introduction

The olive tree has been cultivated for approximately 6000 years in Mediterranean countries, where 95% of olive germplasm is located. Its habitat is determined by the Mediterranean climate, and it stands as the most highly cultivated fruit crop among temperate crops in the world. According to data published by International Olive Council (IOC) (www.international.oliveoil.org (accessed on 10 February 2021)), in the last 25 years, olive oil production and consumption has increased by 1 million tons. The olive crop is mainly located in the Mediterranean Basin (the leading producers being Spain, Italy and Greece). Moreover, the olive is also a crop under increasing cultivation in non-traditional countries such as Argentina, Australia, Chile, China, Japan and the United States.

Both olive oil and fruits have been found to be a rich source of antioxidants and various other secondary metabolites (phenolics, carotenoids, tocopherols, anthocyanins and oleosides). Olive oil in particular has an unique lipid fatty acid composition and health benefits such as defense against chronic degenerative diseases, and reduced cardiovascular risks are attributed to the consumption of olive oils and table olives [1–3]. The increased demand of nutritionally superior olive oil such as extra virgin olive oil (EVOO) and virgin olive oil (VOO) and table olives has also led to increased adulteration of premium quality oils and fruits. Hence, regulations and certifications such as protected designation of origin (PDO) and protected geographical indication (PGI) (EC Regulation no. 510/2006) have been laid out to check product authenticity and traceability.

The exchange of germplasm in ancient times and increased commerce among olive growing nations has established complex genetic relationships among different olive gene pools [4]. The cultivation of cultivars in new climatic conditions and the adoption of local names for new introduced material have led to confusion in the denominations of varieties [5,6]. More than 1200 cultivars of olive spread across the Mediterranean region, with around 600 olive cultivars under cultivation in Italy itself, have been described in the olive germplasm database [7]. The characterization and recognition of many other cultivars and ancient and wild forms is still an ongoing process, and several studies have been undertaken in this direction using morphological as well as molecular tools [8]. Germplasm banks have been established to ensure ex situ conservation of olive genetic resources, and emphasis is being given to the use of microsatellite markers or simple sequence repeats (SSRs) as tools to better inventory these valuable repositories. Molecular characterization or genotypic profiling of available germplasm will not only provide unique identification keys but also help in the development of molecular authentication platforms, wherein these cultivars, wild forms or related species can easily and accurately be identified. Ever since being developed, SSRs or microsatellites are among the most frequently used molecular markers in olives. This is also evident from the large number of publications available pertaining to the use of SSRs in olive research. The characteristic features such as the multiallelic nature, wide genomic distribution, codominant inheritance, locus specificity, high mutation rates, utility as functional markers (present in transcribed regions), cross-transferability, amenability to automation, easy in silico mining and primer design have established SSRs as the markers of choice in most species [9,10]. Detailed reviews are already available, explaining the development, uses and advantages of SSR markers in plants [11–13], and these can be consulted for more elaborate information.

In olives, microsatellite markers have been used in various applications such as cultivar identification, characterization of autochthonous olives (ancient olive trees and oleasters), the management of olive germplasm banks, phylogenetics, diversity analysis and mapping. Moreover, these have also been widely utilized in the authentication and traceability of cultivars in olive agri-food products. Most of the studies involved the use of nuclear genomic SSRs, and recently expressed sequence tag (EST)-based SSRs or the EST-SSRs are also being exploited in several olive genetic studies. Olive SSRs have also been used in combination with other marker systems such as amplified fragment length polymorphisms (AFLPs), inter simple sequence repeats (ISSRs), single nucleotide polymorphisms (SNPs) and random amplified polymorphic DNA (RAPD) in various studies related to mapping, cultivar discrimination and genetic relationships [14–24]. Microsatellites, being so extensively applied in olive germplasm cataloging, authentication and traceability studies, need to be reviewed in detail, and therefore, the present review aims to elaborate on the development of SSRs in olives and specifically targets their use in olive cultivar identification, cataloging of germplasm and the traceability of oils and table olives. Information generated through such studies has been thoroughly compiled and presented in this review through extensive literature searching, mainly using Google (www.google.com accessed on 26 July 2021) and Google Scholar (scholar.google.com, accessed on 26 July 2021). Research articles and reviews covering a wide timeframe and encompassing information about olive distribution, the development of SSR markers and databases on olives and their vast applications were referred. Since the aim of the review is to mainly highlight the utility of SSR markers in the characterization of germplasm banks and local, wild and centennial olive germplasm, thereby leading to proper cultivar identification and cataloging and utilization of such information in olive agri-food authentication and traceability, articles pertaining to these fields were mainly included in this review. The review should be useful to researchers working in the above-mentioned areas. Key factors that affect the applicability and usefulness of microsatellites in olive varietal identification are also emphasized and discussed in the manuscript.

2. The Olive Germplasm

The olive (*Olea europaea* L.) belongs to the family Oleaceae, which comprises around 30 genera and over 600 species. The genus Olea has some 35 species, including both *O. europaea* subsp. *europaea* var. sativa (cultivated olive) and *Olea europaea* subsp. *europaea* var. sylvestris (wild olive or oleasters). In addition, the wild olive includes feral forms which are seedlings of the cultivated olives or the result of hybridizations between the oleasters and cultivars [25,26]. Additionally, five subspecies, namely *laperrinei* (Saharan massifs), *cuspidata* (Afro-Asiatic), *guanchica* (Canary Islands), *maroccana* (Morocco) and *cerasiformis* (Madeira), comprise the *Olea europaea* complex.

The olive was probably domesticated in the Middle East about 6000 years ago [27]. Afterward, commercial shipping spread this crop westward across the Mediterranean Basin, leading to complex genetic relationships among cultivars [4]. The empiric selection of outstanding individuals within wild olives, crosses between the previous selected or introduced cultivars and other local cultivars or wild olives in all growing areas have yielded a huge number of local cultivars. The easy vegetative propagation of the olive cultivars has allowed for maintaining the characteristics by which they were selected, such as greater productivity, fruit size, oil production and environmental adaptation. It is estimated that there are more than 2000 olive varieties worldwide [28]. The denomination of olive cultivars is usually a process synchronous to their diffusion. Initially, olive cultivars were named using generic criteria, like their outstanding morphological traits, utility of production or the locality of origin of the propagated material, or based on other characteristics [8]. Consequently, in olives, the existence of synonymy (different names for the same cultivar) and homonymy (same name for different cultivars) among and within olive-growing countries is very frequent [5–8].

Germplasm banks are facilities that permit us to ensure "ex situ" conservation of genetic resources. Clonally propagated fruit crops such as olives are typically conserved in "live collections", which are suitable selected field plantations where the crop can fulfill its normal biological cycle [29]. Prospecting surveys of olive cultivars in many countries and the exchanges of cultivars between countries have contributed to the high number of conserved accessions in "ex situ" collections. Bartolini et al. [7,30] reviewed for the FAO the accessions conserved in approximately 100 regional and national collections in 54 countries, which include more than 4000 accessions supposedly belonging to 1250 cultivars [31]. Most of these cultivars come from major producer countries like Italy (538 cultivars), Spain (183), France (88) and Greece (52) [32]. Since 1994, the IOC has been promoting a network of banks to preserve the heritage of olive varieties grown in countries around the world. The network presently includes a total of 23 germplasm banks, housing over 1700 varieties andis composed of 3 international banks—Cordoba (Spain), Marrakech (Morocco) and Izmir (Turkey)—and 20 national banks (Albania, Algeria, Argentina, Croatia, Cyprus, Egypt, France, Greece, Iran, Israel, Italy, Jordan, Lebanon, Libya, Montenegro, State of Palestine, Portugal, Slovenia, Tunisia and Uruguay) (https://www.internationaloliveoil.org/the-ioc-network-of-germplasm-banks/, accessed on 5 February 2021). The Olive World Olive Germplasm Bank of Cordoba (Spain) (WOGBC) was established in 1970, and it is one of the largest with more than 1000 accessions from 29 countries [33,34]. The second international bank (WOGBM) was established in 2003 in Marrakech (Morocco) and contains around 560 accessions from 14 countries (mainly from the Mediterranean region) [35]. The third international bank was recently established (2017) in Izmir (Turkey), including 274 accessions [36]. The national olive banks preserve the local as well as important international cultivars.

Despite these efforts, the exploration and conservation of the genetic patrimony of olives is still incomplete. In recent years, numerous initiatives have been promoted to explore, preserve and exploit unknown material, including minority local varieties, centenary trees and wild olive populations (see Section 4.1). It is indeed very clear from the above information that a vast collection of olive cultivars is presently available, but challenges related to correct denominations, geographical origin and proper cataloging of these germplasm still persist, and molecular tools such as SSR markers can be a preferred choice for addressing these aspects, contributing to the proper authentication of agro-food products.

3. Microsatellites in Olives

3.1. Development and Available SSRs

The earliest reports of the development of microsatellites in olives are from the year 2000 by two independent groups. Rallo et al. [37] developed 13 SSR loci (prefixed as IAS-oli) by sequencing 43 clones screened as positive on a GA-enriched olive genomic library of the cultivar "Arbequina". Among these, only five were found to be polymorphic when analyzed for polymorphism in 46 olive cultivars. The occurrence of repeats, other than the enriched "GA" repeats, was found in the form of compound microsatellites and presumed to be common in the olive genome. Sefc et al. [38] screened a size-selected olive genomic library for GA and CA repeats and designed primers (prefixed as ssrOeUA-DCA or DCA) for 28 microsatellite loci. Among these 15 loci, amplified specific products were polymorphic across a set of 47 olive trees from Iberian Peninsula and Italy. In the year 2002, other groups simultaneously reported the genomic library-based development of microsatellites in olives. Carriero et al. [39] screened a $(GA/CT)_n$-enriched genomic library and characterized 20 SSR primer pairs (prefixed as GAPU) in 6 olive cultivars and finally reported 10 polymorphic SSR loci after testing on a set of 20 olive accessions. An average of 5.7 alleles per SSR loci was obtained with these markers. Although enriched for dinucleotide repeats, clones in the library also possessed "CCT" and "TTC" trinucleotide repeat motifs. Cipriani et al. [40] also reported the selection and sequencing of 52 SSRs from (AC/GT) and (AG/CT) repeat-enriched genomic libraries of the olive cultivar "Frantoio". Out of these, a set of 30 SSR primers (prefixed as UDO99) were screened for polymorphism in 13 olive Italian cultivars. GA and CA repeat-enriched libraries were also developed by De La Rosa et al. [41] from the cultivar "Picual" and designed 13 primer pairs (EMO prefixed), out of which only 6 were found to be polymorphic in a set of 23 olive cultivars and were also tested for cross-species transferability.

To further expand the arsenal, the olive cultivar "Arbequina" was used in genomic library preparation and enrichment for GA, GT and ACT repeats by Diaz et al. [42]. However, inserts with the "ACT" repeat motif were not obtained even after the enrichment step. Specific primers (prefixed as IAS-oli) could be designed from 10 of the sequences containing repeats and an additional 14 sequences available from an earlier report. Gil et al. [43] also employed similar techniques of genomic library enrichment, screening and sequencing with the olive cultivar "Lezzo", and they reported 12 polymorphic SSR primers (prefixed as ssrOeIGP) when amplified in a set of 33 olive cultivars. All these genomic SSRs have been extensively used in the characterization of olive cultivars and molecular genetic studies in olives, as reviewed in the sections below. Series DCA-, GAPU- and UDO have been very used; nevertheless, others (e.g., the EMO and IAS-oli series) have been scarcely used. Most of these attempts involved dinucleotide repeat-containing sequences for the primer design, and the GA/CT motif was commonly used. An olive genome is presumed to have a relatively frequent occurrence of compound microsatellite motifs, as found in most of the SSR development studies described above. Multiple amplification products were also reported in some genotyping experiments and probably occurred due to, for example, priming at more than one site, ploidy of the species, the presence of compound microsatellites and genome duplication events [37,40].

EST-SSRs have gained interest in recent years, owing to their easy development through user-friendly bioinformatics tools, higher cross-transferability across species and ability to be used as functional markers in marker-assisted breeding [10]. With the beginning of sequencing projects and advanced sequencing technologies, genomic resources in the form of whole genome sequences and transcriptomes have been made available in public databases for olives. These are a rich source for the in silico development of SSR markers in olives. The availability of different transcriptomes has given researchers the opportunity to screen and design primers for microsatellite repeats present in the coding regions of the genome, thus allowing association of marker variability with phenotypic traits in olives. Data from cDNA libraries sequenced as a part of the OLEAGEN project, an olive genomic project in Spain [44] was used to extract sequences with core hexanucleotide repeats, and a set of eight EST SSR primers were designed (prefixed as OLEAGEN-H) which were successfully tested for genotyping as well as paternity testing in olives and were found to be comparable to dinucleotide-based genomic SSRs reported in earlier studies [45].

Adawy et al. [46] identified 8295 SSR repeat motifs after in silico mining of the EST sequences available in the NCBI database and described 1801 EST SSR primers (prefixed as Oe-ESSR) that could be amplified in different genes. Among the set of ESTs, the highest percentage (77.6%) for mononucleotide repeats and lowest for tetranucleotide repeats (0.29%) were reported, with the AAG/CTT repeat dominating among trinucleotide types and AG/CT dominant in the dinucleotide repeats. Twenty-five primers randomly chosen for amplification in a set of 9 cultivars were able to amplify, and 10 of these were found to be polymorphic. Tissue-specific transcriptomes [47–49] were utilized for the in silico mining of microsatellite repeats in transcripts in [50]. Trinucleotide and longer repeat motifs containing sequences were BLAST aligned to available olive genome data (oleagenome.org), and after screening for locus redundancy, 80 SSR sequences were targeted for primer design. From a prescreening of 5 olive cultivars for amplifiable loci and expected product size, a set of 26 EST SSRs were finalized (prefixed as OLEST). The authors described a set of the 10 best OLEST SSRs after allele sequencing and validation on a larger set of olive cultivars and related species as potential functional markers in olives. EST SSRs (prefixed as OeUP) were also identified in [51] from a transcriptome of developing fruits of the olive variety "Istrska belica" [52]. Dinucleotide repeats appeared to be abundantly present (36%), with "GA" as a common repeat motif and trinucleotides showing a presence of 33% and "GAA" as a common motif. Out of the 110 EST SSRs chosen for primer designing, 46 showed positive amplification and polymorphism when validated on a set of 8 cultivars and analyzed for diversity among 24 olive varieties. A final set of 27 EST SSRs was recommended on the basis of a low null allele frequency and no deviation from the Hardy–Weinberg equilibrium for diversity and population genetics in olives. Dervishi et al. [53] also performed in silico mining of developing fruit transcriptome of the variety "Istrska belica" for tri- and tetranucleotide repeats and reported 12 primers (prefixed SNB and SiBi) out of 35 EST SSRs for olive genetic studies. Gene annotation for sequences carrying microsatellite repeats was also performed, and genes for disease resistance were reported. Similar to earlier reports, the "AAG" motif was found to be most prevalent among the trinucleotide repeats which were found in 0.18% of the sequences. In the case of tetranucleotides, "AAAT" was most frequent, and the number of repeat units in a sequence ranged from 6 to 21 in the case of trinucleotides and 4–14 for the tetranucleotides. SSRs were also found to exist in compound form in a few of the cases.

More recently, genomic SSRs based on trinucleotide repeats (with at least five core repeats) were retrieved from the whole genome sequence information in olives, and SSR primers were developed (prefixed as BFU), covering most of the chromosomes. Twenty-one SSRs were found to be highly polymorphic and effectively discriminated among a panel of 53 accessions of olives [54]. EST SSRs have also been developed by Gómez-Rodríguez et al. [55], where tetra-, penta- and hexa-nucleotide repeats were retrieved from cDNA sequences, and primers were designed (prefixed as Olea). These newly developed markers could successfully discriminate the cultivars present in the core collection of olives available at the Worldwide Olive Germplasm Bank of Cordoba, Spain. Moreover, both the genomic and EST SSRs in olives have shown transferability across oleasters as well as cultivated olives [41,53,56]. Table 1 depicts the key genetic indices as observed while developing different microsatellite resources in olives. These SSRs are a valuable resource and can be utilized in various studies related to germplasm characterization, cataloging, cultivar identification and authetication in food products as discussed in the sections below.

Table 1. Key genetic indices as reported for SSR markers developed in olives using enriched genomic libraries and EST sequences.

Reference	Naming of SSR Loci (Prefixes)	Type of SSR	No. of Polymorphic SSRs Reported	No. of Cultivars Used in Characterization of SSRs	N_a	H_o	H_e
[37]	IAS-oli	Genomic	05	46	3–9	-	0.460–0.710
[38]	DCA	Genomic	15	47	4–15	0.283–0.979	0.357–0.859
[41]	EMO	Genomic	13	23	6–9	0.391–0.913	0.620–0.811
[39]	GAPU	Genomic	10	20	3–9	-	-
[40]	UDO99	Genomic	28	13	1–5	-	0.000–0.770
[43]	ssrOeIGP	Genomic	12	33	2–14	0.188–0.813	0.417–0.895
[42]	IAS-oli	Genomic	12	51	1–13		
[45]	OLEAGEN-H	EST-SSR	08	15	2–7	0.380–1.000	0.490–0.850
[46]	Oe-ESSR	EST-SSR	1801; 25 of these used	09	-	-	-
[51]	OeUP	EST-SSR	46	24	2–8	0.042–1.000	0.042–0.869
[50]	OLEST	EST-SSR	26	32	2–10	0.219–0.813	0.195–0.839
[53]	SNB, SiBi	EST-SSR	-	-	-	0.357–0.932	0.294–0.790
[54]	BFU	EST-SSR	21	53	3–10	0.140–0.910	0.520–0.810
[55]	Olea	EST-SSR	08	36	4–7	0.350–0.710	0.540–0.750

(N_a) Average number of alleles per locus. (H_o) Observed heterozygosity. (H_e) Expected heterozygosity or gene diversity.

3.2. SSR Protocols for Cultivar Genotyping

Allele size discrepancies found while comparing the same set of SSRs across different samples and laboratories make the task of fingerprinting cultivars quite challenging, and thus, the utility of SSRs in cultivar authentication or in food traceability is also hampered. SSR protocols for the genotyping of olive cultivars and consensus sets of microsatellites have been proposed by various research groups for uniform data analysis and comparison. With an aim to standardize a set of SSR markers for olive genotyping, Doveri et al. [57] found that among 17 SSR markers, 6 (DCA3, DCA8, DCA11, DCA13, DCA14 and DCA15) showed maximum concordance between data points scored from all partner laboratories. Emphasis was made toward harmonization of SSR profiles for better resolution of the alleles. Baldoni et al. [58] performed an exhaustive exercise across four independent laboratories and proposed a consensus set of 11 SSRs (UDO-043, DCA9, GAPU103A, DCA18, DCA16, GAPU101, DCA3, GAPU71B, DCA5, DCA14 and EMO90) for olive genetic studies. SSRs were ranked according to the peak intensity, stuttering, null alleles, number of amplified loci and allelic error rate, which were calculated to determine the concordance of the SSRs being tested. Allelic ladders were constructed using a set of genotypes which carried true-sized alleles as confirmed by sequencing to identify the corresponding alleles between labs and to reduce the chance of mistyping alleles. The generation of allelic ladders using known profiled cultivars will allow univocal allele binning and assigning correct sizes to the new alleles. The SSRs present in the consensus list have been used in several genotyping and diversity studies of olives since then.

A protocol was also proposed by Trujillo et al. [8] using a nested set of 5, 10 and 17 SSR markers that allowed for quick characterization, authentication and identification of olive cultivars present in the WOGB in Cordoba, Spain and which could be used for management of germplasm resources in any olive gene banks. A molecular key for the identification of cultivars was also proposed by Aksehirli-Pakyurek et al. [59], where a classification binary tree (CBT) was developed and provided sorting of unknown new material that could be originating from any of the cultivars being analyzed. Hence, well-accepted SSR allelic profiles for specific cultivars are absolutely essential in order to avoid any confusion during molecular genotyping by different laboratories. This will also help in adopting a more uniform and application-worthy traceability and authenticity protocol based on SSRs.

3.3. Genotyping Methods

Over the years, genotyping methods used for SSR analysis have advanced to a great extent. When the aim is to specifically use SSRs for food authenticity and traceability, the genotyping methods being used are of the utmost importance, as any discrepancy in allele identification may lead to wrong cultivar identification and hamper the results. Earlier research mainly involved the use of agarose gel electrophoretic separation of SSR amplification products, and the resolution of alleles with 2–4 bp (base pair) differences in size was quite difficult. Denaturing polyacrylamide gels (4–8%) were also used for fragment separation [39,40,60], as these allow for better resolution compared with agarose gels when small base pair differences are to be identified, but these are more cumbersome to prepare, use toxic chemicals like acrylamide and involve silver staining for visualization of the separated bands. Development of more precise separation matrices in the form of high-resolution agarose have been used in amplicon separation in olive SSR analyses to resolve amplicons that differ in size by as little as 2% [37]. With more and more advancement in amplicon resolution and separation methods, matrices such as polyacrylamide and agarose are becoming obsolete and being replaced with automated capillary electrophoresis techniques and sequencing-based instruments which could achieve more sensitive allele separation and base pair calling. These advanced technologies reduced the separation time; hence, results could be obtained faster, and working with a huge sample size became easier. Moreover, integrated data analysis software, multiplexing, better reproducibility and elimination of staining procedures makes automated sequencers quite advantageous over the conventional methods of genotyping. This becomes very important when SSRs are to be used as a potential tool in olive authentication and traceability. Robust allele separation and detection is very crucial in such cases and thus requires high-throughput techniques. One of the major limitations while using microsatellites is the allele calling differences that may emerge due to polymerase slippage, DNA quantity or quality and the use of different instruments and reagents by different laboratories. Additionally, variations in results may arise due to post-PCR handling of samples in the case of gel-based platforms. These factors may cause problems in accurate determination of cultivar-specific SSR profiles and hence need to be taken into consideration while comparing genotyping results across laboratories and identifying correct cultivars [58].

High-resolution melting (HRM) analysis, an advanced method that compares the melting curve profiles of double-stranded DNA products and detects polymorphism, has recently been used as an alternative to gel-based polymorphism detection methods in olives and other species [61,62]. HRM shows greater resolving power compared with conventional melting curves, which are based on only the value of the melting temperature (Tm) and may not give better discrimination between different genotypes [63,64]. More nucleotide variations associated with the flanking regions of repeat sequences, such as single-nucleotide polymorphisms (SNPs), can be detected through this method and hence expand the applicability and potential of SSR marker systems. Refinements in the method are still going on so as to overcome challenges like specificity of the PCR, multilocus markers, and a high number of alleles [64]. Thus, continuous advancements are being made toward achieving more effective and accurate genotyping of the samples. This would help adopt a uniform method for olive genotyping, and hence information could be easily communicated and transferred between laboratories.

3.4. SSR Databases

It is indeed very clear that large-scale SSR genotyping projects have generated a vast amount of molecular data for different cultivars across the olive-growing regions of world. Nevertheless, this remains unutilized and inaccessible most of the time. A database is a necessary tool to correctly catalog any germplasm bank and optimize its management. Moreover, the database is the keystone to guarantee that a commercial edible product (oils or table olives) matches the cultivar specified on the label. For these reasons, the data from such independent studies need to be available on uniform platforms for easy access and use of the information. Attempts have been made to develop informative databases for olive trees, such as the Istrian olive database (http://old.iptpo.hr/iod, accessed on 20 January 2021), formed by assembling information about the morphological and molecular profiles of Istrian olive cultivars. This was an outcome of the DNA fingerprinting study of olive varieties of Istria conducted by Poljuha et al. [65]. The OLEA database (http://www.oleadb.it/, accessed on 20 January 2021) was yet another olive molecular database established in 2007 by researchers in Italy, and it comprised SSR marker data of a broad set of olive cultivars. Users could search for cultivars corresponding to a particular data type and variety identity and also look for cultivar information across different olive collection facilities.

With the generation of more and more EST information in public databases and the development of EST SSRs in olives, genetic studies have also been conducted using these SSRs. ReprOlive (http://reprolive.eez.csic.es, accessed on 20 January 2021) is a freely available database that gives access to the reproductive transcriptomes of olive trees, where information can be retrieved about tentative transcripts containing SSR units and suitable primers can be designed [66]. Another comprehensive olive database, the Olive Genetics Diversity Database (OGDD) pertaining to SSR molecular data, was generated by Ben Ayed et al. [67], and it is reported to contain morphological, chemical as well as molecular genetic (SSR) information about several olive varieties and oils. However, it is emphasized that the regular addition of newly generated information, updated software and easy access of these databases are required so that users can access the webpages and information smoothly. Public databases would make comparative studies much easier and more useful in the identification and authentication of cultivars and their products, and the information could be used by breeders, population geneticists and researchers across laboratories.

4. Applications of SSRs: Cataloging of Olive Germplasm, Food Authenticity and Traceability Studies

4.1. Cataloging Olive Germplasm

The varietal cataloging process implies (1) characterization or description of the cultivars at different levels (e.g., morphological, molecular or agronomical); (2) identification, a process that allows us to classify or differentiate one cultivar from the rest; (3) authentication, a process that guarantees that a cultivar corresponds to the original cultivar from its natural area of cultivation or origin; and (4) assigning the correct name to the cultivar once identified and authenticated and defining its synonyms and homonyms [68]. Therefore, the cataloging of any bank should be an essential requisite before using plant material for conservation, propagation and breeding purposes. Varietal information is also a key identifier in quality control for high-value virgin olive oils and table olives in the food industry.

In species like the olive, this task becomes particularly challenging. There are several factors that contribute to this, such as the vast number of olive cultivars, the use of generic criteria to name them and the misunderstanding around basic concepts that has led to a confusing scenario. In addition, the heterogeneity of criteria and methodologies applied for cataloging has hampered the completion of varietal cataloging in most traditional olive-growing countries. In this regard, the integration of molecular markers, particularly the microsatellites with the pomological scheme defined by Barranco et al. [69], has allowed for important advancements in the cataloging of olive germplasm [8]. In this work, the challenges of the incorporation of SSR markers both for the cataloging of germplasm and traceability studies in olive oil and table olives are highlighted. Figure 1 summarizes the microsatellites available in olives, the genotyping process and their applications in the cataloging and management of olive germplasm.

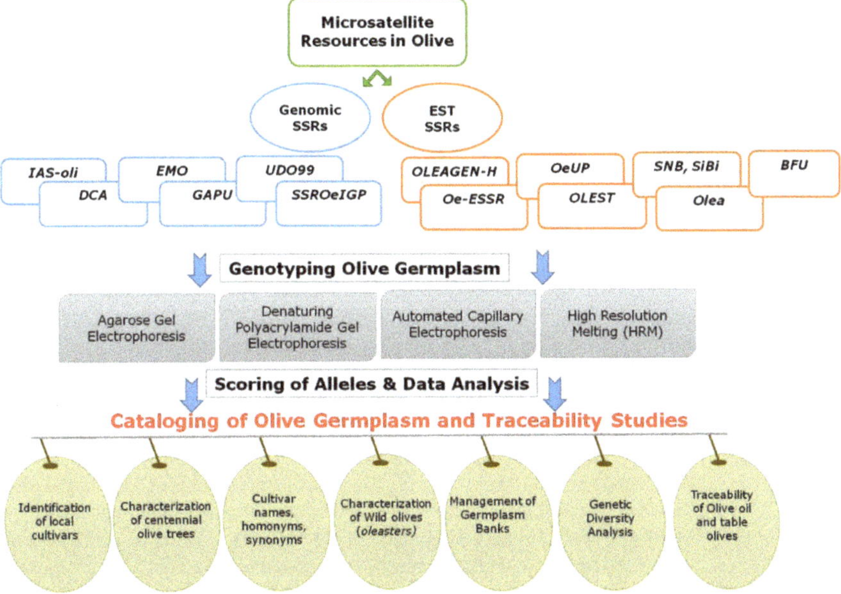

Figure 1. Microsatellite resources and applications in olive germplasm cataloging, authenticity and traceability.

4.1.1. Cataloging of Germplasm Banks

Collections in a germplasm bank are a proper source of confirming the true identity of the cultivar in question. Hence, proper identification and cataloging of plant material becomes a prerequisite for efficient management of germplasm banks. The cataloging (characterization, identification, authentication and naming of the cultivar) of the accessions of any olive germplasm bank should be compulsory before distribution of any plant material from that bank. Only the diffusion of true-to-type cultivars will avoid worldwide confusion between the denominations and cultivars existing in almost any world germplasm collection [30]. Aside from that, the SSR profiles of correctly identified and authenticated material can be used as a reference when dealing with the authenticity and traceability of olive products. In this direction, Trujillo et al. [8] exhaustively characterized, identified and authenticated the 499 accessions (824 trees in total) present in the WOGBC in Córdoba, Spain, representing samples from 21 countries using both phenotypic characters and molecular profiles generated by 33 available SSR markers. Several cases of synonyms and homonyms were detected and rectified, along with the identification of unique genotypes. The WOGBC has now become one of the most characterized olive germplasm banks and has paved way for other worldwide collections to also be well cataloged.

Trujillo et al. [68] also proposed and presented a guide in the international seminar "The IOC Network of Germplasm Banks and The True Healthy Olive Cultivars Project" held in Cordoba (Spain) in 2019. In this guide, the successive necessary steps and methodologies for accomplishing these goals are described, from the arrival of the vegetal material to the bank to the establishment of the plants in the field collection once identified, authenticated and free of pathogens. The molecular protocol is based on a set of 17 previously selected SSRs. All of them are robust and extremely polymorphic, with almost a limitless capability to catalog olive cultivars [8]. Aside from that, in most of the IOC Network collections, there is a considerable amount of information generated with SSR markers. These exhaustive studies establish the potential of microsatellites as robust markers for the characterization and identification of cultivars in rich olive germplasm. Better management of ex situ collections would in turn facilitate the easy exchange of germplasm material even at international levels, eliminate any mislabeling or misinterpretation of cultivars and ensure a reliable supply of cultivars to research labs, breeders and markets. These are indeed very useful resources in developing olive authentication and traceability studies, where the genotypic profile of any cultivar in question can essentially be matched with its true representative maintained in these worldwide collections.

4.1.2. Local Cultivars and Centennial Trees

In the last 25 years, important socioeconomic changes in many Mediterranean countries have driven significant technological improvements in olive cultivation. These changes are increasing the risk of genetic erosion of olive germplasm because local traditional cultivars are being replaced by a few cultivars that are suitable for new mechanically harvested plantations. Therefore, the identification and conservation of traditional olive cultivars are currently high-priority tasks that are needed to ensure the sustainable use of those cultivars in the future [70]. Microsatellite markers have been proven to be immensely useful in describing olive cultivars cultivated locally in certain regions [71–76]. Genotypic data about these local cultivars are useful information when authenticating commercial products coming out of these areas and certifying the origins of cultivars.

In Montenegro, when characterized using 10 SSR markers from the consensus set described by Baldoni et al. [58], the genotypic profile of the oldest olive tree, "Stara Maslina", was found to be quite distinct from other ancient trees and main varieties, including the most diffused "Zutica Bar" variety. In addition, all locally grown and ancient germplasm of Montenegro were grouped together into a separate cluster when analyzed with other foreign cultivars [77]. Similarly, the autochthonous olive germplasm in Crete, Greece, represented by three cultivars ("Koroneiki", "Mastoidis" and "Throubolia") were characterized, along with two cultivars from Turkey and some representative wild genotypes from Crete, using seven informative SSR markers (from the DCA, UDO99 and IAS-oli series). The autochthonous cultivars were grouped into separate clusters showing their distinctness, and the cultivar "Throubolia" was found to be close to Turkish cultivars, indicating possible exchange or movement of the germplasm in the past [59]. Such studies supported by SSR-based genotypic information highlight the uniqueness of local germplasm and point toward more targeted genetic evaluation and conservation of such germplasm in olive-growing regions. Additionally, the information thus generated can also be utilized in developing SSR-based cultivar identification keys to be used in any future authentication of agri-food products based on such cultivars.

Since antiquity, olives have been grown and cultivated in the Mediterranean region of the world, and to date, many such centennial olive trees can be found growing in different regions. Microsatellites have been the molecular marker of choice for the characterization and identification of monumental or centennial olives from different olive-growing regions and proved helpful in generating valuable information with respect to the genotypic identities of trees. These studies supported the hypothesis that ancient olive trees might be unknown traditional cultivars that remained uncharacterized. Rotondi et al. [78] reported that most of the 206 ancient olive trees growing across the Emilia-Romagna region in Italy belonged to 10 cultivars that were already characterized, and the origins of the remaining genotypes remained unknown. In yet another study, 4526 ancient olive trees were surveyed in the "Taula del Sénia" (M-TdS) area (northeast Iberian Peninsula), and a subset of 293 trees was molecularly characterized using eight SSR markers, which revealed 43 different genotypic profiles, with 98% of the trees belonging to the local cultivar "Farga" [79]. Erre et al. [80] genotyped 21 wild and 57 cultivated olives in Sardinia using 13 SSR markers, where novel genotypes were identified and cluster analysis grouped the trees into distinct "wild" and "local" gene pools. Hence, valuable information could be deciphered with reference to the cultivar identity and existence of these trees using molecular as well as phenotypic tools. This would also be helpful in devising strategies for the cataloging, conservation and protection of such a rich ancient resource. The molecular information in the form of SSR profiles generated through such genetic studies can be very useful in cases where local cultivars are blended with commercial ones or when any high-value local cultivars are being adulterated.

4.1.3. Characterization of Wild Olive Populations

Wild germplasm in olives, also known as oleasters, can be exploited in breeding and genetic improvement programs as a rich source of variation in the development of varieties with improved traits, such as biotic and abiotic resistance and increased growth and yields. Microsatellite marker-based diversity studies and the estimation of genetic relationships within wild olive populations and between cultivated and wild forms were reported [80–90]. This would give better insights into the history of the domestication of olives, the regional distribution of genetic diversity and any gene flow between oleasters, feral forms and cultivated types. This topic has been recently reviewed by Belaj et al. [91]. Therefore, to obtain more detailed information, the reader is referred to this review.

In summary, Table 2 provides a list of studies highlighting the various applications of microsatellites in the characterization of olive genetic resources. These studies actually provide useful information about the various microsatellite markers used, and the different genetic indices thus generated can help in the selection of the most appropriate set of SSRs for any future work related to characterization or cultivar authentication. High genetic variability can be utilized in selecting superior genotypes and cultivars for future breeding programs and cultivation. Broad genetic diversity in olive germplasm is also reflected by high heterozygosity levels (both expected and observed heterozygosity) obtained through SSR analysis. For the most part, the expected heterozygosity (H_e) values were lower than the observed heterozygosity (H_o) in olives, as represented in Table 2. Another informative genetic index for SSR usefulness is the polymorphic information content value (PIC value), which in the case of olive SSRs was >0.5 in most of the cases for different SSR loci and reported to be as high as 0.95 by Omrani-Sabbaghi et al. [92].

4.2. Agri-Food Traceability: Olive Oil and Table Olives

Two of the essential components of Mediterranean cuisine, table olives and olive oil, are among the most valuable agri-food products, especially in European markets. Their rich nutritional value and antioxidant properties have also attracted customers from non-olive oil producing areas such as the U.S. and Asia. Increasing demands and market value have tempted certain groups toward fraud and adulteration of high-quality extra virgin olive oils as well as table olives, with the mixing of cheaper low-quality oils such as other vegetable oils [105] and mislabeling of products produced from high-value cultivars or olive-growing regions being among the identified adulterations [106]. To prevent such fraudulent practices, the European Union (EU) has enacted regulations and introduced certifications (European Council Regulation EEC/2081/1992) in the form of "protected designation of origin" (PDO) and "protected geographical indication" (PGI) and launched a consortium-led project called "OLIV-TRACK" to work on olive oil traceability. Additionally, recent projects such as the OLEUM project (http://www.oleumproject.eu/, accessed 20 June 2021) and the Food Integrity Project (https://secure.fera.defra.gov.uk/foodintegrity/index.cfm?sectionid=21, accessed 20 June 2021) have also presented strategies to tackle olive oil fraud. Apart from the geographical origin of the cultivar and processing methods, the cultivar genotype is one of the key determinants in defining these designations. Therefore, methods are needed that can ascertain the authenticity of cultivars present in a particular batch of olive oil. The concept of cultivar authentication has primarily been used in the context of modern food technology to guarantee that the commercial edible product matches the cultivar specified on the label [107–109]. The authenticity of olive oil and table olives has been assessed through conventional methods, including chemical analyses where the presence of the key metabolites responsible for a peculiar flavor and taste is examined. These mainly include the HPLC-based identification of distinct metabolites [110,111]. Several other techniques such as lipid profiling, stable isotope analysis and chromatography-based metabolite analysis have also been used in olive product authentication and traceability [112,113]. Although these methods allow the identification of cultivars and their origins, complex multivariate analyses and statistical procedures are generally needed, which makes these challenging tasks. Additionally, many of these methods are affected by the environment and physiological conditions during the growth of plants, and hence variations in compositions may be seen. Therefore, DNA-based detection methods have gained interest in recent years, as DNA is unaffected by environmental conditions, and thus more specific, accurate and sensitive results could be obtained regarding the origin and identity of a cultivar. Various DNA-based molecular markers have been used in the authentication of olive trees and oil [114,115]. As already detailed in this review, SSRs possess a high power of discrimination and are among the most widely employed molecular marker systems in olives. Difference in SSR profiles between olive oil-producing cultivars can be used to identify their presence in monovarietal oils as well as mixtures of olive oils. The isolation of DNA in adequate amounts and quality from difficult matrices like olive oil is a challenging task, and the success and reproducibility of PCR amplification and marker analysis largely depends on this. Over the last decade, different isolation protocols and kits have been tested and modified for better DNA extraction from fruits and oils, and these studies highlight the importance of DNA quality and its impact on molecular marker-based tests [116–120]. Recently, Piarulli et al. [121] compared four DNA isolation methods referenced in the literature and came up with a modified method based on the work of Consolandi et al. [122] for the extraction of DNA from extra virgin olive oil in a much smaller time frame (4 h as compared with the 30 h reported) and involving low-cost options. A washable and reusable miniaturized device has been developed as well and tested for highly efficient DNA purification from olive oil, providing an increased surface-area-to-volume ratio when compared with other approaches, allowing highly efficient DNA purification and concentration from samples with minute DNA contents [123]. Molecular markers that amplify shorter fragments are supposed to work efficiently with low-quality or fragmented DNA isolated from oil, and SSRs and SNPs are the favored choice in such cases. Here, key achievements in the field of olive oil and table olive traceability using microsatellite or SSR markers are reviewed and summarized in Table 3 with details of the SSR markers and sample types used.

Breton et al. [116] used magnetic beads for DNA purification and amplified SSR alleles from leaves as well oil DNA. The SSR patterns were verified in virgin oil samples of known origins, either in separate cultivars or in mixtures, as well as in commercial virgin oil samples available from markets. Virgin olive oil originating from 10 different olive cultivars were also identified by Pasqualone et al. [124], and a set of three primers (DCA4, DCA17 and GAPU89) was used to describe an identification key for olive cultivars and oil traceability. Testolin and Lain [117] reported DNA extraction from olive oil, comparing different protocols and commercial kits and utilizing conventional and nested SSR-PCR to identify specific cultivar DNA in oil. Similarly, Muzzalupo et al. [118] performed SSR-based authentication of virgin olive oil from "Ogliarola salentina" and Pasqualone et al. [125] identified a PDO-designated extra virgin olive oil (Collina di Brindisi) which contained aminimum of 70% oil from the cultivar "Ogliarola salentina".These studies established the utility of microsatellites in authenticating a cultivar in a mixture of oils as well.

Table 2. List of studies highlighting applications of microsatellites in the characterization of olive genetic resources.

Objective	Cultivars or Accessions	Region	No. of SSRs	H_o	H_e	N_a	PIC	Reference	Key Remarks
Characterization and Identification of Olive Cultivars	19	Slovenia, Italy, France, Spain	14 (DCA-1,3,4,5,7,8,9,10, 11, 13, 14, 15, 16, 17)	0.263–1.000	NA	3–12	NA	[60]	Identification key of 19 olive varieties
Characterization and Identification of Olive Cultivars	87	Iran	16 (DCA 18, 17, 16, 15, 14, 13, 11, 10, GAPU101, 103A, 89, 71B, 72, 90)	NA	NA	NA	0.620–0.950	[92]	Intra-cultivar variation, cultivar denominations and origin investigated using SSRs
Characterization of Autochthonous Olives	44	Croatia	16 (UDO-08, 12, 19, 24, 28, 31, 39, 43, DCA3, 8, 9, 10, 14, 16, EMO2, 3)	0.273–0.932	0.499–0.910	5–20	NA	[93]	SSR-based varietal discrimination achieved
Germplasm Characterization	154	Tuscany	12 (UDO-04, 06, 09, 11, 12, 17, 19, 24, 27, 31, SIU06, 08)	0.278–0.722	0.428–0.855	3–10	NA	[94]	Homonyms and synonyms detected
Database Development and Cultivar Identification	17	Mediterranean Basin	08 (DCA3, 4, 8, 9, 11, 13, 14, 15)	NA	NA	3–12	NA	[57]	Standardization of SSR set for olive cultivar studies
Characterization and Identification of Olive Cultivars	18	Bologna, Italy	17 (DCA3,4,5,7,9,13,14,15,16, 17,18, EMO90, GAPU59,101,103, UDO-24, 43)	0.500–1.000	0.431–0.841	4–10	NA	[95]	Synonyms identified and diversity in germplasm revealed
Characterization of Olive Cultivars	20	Tunisia	10 (GAPU59, 71A, 71B, 103A, UDO-03, 12, 28, 39, DCA9, 18)	0.300–0.950	0.562–0.801	3–6	NA	[96]	Cultivars broadly grouped by their end use and phenotypes
Characterization of Olive Cultivars	38	Southern Marmara region, Turkey	10 (GAPU103A, 101, UDO-06, 07, 09, 11, 12, 14, 15, 35)	NA	NA	2–5	NA	[97]	Cultivar Gemlik revealed as major olive cultivar in the region

Table 2. *Cont.*

Objective	Cultivars or Accessions	Region	No. of SSRs	H_o	H_e	N_a	PIC	Reference	Key Remarks
Characterization of Olive Cultivars	51	Iran	13 (DCA3,9,16,18,11,15, UDO-43, 11, 19, 24, GAPU59, 71B, 101)	0.000–1.000	0.000–0.800	1–8	0.000–0.750	[98]	Synonyms, homonyms and intra-cultivar polymorphisms detected
Characterization and Diversity Analysis	47	Southern Italy	6 (DCA3, 4, 16, 17, UDO-31, GAPU 47)	0.700–0.890	NA	11–17	0.830–0.870	[99]	Genetic distinctness of accessions from Campania region established
Characterization of Olive cultivars	10	Turkey	7 (DCA-4, 9, 11, 16, 17, GAPU-89, UDO-14)	NA	NA	3–6	NA	[100]	Misnamings among cultivars identified
Cultivar Identification Using SSR	53	Australia	7 (9, 3, 16, 18, 5, EMO90, 30)	0.490–0.980	0.480–0.840	7–12	0.467–0.813	[101]	Samples grouped into distinct genotypes
Characterization of Olive Germplasm	561	Marrakech (OWGB collection)	17 (DCA1, 3, 4, 5, 8, 9, 11, 14, 15, 18, UDO-36, GAPU59, 71A, 71B, EMO03, 90,PA(ATT)2*)	0.490–0.928	0.454–0.876	5–32	0.403–0.864	[102]	Construction of two core collections
Cultivar Characterization and Diversity Analysis	26	Algeria	11 (DCA9, 18, GAPU59, 71A, 71B, 101, 103A, UDO-12, 43, 28, 39)	0.135–0.889	0.070–0.510	6–21	NA	[103]	SSR genotyping allowed unambiguous identification of all the cultivars
Cultivar characterization and Diversity Analysis	39	Colombia	10 (DCA3, 5, 9, 16, 17, 18, UDO-43, GAPU101, 103A, EMO90)	0.250–1.000	0.312–0.909	3–15	0.282–0.876	[104]	Synonyms, homonyms identified and 19 genetic profiles discriminated

(N_a) Average number of alleles per locus. (H_o) Observed heterozygosity. (H_e) Expected heterozygosity or gene diversity. (PIC) Polymorphic information content.

Table 3. Applications of microsatellites in olive oil and table olive traceability.

Sample Type	SSR Markers Used	References
Oil samples of different origins	EMO series SSR primers	[116]
VOO (virgin olive oil) from 10 Italian olive cultivars	DCA4, 15,17; GAPU71, 89, 101; UDO03	[124]
Samples of filtered and unfiltered VOO of cv. Carolea	UDO08, 09, 12, 24, 39, 043 and respective shortened internal primers	[117]
VOO of cv. Ogliarola salentina cultivar	GAPU59, 71A 103A and UDO01, 03, 39	[118]
Samples of Collina di Brindisi PDO oil (four unfiltered and two filtered oils); samples of constituent cultivars used for preparation of the PDO mix	UDO09, 19, 25, 35, 044 and GAPU89, 101	[125]
11 monovarietal olive oil samples from Portuguese cultivars; 12 commercial olive oils	DCA1, 3, 5, and 9 along with 11 RAPD and 08 ISSR primers	[126]
Oil from 2 Tunisian olive cultivars: *Chemlali* and *Chetoui*	DCA1, 3; GAPU59, 71A, 71B and UDO12	[119]
Monovarietal oils from 7 Italian cultivars (Coratina, Picholine, Toscanina, Cima di Melfi, Frantoio, Leccino and Cellina di Nardo)	DCA3, 4, 7, 14, 15, 18; GAPU103; EMO90; EMOL and UDO43	[127]
Monovarietal extra virgin olive oil (EVOO) from cultivars Pisciottana, Frantoio and Leccino and their 1:1 mixtures	DCA 3, 4, 16, 17; UDO31 and GAPU 47	[99]
21 monovarietal olive oils from 16 cultivars	EMO30, 90; DCA5, 8, 17, 18; GAPU71B, 89; UDO09; and Shortened DCA14 and EMO30	[128]
VOO from 22 cultivars	DCA1, 3, 4; GAPU59, 71A, 71B; UDO12, 09	[129]
03 Italian PDO table olives and 07 highly diffused cultivars of table olives	16 SSR markers of DCA, GAPU, EMO and UDO99 series.	[130]
14 monovarietal and commercial olive oils	09 cpDNA markers	[131]
3 VOO samples (Frantoio, Italian PDO *Terre di Bari* and other)	GAPU59, 71A, 71B, 103A; UDO01, 03, 12, 28, 39 and DCA9, 18	[132]
"Terra di Bari" PDO EVOO; 9 Apulia region cultivars; experimental mixtures of oils	17 SSR markers of DCA, GAPU, EMO and UDO99 series; HRM analysis of DCA18	[133]
Oil from cv. Coratina	SSR-HRM analysis of DCA3, 16, 18; GAPU103A	[134]
10 monovarietal olive oils from different Portuguese PDO regions; 2 commercial EVOO olive oils	HRM analysis of 15 SSR from DCA and UDO99 series	[135]
VOO and crude olive pomace oil of cv. Coratina	HRM analysis DCA4, 9 and 14	[136]

The use of principal component analysis (PCA) was emphasized in a study with 23 olive oil samples of Portuguese origin (11 monovarietal and 12 commercial oils), which were fingerprinted using 4 SSR loci in combination with 2 RAPD and 4 ISSR markers. No correlation among the common denominations was revealed and commercial samples from the same olive oil brand as well as the samples from the PDO olive oil Tras-os-Montes were found to be distributed in different PCA quadrants. The use of a larger set of markers was therefore required in order to cluster the cultivars and identify each denomination. The study found PCA analysis to be useful in the categorization of samples according to the regions of origin [126]. While dealing with oil traceability through genetic markers, the presence of alleles from pollinators also needs to be distinguished in order to infer denominations correctly, as observed by Ben-Ayed et al. [119] and Alba et al. [127], where parental contributions are assessed while comparing the microsatellite profiles generated from the DNA of the leaves and oil of certain cultivars. The utility of microsatellites in the genetic traceability of oil in agro-food chains was also established when DNA isolated from the drupes or leaves samples of three olive oil cultivars, namely "Pisciottana", "Frantoio" and "Leccino", were genotyped using six SSRs, and similar genetic profiles were obtained with their monovarietal oils. A 1:1 DNA mixture from two extra virgin monovarietal oils was also tested and could detect the expected alleles in the mixture [99].

Microsatellites have also been used in traceability analyses for PDO table olives. Three Italian PDO olives could be reliably identified among a set of 10 olive cultivars using 16 SSR primer pairs. A power of discrimination as high as 0.9 was obtained in the microsatellite set used for analysis [130]. A combination of genetic and biochemical tools in olive oil traceability studies can add to the accuracy of the experiments. Correlation between the SSR genetic data from cultivars and chemical and sensory profiles of nine monovarietal oils was observed by Rotondi et al. [137]. However, no correlation was obtained between genetic and pleasant flavor profiles. A bunch of parameters could play a role in the success of a traceability system based on genetic markers like microsatellites. An evaluation of such parameters was conducted by Vietina et al. [128] through the genotyping of 21 monovarietal oils obtained from 16 cultivars using 11 microsatellite markers. Each marker was assessed for its amplification ability over different oil DNA, reproducibility across a set of replicates in an experiment and correspondence of alleles in oil as well as leaf DNA. Significant correlation was found between the amplification ability and DNA yield, indicating the role of the extraction method. SSR marker GAPU89 gave a total correspondence and amplification ability value of 49.32%, and marker DCA5 was found to have the highest reproducibility, being $71.43 \pm 21.82\%$. The high standard deviation values were attributed to variations within the samples caused by DNA extraction. Microsatellites were also successfully used by Ben-Ayed et al. [129] in the authenticity and traceability of virgin olive oils, and they also reported the non-correspondence of SSR profiles between oil and leaf DNA in some cases, thereby further strengthening the importance of distinguishing the pollinator and maternal alleles. Figure 2 summarizes the process and the main factors that may potentially affect the molecular traceability of olive oils and table olives when using SSR markers. As depicted in the figure, during DNA isolation, DNA that is too fragmented and very low yields may not always provide sufficient target templates and hence do not amplify the correct alleles. Similarly, the presence of inhibitor compounds from DNA extracts may lead to poor PCR amplification. Amplificability of the markers is also required to be checked for different SSRs in DNA isolated by different methods. Only those markers which give a consistent result in one or two methods should be used further. For the genotyping methods, the resolution of alleles needs to be highly precise for using SSRs in traceability and authenticity testing. Methods like capillary electrophoresis and high-resolution melting have proven to be useful. The correspondence of alleles is yet another important factor, where any microsatellite that generate similar profiles in a target oil and corresponding leaf sample of the cultivar in question can be used as a traceability marker. Ideally, the allelic pattern should be similar, but knowledge of the pollinating behavior of the cultivar is beneficial for result interpretation. As for reproducibility, an ideal

SSR used for traceability should be highly reproducible irrespective of the laboratories, instruments and reagents used.

Factors Affecting SSR Based Olive Oil Traceability System

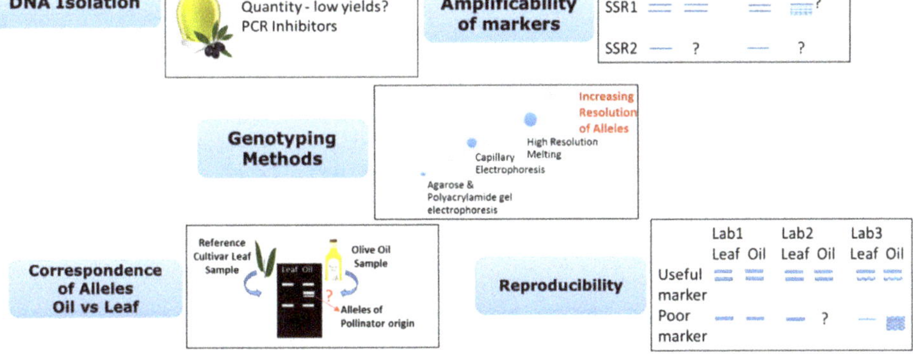

Figure 2. Process and main factors influencing the applicability of a microsatellite marker in the authentication and traceability of olive oils and table olives.

Concerns with respect to the presence of traces of pollinator DNA in extractions made out of oil matrices leading to differences in the allelic profiling of oil and leaf samples also attracted researchers toward the applicability of plastid-based markers. However, chloroplast DNA (cpDNA) among cultivars has shown low levels of variation, which has limited its use in authenticity testing or traceability analyses.

Pérez-Jiménez et al. [131] utilized nine cpDNA loci that consisted of microsatellites and small insertion–deletions (indels) to identify the olive cultivar in leaves and corresponding oil DNA. Six haplotypes could be fingerprinted, and a rare haplotype was identified in genotypes producing regionally high-valued commercial oil. The available olive plastid genome can therefore be analyzed for the presence of more such microsatellite regions. In order to overcome the challenges of DNA isolation from oil matrices, Muzzalupo et al. [132] reported a direct DNA amplification method which avoided the routine extraction step and instead used KAPA3G plant DNA polymerase (an engineered DNA polymerase which could tolerate plant PCR inhibitors) for SSR amplification of membrane-filtered DNA molecules. DNA isolated from this method was used to check the traceability of three distinct types of virgin olive oil. The diagnostics power of microsatellite markers was further proven in the analysis of processed olives by Crawford et al. [138], where a panel of 5 SSRs was selected out of the 15 tested to authenticate California-style olive cultivars, widely marketed as packed forms. Based on the differences in allele combinations generated through these markers, any two samples could be differentiated. While comparing the genotyping method based on SSR alongside fatty acid analysis, phenolic content and nuclear magnetic resonance (NMR) analysis, Crawford et al. [139] found NMR to be able to discriminate all four tested cultivars in their processed forms. However, the five SSR markers could still detect genetic similarity between Sevillano and Gordal cultivars and indicated possible synonymy between the two.

More recently, techniques like high-resolution melting (HRM) have been reported to be coupled to SSR genotyping for the identification of target cultivars in commercial olive oil samples. HRM gives an additional advantage of closed-tube analysis post-PCR and is a sensitive and cost-effective method. Montemurro et al. [133] identified the constituent cultivars of PDO, designated "Terra di Bari" extra virgin olive oil, using HRM curve analysis of the SSR marker DCA18, and Gomes et al. [135] also applied this method for varietal identification in monovarietal PDO as well as blended olive oils using three SSRs from the UDO99 series (UDO99-011, UDO99-039 and UDO99-024) and one SSR from the DCA series (ssrOeUA-DCA16). In addition, Pasqualone et al. [134] evaluated the effect of talc addition during olive oil processing on DNA by comparing the SSR-HRM profiles of treated as well as control samples. Similarly, Pasqualone et al. [136] carried out varietal authentication in samples from crude olive pomace oil and corresponding virgin olive oil. Chedid et al. [140] performed both SSR-HRM and SNP-HRM for authentication and trace adulteration in olive oils and found that the discrimination power of SSRs was greater in the case of monovarietal olive oils, while SNPs were the marker of choice when the oils were blended together or adulterated.

Overall, microsatellites present a desirable system for formulating olive oil and table olive traceability studies, and key parameters like DNA extraction efficiency, reproducibility of the SSR profiles, knowledge about the breeding and pollinating behavior of the cultivars in question and correspondence levels between the oil and reference leaf SSR profiles should be focused on in order to utilize the method as a successful detection tool.

5. Concluding Remarks

A vast amount of genetic information about olive populations, wild relatives, local cultivars and germplasm banks around the world is now available to researchers, which can be utilized for developing cultivar breeding programs and better management of global olive genetic resources. However, organizing this valuable information in the form of easy-to-access and routinely updated databases is essential for the smooth transfer and sharing of scientific knowledge to the olive research community and control laboratories for the olive industry. Olives and olive oil have been an essential part of the diets for many populations, especially the Mediterranean region, with a notable presence nowadays in the non-olive growing nations of the world as well. Therefore, genetic characterization of the available unexplored germplasm is an important step for the introduction of new and improved cultivars. There are challenges associated with use of SSRs as tools to identify olive cultivars and obtain reproducible DNA profiles extracted from its oils.

One of the main limitations in implementing a traceability system based on microsatellites or any of the marker systems is the reproducibility of genotypic profiles across different laboratories. There can be variations due to the quality of the DNA extracted and the genotyping method used, and therefore, results need to be carefully interpreted while using the same set of cultivars and markers under different conditions. Additionally, identifying pollinator origin alleles while comparing olive oil and corresponding leaf DNA is crucial for correct result interpretation. A set of reference cultivars and their respective SSR profiles should be defined globally, and this can be used as a set of controls during experiments by all the laboratories working in cultivar identification and traceability of oil and table olives in order to maintain the authenticity of the data. Olive oil and table olive quality and authenticity is a topic of concern nowadays, and continuous efforts are being made to develop traceability tools based on chemical as well as molecular methods. The available literature indicates that microsatellites are a potential marker system with excellent utility in cultivar identification and coupling with high-throughput platforms, like automated sequencers, and high-resolution melting provides much faster and more sensitive and accurate results. As developments are being made in sophisticated techniques of genotyping, the problems associated with microsatellite profiling, such as mis-scoring of alleles or poor resolution of the electrophoresis gels, are being overcome, allowing users to obtain robust and reliable molecular profiles from samples of commercial olive oil and table olives.

With the use of next-generation sequencing (NGS) technologies in olive trees, more and more genomic information is being added and can be used as a rich source for the development of new sets of long core repeats containing microsatellite markers to overcome limitations while using dinucleotide repeat-rich SSRs. The increasing number of available genomic as well as EST SSRs will not only escalate the existing molecular arsenal but also pave the way for their application in the development of functional markers and linkage, as well as association mapping, map-based cloning and marker-assisted selection in the future, in addition to variety identification in high-quality food products such as table olives and olive oil. The use of techniques like HRM has opened new ways of analyzing microsatellites and exploring their potential beyond length polymorphisms. The development and applications of SNP markers in olives have also gained attention in recent years, but SSRs still remain a marker of choice to initiate preliminary genetic studies in a collection of cultivars, especially in resource-limited laboratories.

Author Contributions: S.Y. and M.P. conceptualized the manuscript design and contents; S.Y. and J.C. wrote the manuscript and carried out the detailed literature search and compilation; M.P. and I.T. contributed to meticulous inputs for refinement and editing of the manuscript. All authors have read and agreed to the published version of the manuscript.

Funding: This research was funded by NANOEATERS: Valorization and transfer of NANOtechnologies to EArly adopTERS of the Euroregion Galicia-Norte Portugal (200005902—NANOEATERS—Aceite de oliva), supported by INTERREG V-A España-Portugal (POCTEP) 2014–2020, and by the project Nanotechnology-Based Functional Solutions (NORTE-01-0145-FEDER-000019), supported by the Norte Portugal Regional Operational Programme (NORTE2020) under the PORTUGAL 2020 Partnership Agreement through the European Regional Development Fund.

Institutional Review Board Statement: Not applicable.

Informed Consent Statement: Not applicable.

Acknowledgments: The author (S.Y.) acknowledges the Indian Council of Forestry Research and Education (ICFRE) in Dehradun, India for administrative support to work as a post-doctoral visiting researcher at the Food Quality and Safety Research Group of the International Iberian Nanotechnology Laboratory (INL) in Braga, Portugal.

Conflicts of Interest: The authors declare no conflict of interest.

References

1. Pérez-Jiménez, F.; Ruano, J.; Perez-Martinez, P.; Lopez-Segura, F.; Lopez-Miranda, J. The influence of olive oil on human health: Not a question of fat alone. *Mol. Nutr. Food Res.* **2007**, *51*, 1199–1208. [CrossRef] [PubMed]
2. Cicerale, S.; Lucas, L.; Keast, R. Biological activities of phenolic compounds present in virgin olive oil. *Int. J. Mol. Sci.* **2010**, *11*, 458–479. [CrossRef]
3. Guasch-Ferré, M.; Hu, F.B.; Martínez-González, M.A.; Fitó, M.; Bulló, M.; Estruch, R.; Ros, E.; Corella, D.; Recondo, J.; Gómez-Gracia, E.; et al. Olive oil intake and risk of cardiovascular disease and mortality in the PREDIMED Study. *BMC Med.* **2014**, *12*, 78. [CrossRef]
4. Besnard, G.; Bakkali, A.E.; Haouane, H.; Baali-Cherif, D.; Moukhli, A.; Khadari, B. Population genetics of Mediterranean and Saharan olives: Geographic patterns of differentiation and evidence for early generations of admixture. *Ann. Bot.* **2013**, *112*, 1293–1302. [CrossRef]
5. Barranco, D.; Trujillo, I.; Rallo, P. Are "Oblonga" and "Frantoio" olives the same cultivar? *HortScience* **2000**, *35*, 1323–1325. [CrossRef]
6. Corrado, G.; La Mura, M.; Ambrosino, O.; Pugliano, G.; Varricchio, P.; Rao, R. Relationships of Campanian olive cultivars: Comparative analysis of molecular and phenotypic data. *Genome* **2009**, *52*, 692–700. [CrossRef]
7. Bartolini, G.; Cerreti, S. Olive Germplasm (*Olea europaea* L.) Cultivars, Synonyms, Cultivation Area, Collections, Descriptors. Available online: www.oleadb.it/olivodb.html (accessed on 10 February 2021).
8. Trujillo, I.; Ojeda, M.A.; Urdiroz, N.M.; Potter, D.; Barranco, D.; Rallo, L.; Diez, C.M. Identification of the Worldwide Olive Germplasm Bank of Córdoba (Spain) using SSR and morphological markers. *Tree Genet. Genomes* **2014**, *10*, 141–155. [CrossRef]
9. Gupta, P.K.; Varshney, R.K. The development and use of microsatellite markers for genetic analysis and plant breeding with emphasis on bread wheat. *Euphytica* **2000**, *113*, 163–185. [CrossRef]
10. Varshney, R.K.; Graner, A.; Sorrells, M.E. Genic microsatellite markers in plants: Features and applications. *Trends Biotechnol.* **2005**, *23*, 48–55. [CrossRef]
11. Morgante, M.; Olivieri, A.M. PCR-amplified microsatellites as markers in plant genetics. *Plant J.* **1993**. [CrossRef]
12. Kalia, R.K.; Rai, M.K.; Kalia, S.; Singh, R.; Dhawan, A.K. Microsatellite markers: An overview of the recent progress in plants. *Euphytica* **2011**, *177*, 309–334. [CrossRef]
13. Vieira, M.L.C.; Santini, L.; Diniz, A.L.; Munhoz, C. de F. Microsatellite markers: What they mean and why they are so useful. *Genet. Mol. Biol.* **2016**, *39*, 312–328. [CrossRef]
14. Belaj, A.; Satovic, Z.; Cipriani, G.; Baldoni, L.; Testolin, R.; Rallo, L.; Trujillo, I. Comparative study of the discriminating capacity of RAPD, AFLP and SSR markers and of their effectiveness in establishing genetic relationships in olive. *Theor. Appl. Genet.* **2003**, *107*, 736–744. [CrossRef]

15. De la Rosa, R.; Angiolillo, A.; Guerrero, C.; Pellegrini, M.; Rallo, L.; Besnard, G.; Bervillé, A.; Martin, A.; Baldoni, L. A first linkage map of olive (*Olea europaea* L.) cultivars using RAPD, AFLP, RFLP and SSR markers. *Theor. Appl. Genet.* **2003**, *106*, 1273–1282. [CrossRef]

16. Bandelj, D.; Jakše, J.; Javornik, B. Assessment of genetic variability of olive varieties by microsatellite and AFLP markers. *Euphytica* **2004**, *136*, 93–102. [CrossRef]

17. Wu, S.; Collins, G.; Sedgley, M. A molecular linkage map of olive (*Olea europaea* L.) based on RAPD, microsatellite, and SCAR markers. *Genome* **2004**, *47*, 26–35. [CrossRef] [PubMed]

18. Montemurro, C.; Simeone, R.; Pasqualone, A.; Ferrara, E.; Blanco, A. Genetic relationships and cultivar identification among 112 olive accessions using AFLP and SSR markers. *J. Hortic. Sci. Biotechnol.* **2005**, *80*, 105–110. [CrossRef]

19. Ganino, T.; Beghè, D.; Valenti, S.; Nisi, R.; Fabbri, A. RAPD and SSR markers for characterization and identification of ancient cultivars of *Olea europaea* L. in the Emilia region, Northern Italy. *Genet. Resour. Crop Evol.* **2007**, *54*, 1531–1540. [CrossRef]

20. Gomes, S.; Martins-Lopes, P.; Lima-Brito, J.; Meirinhos, J.; Lopes, J.; Martins, A.; Guedes-Pinto, H. Evidence for clonal variation in "Verdeal-Transmontana" olive using RAPD, ISSR and SSR markers. *J. Hortic. Sci. Biotechnol.* **2008**, *83*, 395–400. [CrossRef]

21. Khadari, B.; El Aabidine, A.Z.; Grout, C.; Ben Sadok, I.; Doligez, A.; Moutier, N.; Santoni, S.; Costes, E. A Genetic Linkage Map of Olive Based on Amplified Fragment Length Polymorphism, Intersimple Sequence Repeat and Simple Sequence Repeat Markers. *J. Am. Soc. Hortic. Sci.* **2010**, *135*, 548–555. [CrossRef]

22. Linos, A.; Nikoloudakis, N.; Katsiotis, A.; Hagidimitriou, M. Genetic structure of the Greek olive germplasm revealed by RAPD, ISSR and SSR markers. *Sci. Hortic.* **2014**, *175*, 33–43. [CrossRef]

23. Kaya, H.B.; Cetin, O.; Kaya, H.; Sahin, M.; Sefer, F.; Kahraman, A.; Tanyolac, B. SNP Discovery by Illumina-Based Transcriptome Sequencing of the Olive and the Genetic Characterization of Turkish Olive Genotypes Revealed by AFLP, SSR and SNP Markers. *PLoS ONE* **2013**, *8*, e93146. [CrossRef] [PubMed]

24. Kaya, H.B.; Cetin, O.; Kaya, H.S.; Sahin, M.; Sefer, F.; Tanyolac, B. Association Mapping in Turkish Olive Cultivars Revealed Significant Markers Related to Some Important Agronomic Traits. *Biochem. Genet.* **2016**, *54*, 506–533. [CrossRef]

25. Angiolillo, A.; Mencuccini, M.; Baldoni, L. Olive genetic diversity assessed using amplified fragment length polymorphisms. *Theor. Appl. Genet.* **1999**, *98*, 411–421. [CrossRef]

26. Lumaret, R.; Ouazzani, N.; Michaud, H.; Vivier, G.; Deguilloux, M.-F.; Di Giusto, F. Allozyme variation of oleaster populations (wild olive tree) (*Olea europaea* L.) in the Mediterranean Basin. *Heredity* **2004**, *92*, 343–351. [CrossRef]

27. Zohary, D.; Spiegel-Roy, P. Beginnings of Fruit Growing in the Old World. *Science* **1975**, *187*, 319–327. [CrossRef]

28. Lavee, S. ¿Por qué la necesidad de nuevas variedades de olivos? *Fruticultura* **1994**, *62*, 29–37.

29. Van Hintum, T.J.L.; Brown, A.H.D.; Spillane, C.; Hodgkin, T. *Core Collections of Plant Genetic Resources*; IPGRI Technical Bulletin No. 3; International Plant Genetic Resources Institute: Rome, Italy, 2000; ISBN 92-9043-454-6.

30. Bartolini, G.; Prevost, G.; Messeri, C.; Carignani, G. *Olive Germplasm: Cultivars and World-Wide Collections*; FAO: Rome, Italy, 1998.

31. Rallo, L.; Barranco, D.; Díez, C.M.; Rallo, P.; Suárez, M.P.; Trapero, C.; Pliego-Alfaro, F. Strategies for Olive (*Olea europaea* L.) Breeding: Cultivated Genetic Resources and Crossbreeding. In *Advances in Plant Breeding Strategies: Fruits*; Al-Khayri, J.M., Jain, S.M., Johnson, D.V., Eds.; Springer International Publishing: Cham, Switzerland, 2018; Volume 3, pp. 535–600. ISBN 978-3-319-91944-7.

32. Baldoni, L.; Belaj, A. Olive. In *Oil Crops*; Vollmann, J., Rajcan, I., Eds.; Springer: New York, NY, USA, 2009; pp. 397–421. ISBN 978-0-387-77594-4.

33. Belaj, A. Germplasm bank: WOGBC-IFAPA (ESP046). In Proceedings of the International Olive Council (IOC) Network of Germplasm Banks and Phytosanitary Management II Workshop True Healthy Olive Cultivars, Videoconference, 9–11 December 2020.

34. Morello, P. Germplasm Bank WOGBC-UCO Spain. In Proceedings of the International Olive Council (IOC) Network of Germplasm Banks and Phytosanitary Management II Workshop True Healthy Olive Cultivars, Videoconference, 9–11 December 2020.

35. Haouane, H.; El Bakkali, A.; Moukhli, A.; Tollon, C.; Santoni, S.; Oukabli, A.; El Modafar, C.; Khadari, B. Genetic structure and core collection of the World Olive Germplasm Bank of Marrakech: Towards the optimised management and use of Mediterranean olive genetic resources. *Genetica* **2011**, *139*, 1083–1094. [CrossRef] [PubMed]

36. Veral, G.M. Germplasm bank: Izmir World Olive and Collection Turkish National Olive Collection. In Proceedings of the International Olive Council (IOC) Network of Germplasm Banks and Phytosanitary Management II Workshop True Healthy Olive Cultivars, Videoconference, 9–11 December 2020.

37. Rallo, P.; Dorado, G.; Martín, A. Development of simple sequence repeats (SSRs) in olive tree (*Olea europaea* L.). *Theor. Appl. Genet.* **2000**, *101*, 984–989. [CrossRef]

38. Sefc, K.M.; Lopes, M.S.; Mendonça, D.; Dos Santos, M.R.; Da Câmara Machado, M.L.; Da Câmara Machado, A. Identification of microsatellite loci in olive (*Olea europaea*) and their characaterization in Italian and Iberian olive trees. *Mol. Ecol.* **2000**, *9*, 1171–1173. [CrossRef] [PubMed]

39. Carriero, F.; Fontanazza, G.; Cellini, F.; Giorio, G. Identification of simple sequence repeats (SSRs) in olive (*Olea europaea* L.). *TAG Theor. Appl. Genet.* **2002**, *104*, 301–307. [CrossRef]

40. Cipriani, G.; Marrazzo, M.T.; Marconi, R.; Cimato, A.; Testolin, R. Microsatellite markers isolated in olive (*Olea europaea* L.) are suitable for individual fingerprinting and reveal polymorphism within ancient cultivars. *Theor. Appl. Genet.* **2002**, *104*, 223–228. [CrossRef]

41. De La Rosa, R.; James, C.M.; Tobutt, K.R. Isolation and characterization of polymorphic microsatellites in olive (*Olea europaea* L.) and their transferability to other genera in the *Oleaceae*. *Mol. Ecol. Notes* **2002**, *2*, 265–267. [CrossRef]

42. Díaz, A.; De La Rosa, R.; Martín, A.; Rallo, P. Development, characterization and inheritance of new microsatellites in olive (*Olea europaea* L.) and evaluation of their usefulness in cultivar identification and genetic relationship studies. *Tree Genet. Genomes* **2006**, *2*, 165–175. [CrossRef]

43. Gil, F.S.; Busconi, M.; Da Câmara Machado, A.; Fogher, C. Development and characterization of microsatellite loci from *Olea europaea*. *Mol. Ecol. Notes* **2006**, *6*, 1275–1277. [CrossRef]

44. Munoz-Merida, A.; Gonzalez-Plaza, J.J.; Canada, A.; Blanco, A.M.; Garcia-Lopez, M.d.C.; Rodriguez, J.M.; Pedrola, L.; Sicardo, M.D.; Hernandez, M.L.; De la Rosa, R.; et al. De Novo Assembly and Functional Annotation of the Olive (*Olea europaea*) Transcriptome. *DNA Res.* **2013**, *20*, 93–108. [CrossRef]

45. De la Rosa, R.; Belaj, A.; Munoz-Merida, A.; Trelles, O.; Ortiz-Martin, I.; Gonzalez-Plaza, J.J.; Valpuesta, V.; Beuzon, C.R. Development of EST-derived SSR Markers with Long-core Repeat in Olive and Their Use for Paternity Testing. *J. Am. Soc. Hort. Sci.* **2013**, *138*, 290–296. [CrossRef]

46. Adawy, S.S.; Mokhtar, M.M.; Alsamman, A.M.; Sakr, M.M. Development of Annotated EST-SSR Database in Olive (*Olea europaea*). *Int. J. Sci. Res.* **2015**, *14*, 1063–1073.

47. Alagna, F.; D'Agostino, N.; Torchia, L.; Servili, M.; Rao, R.; Pietrella, M.; Giuliano, G.; Chiusano, M.L.; Baldoni, L.; Perrotta, G. Comparative 454 pyrosequencing of transcripts from two olive genotypes during fruit development. *BMC Genom.* **2009**. [CrossRef]

48. Alagna, F.; Cirilli, M.; Galla, G.; Carbone, F.; Daddiego, L.; Facella, P.; Lopez, L.; Colao, C.; Mariotti, R.; Cultrera, N.; et al. Transcript analysis and regulative events during flower development in olive (*Olea europaea* L.). *PLoS ONE* **2016**. [CrossRef]

49. Corrado, G.; Alagna, F.; Rocco, M.; Renzone, G.; Varricchio, P.; Coppola, V.; Coppola, M.; Garonna, A.; Baldoni, L.; Scaloni, A.; et al. Molecular interactions between the olive and the fruit fly *Bactrocera oleae*. *BMC Plant Biol.* **2012**, *12*, 86. [CrossRef] [PubMed]

50. Mariotti, R.; Cultrera, N.G.M.; Mousavi, S.; Baglivo, F.; Rossi, M.; Albertini, E.; Alagna, F.; Carbone, F.; Perrotta, G.; Baldoni, L. Development, evaluation, and validation of new EST-SSR markers in olive (*Olea europaea* L.). *Tree Genet. Genomes* **2016**, *12*, 120. [CrossRef]

51. Arbeiter, A.B.; Hladnik, M.; Jakše, J.; Bandelj, D.; Arbeiter, A.B.; Hladnik, M.; Jakše, J.; Bandelj, D. Identification and validation of novel EST-SSR markers in olives. *Sci. Agric.* **2017**, *74*, 215–225. [CrossRef]

52. Resetic, T.; Stajner, N.; Bandelj, D.; Javornik, B.; Jakse, J. Validation of candidate reference genes in RT-qPCR studies of developing olive fruit and expression analysis of four genes involved in fatty acids metabolism. *Mol. Breed.* **2013**, *32*, 211–222. [CrossRef]

53. Dervishi, A.; Jakše, J.; Ismaili, H.; Javornik, B.; Štajner, N. Comparative assessment of genetic diversity in Albanian olive (*Olea europaea* L.) using SSRs from anonymous and transcribed genomic regions. *Tree Genet. Genomes* **2018**, *14*, 53. [CrossRef]

54. Li, D.; Long, C.; Pang, X.; Ning, D.; Wu, T.; Dong, M.; Han, X.; Guo, H. The newly developed genomic-SSR markers uncover the genetic characteristics and relationships of olive accessions. *PeerJ* **2020**, *8*, e8573. [CrossRef] [PubMed]

55. Gómez-Rodríguez, M.V.; Beuzon, C.; González-Plaza, J.J.; Fernández-Ocaña, A.M. Identification of an olive (*Olea europaea* L.) core collection with a new set of SSR markers. *Genet. Resour. Crop Evol.* **2021**, *68*, 117–133. [CrossRef]

56. Rallo, P.; Tenzer, I.; Gessler, C.; Baldoni, L.; Dorado, G.; Martín, A. Transferability of olive microsatellite loci across the genus Olea. *Theor. Appl. Genet.* **2003**, *107*, 940–946. [CrossRef] [PubMed]

57. Doveri, S.; Sabino Gil, F.; Díaz, A.; Reale, S.; Busconi, M.; da Câmara Machado, A.; Martín, A.; Fogher, C.; Donini, P.; Lee, D. Standardization of a set of microsatellite markers for use in cultivar identification studies in olive (*Olea europaea* L.). *Sci. Hortic.* **2008**, *116*, 367–373. [CrossRef]

58. Baldoni, L.; Cultrera, N.G.; Mariotti, R.; Ricciolini, C.; Arcioni, S.; Vendramin, G.G.; Buonamici, A.; Porceddu, A.; Sarri, V.; Ojeda, M.A.; et al. A consensus list of microsatellite markers for olive genotyping. *Mol. Breed.* **2009**, *24*, 213–231. [CrossRef]

59. Aksehirli-Pakyurek, M.; Koubouris, G.C.; Petrakis, P.V.; Hepaksoy, S.; Metzidakis, I.T.; Yalcinkaya, E.; Doulis, A.G. Cultivated and Wild Olives in Crete, Greece—Genetic Diversity and Relationships with Major Turkish Cultivars Revealed by SSR Markers. *Plant Mol. Biol. Report.* **2017**, *35*, 575–585. [CrossRef]

60. Bandelj, D.; Jakse, J.; Javornik, B. DNA fingerprinting of olive varieties by microsatellite markers. *Food Technol. Biotechnol.* **2002**, *40*, 185–190.

61. Mackay, J.F.; Wright, C.D.; Bonfiglioli, R.G. A new approach to varietal identification in plants by microsatellite high resolution melting analysis: Application to the verification of grapevine and olive cultivars. *Plant Methods* **2008**, *4*. [CrossRef] [PubMed]

62. Xanthopoulou, A.; Ganopoulos, I.; Koubouris, G.; Tsaftaris, A.; Sergendani, C.; Kalivas, A.; Madesis, P. Microsatellite high-resolution melting (SSR-HRM) analysis for genotyping and molecular characterization of an *Olea europaea* germplasm collection. *Plant Genet. Resour.* **2014**, *12*, 273–277. [CrossRef]

63. Ganopoulos, I.; Argiriou, A.; Tsaftaris, A. Microsatellite high resolution melting (SSR-HRM) analysis for authenticity testing of protected designation of origin (PDO) sweet cherry products. *Food Control* **2011**, *22*, 532–541. [CrossRef]

64. Distefano, G.; Caruso, M.; La Malfa, S.; Gentile, A.; Wu, S.B. High Resolution Melting Analysis Is a More Sensitive and Effective Alternative to Gel-Based Platforms in Analysis of SSR—An Example in Citrus. *PLoS ONE* **2012**, *7*. [CrossRef] [PubMed]

65. Poljuha, D.; Sladonja, B.; Šetić, E.; Milotić, A.; Bandelj, D.; Jakše, J.; Javornik, B. DNA fingerprinting of olive varieties in Istria (Croatia) by microsatellite markers. *Sci. Hortic.* **2008**, *115*, 223–230. [CrossRef]

66. Carmona, R.; Zafra, A.; Seoane, P.; Castro, A.J.; Guerrero-Fernández, D.; Castillo-Castillo, T.; Medina-García, A.; Cánovas, F.M.; Aldana-Montes, J.F.; Navas-Delgado, I.; et al. ReprOlive: A database with linked data for the olive tree (*Olea europaea* L.) reproductive transcriptome. *Front. Plant Sci.* **2015**, *6*, 1–14. [CrossRef]

67. Ben Ayed, R.; Ben Hassen, H.; Ennouri, K.; Ben Marzoug, R.; Rebai, A. OGDD (Olive Genetic Diversity Database): A microsatellite markers' genotypes database of worldwide olive trees for cultivar identification and virgin olive oil traceability. *Database* **2016**, *2016*, 1–9. [CrossRef] [PubMed]

68. Trujillo, I.; Barranco, D.; Cabello, A.; Gordon, P.; Morello, C.; Díez, M.; Rallo, L. Proposal of a Guide for the cataloging, sanitation and management of the Germplasm Banks of the IOC Network (UCOLIVO). In Proceedings of the International Seminar. The IOC Network of Germplasm Banks and the True Healthy Olive Cultivars Project, Córdoba, Spain, 21–24 October 2019.

69. Barranco, D.; Cimato, A.; Fiorino, P.; Rallo, L.; Touzani, A.; Castañeda, C.; Serafini, F.; Trujillo, I. *World Catalogue of Olive Cultivars*; International Olive Oil Council: Madrid, Spain, 2000.

70. Rallo, L.; Barranco, D.; Castro-García, S.; Connor, D.J.; Gómez del Campo, M.; Rallo, P. High-Density Olive Plantations. In *Horticultural Reviews*; Janick, J., Ed.; John Wiley & Sons, Inc.: Hoboken, NJ, USA, 2014; Volume 41, pp. 303–384.

71. Fernández i Martí, A.; Font i Forcada, C.; Socias i Company, R.; Rubio-Cabetas, M.J. Genetic relationships and population structure of local olive tree accessions from Northeastern Spain revealed by SSR markers. *Acta Physiol. Plant.* **2015**, *37*. [CrossRef]

72. Sakar, E.; Unver, H.; Bakir, M.; Ulas, M.; Sakar, Z.M. Genetic Relationships Among Olive (*Olea europaea* L.) Cultivars Native to Turkey. *Biochem. Genet.* **2016**, *54*, 348–359. [CrossRef]

73. Sakar, E.; Unver, H.; Ulas, M.; Lazovic, B.; Ercisli, S. Genetic Diversity and Relationships among Local Olive (*Olea europaea* L.) Genotypes from Gaziantep Province and Notable Cultivars in Turkey, Based on SSR Markers. *Not. Bot. Horti Agrobot. Cluj-Napoca* **2016**, *44*, 557–562. [CrossRef]

74. Unver, H.; Sakar, E.; Ulas, M.; Ercisli, S.; Ak, B. Molecular characterization of indigenous olive genotypes based on SSR analysis. *Genetika* **2016**, *48*, 1017–1025. [CrossRef]

75. Hmmam, I.; Mariotti, R.; Ruperti, B.; Cultrera, N.; Baldoni, L.; Barcaccia, G. Venetian olive (*Olea europaea*) germplasm: Disclosing the genetic identity of locally grown cultivars suited for typical extra virgin oil productions. *Genet. Resour. Crop Evol.* **2018**, *65*, 1733–1750. [CrossRef]

76. Reboredo-Rodríguez, P.; González-Barreiro, C.; Cancho-Grande, B.; Simal-Gándara, J.; Trujillo, I. Genotypic and phenotypic identification of olive cultivars from north-western Spain and characterization of their extra virgin olive oils in terms of fatty acid composition and minor compounds. *Sci. Hortic.* **2018**, *232*, 269–279. [CrossRef]

77. Lazović, B.; Adakalić, M.; Pucci, C.; Perović, T.; Bandelj, D.; Belaj, A.; Mariotti, R.; Baldoni, L. Characterizing ancient and local olive germplasm from Montenegro. *Sci. Hortic.* **2016**, *209*, 117–123. [CrossRef]

78. Rotondi, A.; Ganino, T.; Beghè, D.; Di Virgilio, N.; Morrone, L.; Fabbri, A.; Neri, L. Genetic and landscape characterization of ancient autochthonous olive trees in northern Italy. *Plant Biosyst. Int. J. Deal. Asp. Plant Biol.* **2018**, *152*, 1067–1074. [CrossRef]

79. Ninot, A.; Howad, W.; Aranzana, M.J.; Senar, R.; Romero, A.; Mariotti, R.; Baldoni, L.; Belaj, A. Survey of over 4, 500 monumental olive trees preserved on-farm in the northeast Iberian Peninsula, their genotyping and characterization. *Sci. Hortic.* **2018**, *231*, 253–264. [CrossRef]

80. Erre, P.; Chessa, I.; Muñoz-Diez, C.; Belaj, A.; Rallo, L.; Trujillo, I. Genetic diversity and relationships between wild and cultivated olives (*Olea europaea* L.) in Sardinia as assessed by SSR markers. *Genet. Resour. Crop Evol.* **2010**, *57*, 41–54. [CrossRef]

81. Belaj, A.; Muñoz-Diez, C.; Baldoni, L.; Porceddu, A.; Barranco, D.; Satovic, Z. Genetic diversity and population structure of wild olives from the north-western Mediterranean assessed by SSR markers. *Ann. Bot.* **2007**, *100*, 449–458. [CrossRef] [PubMed]

82. Belaj, A.; León, L.; Satovic, Z.; De la Rosa, R. Variability of wild olives (*Olea europaea* subsp. *europaea* var. *sylvestris*) analyzed by agro-morphological traits and SSR markers. *Sci. Hortic.* **2011**, *129*, 561–569. [CrossRef]

83. Breton, C.; Tersac, M.; Bervillé, A. Genetic diversity and gene flow between the wild olive (oleaster, *Olea europaea* L.) and the olive: Several Plio-Pleistocene refuge zones in the Mediterranean basin suggested by simple sequence repeats analysis. *J. Biogeogr.* **2006**, *33*, 1916–1928. [CrossRef]

84. Hannachi, H.; Breton, C.; Msallem, M.; Ben El Hadj, S.; El Gazzah, M.; Bervillé, A. Are olive cultivars distinguishable from oleaster trees based on morphology of drupes and pits, oil composition and microsatellite polymorphisms? *Acta Bot. Gall.* **2008**, *155*, 531–545. [CrossRef]

85. Hannachi, H.; Breton, C.; Msallem, M.; Ben El Hadj, S.; El Gazzah, M.; Bervillé, A. Differences between native and introduced olive cultivars as revealed by morphology of drupes, oil composition and SSR polymorphisms: A case study in Tunisia. *Sci. Hortic.* **2008**, *116*, 280–290. [CrossRef]

86. Hannachi, H.; Breton, C.; Msallem, M.; Hadj, S.B.E.; Gazzah, M.E.; Berville, A. Genetic Relationships between Cultivated and Wild Olive Trees (*Olea europaea* L. var. *europaea* and var. *sylvestris*) Based on Nuclear and Chloroplast SSR Markers. *Nat. Resour.* **2010**, *1*, 95–103. [CrossRef]

87. Belaj, A.; Muñoz-Diez, C.; Baldoni, L.; Satovic, Z.; Barranco, D. Genetic diversity and relationships of wild and cultivated olives at regional level in Spain. *Sci. Hortic.* **2010**, *124*, 323–330. [CrossRef]

88. Yoruk, B.; Taskin, V. Genetic diversity and relationships of wild and cultivated olives in Turkey. *Plant Syst. Evol.* **2014**, *300*, 1247–1258. [CrossRef]

89. Boucheffa, S.; Miazzi, M.M.; di Rienzo, V.; Mangini, G.; Fanelli, V.; Tamendjari, A.; Pignone, D.; Montemurro, C. The coexistence of oleaster and traditional varieties affects genetic diversity and population structure in Algerian olive (*Olea europaea*) germplasm. *Genet. Resour. Crop Evol.* **2017**, *64*, 379–390. [CrossRef]

90. Chiappetta, A.; Muto, A.; Muzzalupo, R.; Muzzalupo, I. New rapid procedure for genetic characterization of Italian wild olive (*Olea europaea*) and traceability of virgin olive oils by means of SSR markers. *Sci. Hortic.* **2017**, *226*, 42–49. [CrossRef]

91. Belaj, A.; Veral, M.G.; Sikaoui, H.; Moukhli, A.; Khadari, B.; Mariotti, R.; Baldoni, L. Olive Genetic Resources. In *The Olive Tree Genome*; Rugini, E., Baldoni, L., Muleo, R., Sebastiani, L., Eds.; Springer: Cham, Switzerland, 2016; pp. 27–54.
92. Omrani-Sabbaghi, A.; Shahriari, M.; Falahati-Anbaran, M.; Mohammadi, S.A.; Nankali, A.; Mardi, M.; Ghareyazie, B. Microsatellite markers based assessment of genetic diversity in Iranian olive (*Olea europaea* L.) collections. *Sci. Hortic.* **2007**, *112*, 439–447. [CrossRef]
93. Stambuk, S.; Sutlović, D.; Bakarić, P.; Petricević, S.; Andelinović, S. Forensic botany: Potential usefulness of microsatellite-based genotyping of Croatian olive (*Olea europaea* L.) in forensic casework. *Croat. Med. J.* **2007**, *48*, 556–562.
94. Cantini, C.; Cimato, A.; Autino, A.; Redi, A.; Cresti, M. Assessment of the Tuscan Olive Germplasm by Microsatellite Markers Reveals Genetic Identities and Different Discrimination Capacity among and within Cultivars. *J. Am. Soc. Hortic. Sci.* **2008**, *133*, 598–604. [CrossRef]
95. Ganino, T.; Beghè, D.; Rotondi, A.; Fabbri, A. Genetic resources of *Olea europaea* L. in the Bologna province (Italy): SSR analysis and identification of local germplasm. *Adv. Hortic. Sci.* **2008**, *22*, 149–155. [CrossRef]
96. Rekik, I.; Salimonti, A.; Grati Kamoun, N.; Muzzalupo, I.; Lepais, O.; Gerber, S.; Perri, E.; Rebai, A. Characterization and Identification of Tunisian Olive Tree Varieties by Microsatellite Markers. *HortScience* **2008**, *43*, 1371–1376. [CrossRef]
97. Ipek, A.; Barut, E.; Gulen, H.; Oz, A.T.; Tangu, N.A.; Ipek, M. SSR analysis demonstrates that olive production in the southern Marmara region in Turkey uses a single genotype. *Genet. Mol. Res.* **2009**, *8*, 1264–1272. [CrossRef] [PubMed]
98. Noormohammadi, Z.; Hosseini-Mazinani, M.; Trujillo, I.; Belaj, A. Study of intracultivar variation among main Iranian olive cultivars using SSR markers. *Acta Biol. Szeged.* **2009**, *53*, 27–32.
99. Corrado, G.; Imperato, A.; la Mura, M.; Perri, E.; Rao, R. Genetic diversity among olive varieties of southern italy and the traceability of olive oil using SSR markers. *J. Hortic. Sci. Biotechnol.* **2011**, *86*, 461–466. [CrossRef]
100. Ercisli, S.; Ipek, A.; Barut, E. SSR Marker-Based DNA Fingerprinting and Cultivar Identification of Olives (*Olea europaea*). *Biochem. Genet.* **2011**, *49*, 555–561. [CrossRef] [PubMed]
101. Rehman, A.U.; Mailer, R.J.; Belaj, A.; De La Rosa, R.; Raman, H. Microsatellite marker-based identification of mother plants for the reliable propagation of olive (*Olea europaea* L.) cultivars in Australia. *J. Hortic. Sci. Biotechnol.* **2012**, *87*, 647–653. [CrossRef]
102. El Bakkali, A.; Haouane, H.; Moukhli, A.; Costes, E.; van Damme, P.; Khadari, B. Construction of Core Collections Suitable for Association Mapping to Optimize Use of Mediterranean Olive (*Olea europaea* L.) Genetic Resources. *PLoS ONE* **2013**, *8*, e61265. [CrossRef]
103. Abdessemed, S.; Muzzalupo, I.; Benbouza, H. Assessment of genetic diversity among Algerian olive (*Olea europaea* L.) cultivars using SSR marker. *Sci. Hortic.* **2015**, *192*, 10–20. [CrossRef]
104. Beghè, D.; Molano, J.F.G.; Fabbri, A.; Ganino, T. Olive biodiversity in Colombia. A molecular study of local germplasm. *Sci. Hortic.* **2015**, *189*, 122131. [CrossRef]
105. Casadei, E.; Valli, E.; Panni, F.; Donarski, J.; Farrús Gubern, J.; Lucci, P.; Conte, L.; Lacoste, F.; Maquet, A.; Brereton, P.; et al. Emerging trends in olive oil fraud and possible countermeasures. *Food Control* **2021**, *124*, 107902. [CrossRef]
106. Kiritsakis, A.; Christie, W.W. Analysis of Edible Oils. In *Handbook of Olive Oil*; Harwood, J., Aparicio, R., Eds.; Springer US: Boston, MA, USA, 2000; pp. 129–158. ISBN 978-1-4757-5371-4.
107. Downey, G.; Boussion, J. Authentication of Coffee Bean Variety by Near-infrared Reflectance Spectroscopy of Dried Extract. *J. Sci. Food Agric.* **1996**, *71*, 41–49. [CrossRef]
108. Melchiade, D.; Foroni, I.; Corrado, G.; Santangelo, I.; Rao, R. Authentication of the 'Annurca' apple in agro-food chain by ampli-fication of microsatellite loci. *Food Biotechnol.* **2007**, *21*, 33–43. [CrossRef]
109. Mouly, P.P.; Gaydou, E.M.; Faure, R.; Estienne, J.M. Blood Orange Juice Authentication Using Cinnamic Acid Derivatives. Variety Differentiations Associated with Flavanone Glycoside Content. *J. Agric. Food Chem.* **1997**, *45*, 373–377. [CrossRef]
110. Montealegre, C.; Marina Alegre, M.L.; García-Ruiz, C. Traceability Markers to the Botanical Origin in Olive Oils. *J. Agric. Food Chem.* **2010**, *58*, 28–38. [CrossRef]
111. Bajoub, A.; Bendini, A.; Fernández-Gutiérrez, A.; Carrasco-Pancorbo, A. Olive oil authentication: A comparative analysis of regulatory frameworks with especial emphasis on quality and authenticity indices, and recent analytical techniques developed for their assessment. A review. *Crit. Rev. Food Sci. Nutr.* **2018**, *58*, 832–857. [CrossRef] [PubMed]
112. Lukić, I.; Ros, A.D.; Guella, G.; Camin, F.; Masuero, D.; Mulinacci, N.; Vrhovsek, U.; Mattivi, F. Lipid profiling and stable isotopic data analysis for differentiation of extra virgin olive oils based on their origin. *Molecules* **2020**, *25*, 4. [CrossRef]
113. Jiménez-Morillo, N.T.; Palma, V.; Garcia, R.; Dias, C.B.; Cabrita, M.J. Combination of Stable Isotope Analysis and Chemometrics to Discriminate Geoclimatically and Temporally the Virgin Olive Oils from Three Mediterranean Countries. *Foods* **2020**, *9*, 1855. [CrossRef] [PubMed]
114. Ben-Ayed, R.; Kamoun-Grati, N.; Rebai, A. An overview of the authentication of olive tree and oil. *Compr. Rev. Food Sci. Food Saf.* **2013**, *12*, 218–227. [CrossRef]
115. Costa, J.; Mafra, I.; Oliveira, M.B.P.P. Advances in vegetable oil authentication by DNA-based markers. *Trends Food Sci. Technol.* **2012**, *26*, 43–55. [CrossRef]
116. Breton, C.; Claux, D.; Metton, I.; Skorski, G.; Bervillé, A. Comparative Study of Methods for DNA Preparation from Olive Oil Samples to Identify Cultivar SSR Alleles in Commercial Oil Samples: Possible Forensic Applications. *J. Agric. Food Chem.* **2004**, *52*, 531–537. [CrossRef]

117. Testolin, R.; Lain, O. DNA Extraction from Olive Oil and PCR Amplification of Microsatellite Markers. *J. Food Sci.* **2005**, *70*, C108–C112. [CrossRef]
118. Muzzalupo, I.; Pellegrino, M.; Perri, E. Detection of DNA in virgin olive oils extracted from destoned fruits. *Eur. Food Res. Technol.* **2007**, *224*, 469–475. [CrossRef]
119. Ben Ayed, R.; Grati-Kamoun, N.; Moreau, F.; Rebaï, A. Comparative study of microsatellite profiles of DNA from oil and leaves of two Tunisian olive cultivars. *Eur. Food Res. Technol.* **2009**, *229*, 757–762. [CrossRef]
120. Raieta, K.; Muccillo, L.; Colantuoni, V. A novel reliable method of DNA extraction from olive oil suitable for molecular traceability. *Food Chem.* **2015**, *172*, 596–602. [CrossRef]
121. Piarulli, L.; Savoia, M.A.; Taranto, F.; D'Agostino, N.; Sardaro, R.; Girone, S.; Gadaleta, S.; Fucili, V.; De Giovanni, C.; Montemurro, C.; et al. A Robust DNA Isolation Protocol from Filtered Commercial Olive Oil for PCR-Based Fingerprinting. *Foods* **2019**, *8*, 462. [CrossRef]
122. Consolandi, C.; Palmieri, L.; Severgnini, M.; Maestri, E.; Marmiroli, N.; Agrimonti, C.; Baldoni, L.; Donini, P.; De Bellis, G.; Castiglioni, B. A procedure for olive oil traceability and authenticity: DNA extraction, multiplex PCR and LDR-universal array analysis. *Eur. Food Res. Technol.* **2008**, *227*, 1429–1438. [CrossRef]
123. Carvalho, J.; Puertas, G.; Gaspar, J.; Azinheiro, S.; Diéguez, L.; Garrido-Maestu, A.; Vázquez, M.; Barros-Velázquez, J.; Cardoso, S.; Prado, M. Highly efficient DNA extraction and purification from olive oil on a washable and reusable miniaturized device. *Anal. Chim. Acta* **2018**, *1020*, 30–40. [CrossRef] [PubMed]
124. Pasqualone, A.; Montemurro, C.; Caponio, F.; Blanco, A. Identification of Virgin Olive Oil from Different Cultivars by Analysis of DNA Microsatellites. *J. Agric. Food Chem.* **2004**, *52*, 1068–1071. [CrossRef] [PubMed]
125. Pasqualone, A.; Montemurro, C.; Summo, C.; Sabetta, W.; Caponio, F.; Blanco, A. Effectiveness of Microsatellite DNA Markers in Checking the Identity of Protected Designation of Origin Extra Virgin Olive Oil. *J. Agric. Food Chem.* **2007**, *55*, 3857–3862. [CrossRef]
126. Martins-Lopes, P.; Gomes, S.; Santos, E.; Guedes-Pinto, H. DNA Markers for Portuguese Olive Oil Fingerprinting. *J. Agric. Food Chem.* **2008**, *56*, 11786–11791. [CrossRef] [PubMed]
127. Alba, V.; Sabetta, W.; Blanco, A.; Pasqualone, A.; Montemurro, C. Microsatellite markers to identify specific alleles in DNA extracted from monovarietal virgin olive oils. *Eur. Food Res. Technol.* **2009**, *229*, 375–382. [CrossRef]
128. Vietina, M.; Agrimonti, C.; Marmiroli, M.; Bonas, U.; Marmiroli, N. Applicability of SSR markers to the traceability of monovarietal olive oils. *J. Sci. Food Agric.* **2011**, *91*, 1381–1391. [CrossRef]
129. Ben-Ayed, R.; Grati-Kamoun, N.; Sans-Grout, C.; Moreau, F.; Rebai, A. Characterization and authenticity of virgin olive oil (*Olea europaea* L.) cultivars by microsatellite markers. *Eur. Food Res. Technol.* **2012**, *234*, 263–271. [CrossRef]
130. Pasqualone, A.; Di Rienzo, V.; Nasti, R.; Blanco, A.; Gomes, T.; Montemurro, C. Traceability of Italian Protected Designation of Origin (PDO) table olives by means of microsatellite molecular markers. *J. Agric. Food Chem.* **2013**, *61*, 3068–3073. [CrossRef]
131. Pérez-Jiménez, M.; Besnard, G.; Dorado, G.; Hernandez, P. Varietal tracing of virgin olive oils based on plastid DNA variation profiling. *PLoS ONE* **2013**, *8*, e70507. [CrossRef]
132. Muzzalupo, I.; Pisani, F.; Greco, F.; Chiappetta, A. Direct DNA amplification from virgin olive oil for traceability and authenticity. *Eur. Food Res. Technol.* **2015**, *241*, 151–155. [CrossRef]
133. Montemurro, C.; Miazzi, M.M.; Pasqualone, A.; Fanelli, V.; Sabetta, W.; Di Rienzo, V. Traceability of PDO Olive Oil "Terra di Bari" Using High Resolution Melting. *J. Chem.* **2015**, *2015*, 1–7. [CrossRef]
134. Pasqualone, A.; Rienzo, V.D.; Miazzi, M.M.; Fanelli, V.; Caponio, F.; Montemurro, C. High resolution melting analysis of DNA microsatellites in olive pastes and virgin olive oils obtained by talc addition. *Eur. J. Lipid Sci. Technol.* **2015**, *117*, 2044–2048. [CrossRef]
135. Gomes, S.; Breia, R.; Carvalho, T.; Carnide, V.; Martins-Lopes, P. Microsatellite High-Resolution Melting (SSR-HRM) to Track Olive Genotypes: From Field to Olive Oil. *J. Food Sci.* **2018**, *83*, 2415–2423. [CrossRef] [PubMed]
136. Pasqualone, A.; Di Rienzo, V.; Sabetta, W.; Fanelli, V.; Summo, C.; Paradiso, V.M.; Montemurro, C.; Caponio, F. Chemical and Molecular Characterization of Crude Oil Obtained by Olive-Pomace Recentrifugation. *J. Chem.* **2016**, *2016*, 1–7. [CrossRef]
137. Rotondi, A.; Beghè, D.; Fabbri, A.; Ganino, T. Olive oil traceability by means of chemical and sensory analyses: A comparison with SSR biomolecular profiles. *Food Chem.* **2011**, *129*, 1825–1831. [CrossRef]
138. Crawford, L.M.; Carrasquilla-Garcia, N.; Cook, D.; Wang, S.C. Analysis of Microsatellites (SSRs) in Processed Olives as a Means of Cultivar Traceability and Authentication. *J. Agric. Food Chem.* **2020**, *68*, 1110–1117. [CrossRef] [PubMed]
139. Crawford, L.M.; Janovick, J.L.; Carrasquilla-Garcia, N.; Hatzakis, E.; Wang, S.C. Comparison of DNA analysis, targeted metabolite profiling, and non-targeted NMR fingerprinting for differentiating cultivars of processed olives. *Food Control* **2020**, *114*, 107264. [CrossRef]
140. Chedid, E.; Rizou, M.; Kalaitzis, P. Application of high resolution melting combined with DNA-based markers for quantitative analysis of olive oil authenticity and adulteration. *Food Chem. X* **2020**, *6*, 100082. [CrossRef] [PubMed]

Article

Origin Determination of Walnuts (*Juglans regia* L.) on a Worldwide and Regional Level by Inductively Coupled Plasma Mass Spectrometry and Chemometrics

Torben Segelke, Kristian von Wuthenau, Anita Kuschnereit, Marie-Sophie Müller and Markus Fischer *

Hamburg School of Food Science, Institute of Food Chemistry, University of Hamburg, Grindelallee 117, 20146 Hamburg, Germany; torben.segelke@chemie.uni-hamburg.de (T.S.); kristian.wuthenau@chemie.uni-hamburg.de (K.v.W.); anita.kuschnereit@studium.uni-hamburg.de (A.K.); marie-sophie.mueller@studium.uni-hamburg.de (M.-S.M.)
* Correspondence: Markus.Fischer@uni-hamburg.de; Tel.: +49-40-42838-4357/59

Received: 25 October 2020; Accepted: 18 November 2020; Published: 20 November 2020

Abstract: To counteract food fraud, this study aimed at the differentiation of walnuts on a global and regional level using an isotopolomics approach. Thus, the multi-elemental profiles of 237 walnut samples from ten countries and three years of harvest were analyzed with inductively coupled plasma mass spectrometry (ICP-MS), and the resulting element profiles were evaluated with chemometrics. Using support vector machine (SVM) for classification, validated by stratified nested cross validation, a prediction accuracy of 73% could be achieved. Leave-one-out cross validation was also applied for comparison and led to less satisfactory results because of the higher variations in sensitivity for distinct classes. Prediction was still possible using only elemental ratios instead of the absolute element concentrations; consequently, a drying step is not mandatory. In addition, the isotopolomics approach provided the classification of walnut samples on a regional level in France, Germany, and Italy, with accuracies of 91%, 77%, and 94%, respectively. The ratio of the model's accuracy to a random sample distribution was calculated, providing a new parameter with which to evaluate and compare the performance of classification models. The walnut cultivar and harvest year had no observable influence on the origin differentiation. Our results show the high potential of element profiling for the origin authentication of walnuts.

Keywords: walnut; *Juglans regia*; origin authentication; element profiling; inductively coupled plasma mass spectrometry; ICP-MS; chemometrics; isotopolomics

---

1. Introduction

Walnuts are the seeds of the *Juglans* tree, particularly of the English or Persian walnut tree *Juglans regia*. They are appreciated for their high level of polyunsaturated fatty acids as well as their high tocopherol and potassium content. There are different walnut cultivars that are commercially grown—e.g., "Lara" and "Chandler" [1–3]. Nowadays, consumers are increasingly interested in products made with selected ingredients. The interest in sustainable and regional food is growing correspondingly. As a result, consumers are willing to accept higher prices for products with a specific geographical origin [4]. The annual financial damage by food fraud is estimated at 40 billion dollars, and there are also health risks with lethal consequences [5]. The omics disciplines are suitable for authentication by creating a fingerprint of the examined food to prevent food fraud [4]. DNA-based methods for food authentication, which are only able to determine the genotype, are inevitably limited to the determination of the biological identity [6,7]. Only in exceptional cases where only certain varieties are grown in certain regions can indications of geographical origin be obtained. Furthermore, the presence of the analyte DNA is the essential prerequisite for carrying out molecular biological analyses. This is usually not the case with fats and oils [4,8].

Isotopolomics is particularly applicable for origin analysis, as it reflects the influence of the soil and thus the geographical origin [9,10]. Inductively coupled plasma mass spectrometry (ICP-MS) has become a routine application in the field of isotopolomics for the generation of the elemental profiles of food by the quantitative determination of the elemental composition of the sample in a wide dynamic range (ng/kg to mg/kg) [11–14].

Walnuts have been analyzed before with regard to their origin, partly by determining the elemental content. Esteki et al. used chromatographic fatty acid fingerprint analysis to differentiate walnuts from six Iranian regions [15]. Gu et al. analyzed Chinese walnut samples with inductively coupled plasma optical emission spectrometry and near-infrared and mid-infrared spectroscopy from three production areas in Xinjiang [16]. Krauß et al. evaluated stable isotope signatures from different regions in Germany [17]. In a preliminary but international study by Popescu et al., the authors used nuclear magnetic resonance spectroscopy to investigate the differences between walnuts varieties from five countries and two years of harvest [18].

To our knowledge, however, there has been no large-scale international walnut study based on the analysis of elementary patterns comparing several geographical origins from at least three harvesting years [19]. The element pattern is considered particularly suitable for determining the geographical origin of walnuts, since walnut kernels grow inside the shell and are therefore protected from the environment—i.e., practically unaffected by perturbations such as anthropogenic aerosols and soil dust [2,20]. As a consequence, the exclusive elemental characteristics of the soil should be recognized in the walnut kernels. Still, the analysis may be challenging from an analytical point of view, since only small quantities of elements are to be expected in the fat-rich walnut kernels [20], requiring sensitive analytical methods. ICP-MS offers limits of detection into the parts per trillion (ng/L) range [10,13], and is therefore even more sensitive than inductively coupled plasma optical emission spectrometry (ICP-OES), as applied in previous analyses of walnuts [3,21–23].

Therefore, the aim of this study was to develop a reliable chemometric model using ICP-MS in combination with machine learning methods for the worldwide and regional origin authentication of walnuts, independent from harvest year and cultivar.

In this context, we focused on the *Juglans regia* walnut species, as it is grown and consumed all over the world and is generally considered to be of the highest quality and has the highest demand [15,17].

In 2018, about 3.6 million tons of walnuts were harvested. China (approx. 1.6 million tons) and the USA (approx. 0.6 million tons) have the largest contribution to the worldwide walnut harvest. In the context of walnut authentication for the west European market, however, not only is the total harvest quantity important but also the quantity of exported and imported goods. Chile, for example, is a significant contributor, exporting more than 10,000 tons to Germany, France, Italy, and Switzerland combined. Turkey, Hungary, and Pakistan also play an important role [1,17,24].

Consequently, 237 walnut reference samples from three harvest years and originating from ten countries were analyzed with high-resolution ICP-MS. Principal component analysis (PCA) [25] and *t*-distributed stochastic neighbor embedding (*t*-SNE) [26] were carried out to visualize the data. Then, machine learning methods were applied to develop classification models for the authentication of walnuts from ten countries on a worldwide scale. Since a high number of authentic walnut samples (>30) could be obtained from France, Germany, and Italy, classification models were also developed on a regional level.

2. Materials and Methods

2.1. Reagents and Materials

The elemental analyses of walnuts were based on a previous study [13]. In Table S1 (supporting information), the reagents and materials used in this study are listed.

2.2. Sample Preparation

A total of 237 reference samples of relevant, market-available walnuts from three years of harvest (2017, 2018, and 2019) were collected and analyzed in this study. The walnut samples originated from ten countries and were purchased as shelled or in-shell goods. Thanks to the cooperation with regional producers and project partners who work according to our internal guidelines to ensure the authenticity of the reference material (e.g., by applying the HACCP guidelines, FSSC 22,000, or providing the structure meta date), authentic walnut samples could be acquired. See Table S2 (Supporting Information) for detailed information and Figure 1 for a visual illustration. On arrival, walnut samples were frozen and stored at −80 °C until further processing could take place. Element patterns are less influenced by storage than other profiling levels, such as the metabolome [27]. At the applied storage conditions of −80 °C, enzyme activities are inhibited—i.e., cell lysis is also inhibited [28]. One German walnut sample (harvest year 2018, Hesse) was selected as a quality control (QC) sample.

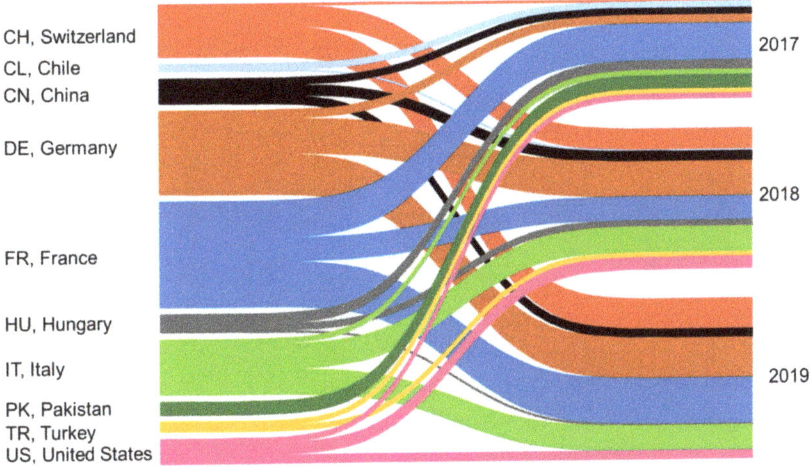

Figure 1. Overview of the 237 walnut samples with regard to their origin and harvest year.

2.3. Sample Preparation and Digestion

For each walnut sample, 100 g of walnut kernels were milled using a knife mill (Grindomix GM 300, Retsch, Haan, Germany) with the addition of dry ice. If necessary, in-shell walnuts were shelled before. Homogenized samples were freeze-dried for 48 h (Beta 1–8 LDplus, Martin Christ Gefriertrocknungsanlagen GmbH, Osterode am Harz, Germany), including a stirring step after 24 h.

The sample digestion of 500 mg of homogenized and lyophilized walnut material was performed using an Ethos.lab microwave (MLS GmbH, Leutkirch, Germany), as described in reference [13] and in Table S3 (Supporting Information). For each digestion run, one vessel was selected for the QC sample and one vessel for a blank. The QC sample was later used for quality assurance and the calculation of the method's precision (see Section 3.2).

2.4. Analytical Procedure and Instrumentation

Multi-elemental analyses were performed on an HR-ICP-MS Element2 (ThermoFisher Inc., Waltham, MA, USA) coupled with an SC-E4 Autosampler (Elemental Scientific Inc., Omaha, NE, USA), following a method validated in a previous study [13].

The multi-element method included 47 isotopes: Li, Be, B, Na, Mg, Al, K, Ca, Sc, V, Cr, Mn, Fe, Co, Ni, Cu, Zn, Ga, As, Se, Rb, Sr, Y, Mo, Ag, Cd, Te, Ba, La, Ce, Pr, Nd, Sm, Eu, Gd, Tb, Dy, Ho, Er, Tm, Yb, Lu, Tl, Pb, Bi, Th, U. The calculated limit of detection and limit of quantitation are given in Table S4 (Supporting Information). Here, also the respective element used for internal standardization is given.

Quantitation was conducted by external calibration. Instrument optimization, mass calibration, and mass offset were performed daily. For further instrumental conditions and method validation, refer to Table S5 (Supplementary Materials) and reference no. [13]. Tubes and pipette tips were pre-cleaned by soaking in 3% (*v/v*) nitric acid overnight and subsequently rinsed with ultrapure water and dried.

2.5. Multivariate Data Analysis and Classification Models

Multi-element data were visualized and interpreted with one-way analysis of variance [29] (ANOVA) tests using Matlab R2019a (The Mathworks Inc., Natick, MA, USA). Bonferroni post-hoc tests [30] were calculated to determine inter-class differences using Microsoft Excel 2016 (Microsoft Corporation, Redmond, WA, USA).

Boxplots for the data visualization of certain elements, including outlier detection, were created using Matlab's boxplot-function [31]. Furthermore, *t*-SNE (Barnes–Hut algorithm, cosine distance, perplexity = 18) and PCA plots were calculated using Matlab. For interpretation, 95% confidence ellipses were added to the score plots [32].

For the calculation of classification models, Matlab was applied; different settings of the data pre-treatment, classification method, and validation were compared for each sub-issue (the differentiation of all walnuts or the differentiation of only French walnuts, etc., as described in the following sections). The settings for the data pre-treatments, the classification methods, and the validation are stated in Table 1. For classification methods, we chose linear discriminant analysis, support vector machine [33], subspace discriminant [34,35], and random forest [36]. Following a design of experiments approach, the results of every combination were compared and the settings of the best results were chosen [37].

Table 1. Overview of the settings for data pre-treatment, classification methods with hyperparameters, and validation for the calculation of the classification models.

Data Pre-Treatment	Classification Method	Validation
(i) no pre-treatment	(1) linear discriminant analysis, LDA $\gamma = 0$	(a) stratified nested cross validation
(ii) log10	(2) support vector machine, SVM polynomial order = 2 box constraint level = 1 coding: one vs. one	(b) leave-one-out cross validation
(iii) center (mean) and scale (standard deviation)	(3) subspace discriminant, SSD number learning cycles = 30	
(iv) center (median) and scale (standard deviation)	(4) random forest, RF split criterion: Gini's diversity index max. number of splits = 100 min leaf size = 1 surrogate: off	
(v) center (median) and scale (range)		
(vi) center (median) and scale (interquartile range)		

For obtaining an unbiased estimate of the model's performance, the models were validated using (a) leave-one-out cross validation or (b) stratified nested cross validation [38,39]. Since the former validation method is more widespread in the scientific community, we would like to explain the procedure of the latter validation method briefly: The whole data set was split into five parts, whereby the samples were not fully randomly divided but stratified by the origins. Hence, all five parts contained a preferably equal number of samples with respect to the walnuts' origins, ensuring a representative and balanced training (four fifth) and test (one fifth) set. For the training set, 10-fold cross validation was applied to select the optimal model parameters (i.e., inner cross validation). The performance of the calculated model was evaluated by predicting the independent test set. The described procedure was repeated for all five parts, so every part of the 5-fold outer cross validation was once used as the test set (i.e., outer cross validation). Finally, since the results by a single nested cross validation can vary, the entire cross validation was repeated 20 times. By repeating this process, a standard deviation of the accuracy was calculated [13].

For the geographical origin authentication, 17 elemental concentrations and 78 ratios of element concentrations were considered (cp. Section 3.3), resulting in a total of 95 variables:

Al, B, Ba, Ca, Co, Cu, Fe, Ga, Mg, Mn, Mo, Ni, Rb, Sr, Te, Tl, Zn, Rb/B, Sr/B, Mo/B, Ba/B, Co/B, Ni/B, Sr/Rb, Mo/Rb, Ba/Rb, Co/Rb, Ni/Rb, Mo/Sr, Ba/Sr, Co/Sr, Ni/Sr, Co/Mo, Mo/Ba, Co/Ba, B/Mg, Rb/Mg, Sr/Mg, Mo/Mg, Ba/Mg, Ca/Mg, Mn/Mg, Fe/Mg, Co/Mg, Ni/Mg, Cu/Mg, Zn/Mg, B/Ca, Rb/Ca, Sr/Ca, Mo/Ca, Ba/Ca, Mn/Ca, Fe/Ca, Co/Ca, Ni/Ca, Cu/Ca, Zn/Ca, B/Mn, Rb/Mn, Sr/Mn, Mo/Mn, Ba/Mn, Fe/Mn, Co/Mn, Ni/Mn, Cu/Mn, Zn/Mn, B/Fe, Rb/Fe, Sr/Fe, Mo/Fe, Ba/Fe, Co/Fe, Ni/Fe, Cu/Fe, Zn/Fe, Mo/Ni, Ba/Ni, Co/Ni, Cu/B, Rb/Cu, Sr/Cu, Mo/Cu, Ba/Cu, Co/Cu, Ni/Cu, B/Zn, Rb/Zn, Sr/Zn, Mo/Zn, Ba/Zn, Co/Zn, Ni/Zn, and Cu/Zn.

3. Results and Discussion

3.1. Explanation for the Usage of Walnut Kernels

Theoretically, both walnut shell and walnut kernel could be usable for an authentication study. It is reasonable to assume that both parts reflect the elemental pattern of the soil and, thus, the origin.

However, walnuts are mostly traded as shelled goods: in 2018, the percentages of shelled walnuts imported to Europe and Germany were 62% and 81%, respectively [40]. An analytical method developed solely based on the shell would only be applicable to 38% or 19% of potential food fraud, respectively. Additionally, as stated earlier, walnut kernels grow inside the shell and are therefore virtually unaffected by perturbations such as anthropogenic aerosols and soil dust [20]. For these reasons, we decided to use the walnut kernels, as is also practiced in other studies [15–17]. Whenever mentioning walnuts samples henceforth, we are referring to walnut kernels.

3.2. Selection of Variables for the Chemometric Analysis of 237 Walnut Kernel Samples

From the 47 elements acquired, not all elements were considered for chemometric analysis. Concentrations below the LOQ were obtained for some elements: if the content of those samples exceeded 20%, the respective element was not used for evaluation (this was the case for Ag (22%), Pr (25%), Sm (27%), Dy (32%), Na (36%), Y (42%), Er (53%), Yb (56%), Pb (64%), As (73%), V (75%), Th (78%), Se (78%), Eu (81%), U (85%), Tb (89%), Ho (89%), Cd (91%), Bi (95%), Sc (97%), Li (99%), Tm (99%), Lu (99%), and Be (100%)). Otherwise, the concentration was set to the LOQ level instead of zero, ensuring logarithmic functions to be applicable. For K, the concentrations were over the calibration range for all of the samples; thus, K concentrations were not used for chemometric modeling.

The long-term stability (reproducibility) evaluated by the QC sample deviated between 5.2% (Co) and 27% (Al) (median: 9.2%), except for Te, Tl, Gd, Nd, Ce, and La (54–107%), which were at very low concentrations in the chosen QC sample.

One-way ANOVA tests indicated that Al, Ba, Co, Cu, Fe, Mo, Ni, and Sr were highly significant (99% confidence level) for the walnut origins. The corresponding boxplots are shown in Figure 2.

B, Ga, Mg, Mn, Rb, Te, Tl, and Zn were significant (95% confidence level). Ce, Cr, Gd, La, and Nd, however, showed no significance and were not used for chemometric modeling, also because of the high deviations for the QC sample.

One can observe an increasing tendency where elemental concentration ratios, especially the rare earth elements (REE) among these, are considered for chemometric evaluation besides absolute concentrations [13,14,41,42]. In this way, the model can become more robust [41]. We recently evaluated the benefit of considering concentration ratios in addition to elemental concentrations. Furthermore, we see the possibility of foregoing a drying step for the samples; the water content is no longer important if only elemental ratios are used for chemometric modeling [13]. The concentrations of B, Ba, Ca, Co, Cu, Fe, Mg, Mn, Mo, Ni, Rb, Sr, and Zn were >LOQ for all 237 samples, and for these 13 elements the concentration ratios were calculated. In order to reduce redundant data, duplicate ratios with an interchangeable nominator and denominator were rejected, resulting in 78 element ratios (see Section 2.5).

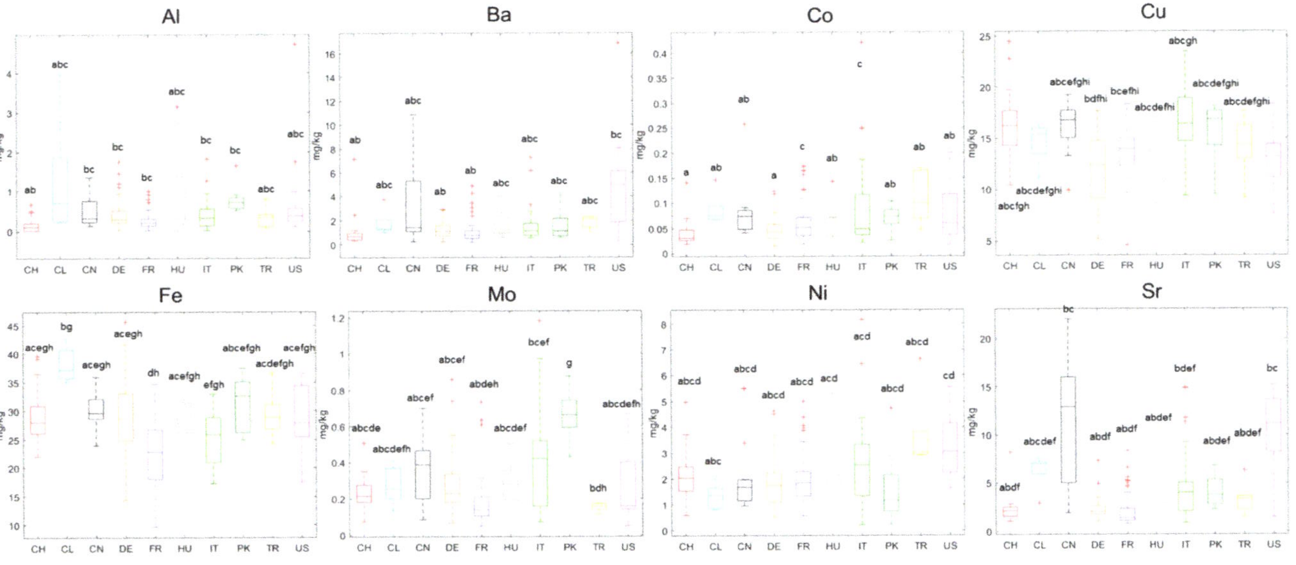

Figure 2. Boxplots for the highly significant elements for the walnuts' origins after one-way ANOVA testing. Different small letters indicate significant inter-class differences, as determined by Bonferroni post-hoc tests. The "+"-symbol indicates outliers as calculated by Matlab's boxplot-function (cf. Section 2.5). Data are expressed as mg/kg in walnut lyophilizate.

3.3. Chemometric Analysis of the Walnut Samples

3.3.1. Data Investigation and Visualization

PCA plots after centering (mean) and scaling (standard deviation) are shown in Figure 3. In the principal component 1 (PC1) vs. PC2 plane, samples from the USA tend to have higher PC1 values and differ most strongly compared to the rest of the samples. In the PC2 vs. PC4 plane, a better visual differentiation can be achieved: Chinese and Pakistani samples are located in the lower right of the scores plot. Swiss and German samples are located in the upper half, while Italian samples tend to have lower PC4 values and French samples tend to have lower PC2 values.

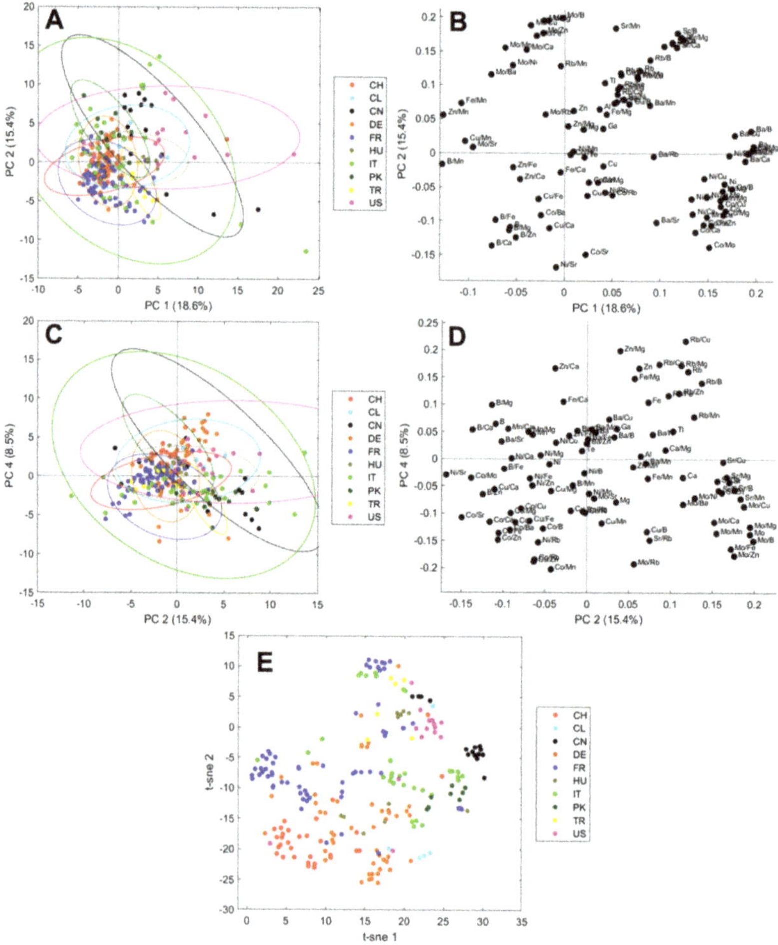

Figure 3. Unsupervised visualization of the multielement data of 237 walnut samples after mean centering and standard deviation scaling using PCA and *t*-SNE. Scores are colored by the origin in the PC1 vs. PC2 plane (**A**) with the corresponding loadings plot (**B**), and the PC2 vs. PC4 plane (**C**) with the corresponding loadings plot (**D**). 95% confidence ellipses were added to the scores plots in (**A**) and (**C**). *t*-SNE plot colored by the origin (**E**).

The usage of PCA models for data investigation and visualization is very common for authentication studies; however, it is not always the best choice for visualizing big data sets.

The non-linear iterative partial least squares (NIPALS) algorithm focuses on the largest differences in the data set and sets the axis towards the greatest variance [25,43]. Few samples of a minor sample population may get lost in the shuffle [44]. Therefore, we chose *t*-SNE as an additional approach to visualize the data. This can be described as a complementary method, since it focuses on the similarities between two data points rather than the differences [26]. Like PCA, it is an unsupervised model, and the closer two data points the more similar they are. As seen from Figure 3E, the samples tend to cluster according to their origins. In this plot, the clusters seem to be more distinguishable for all sample populations, though the clusters still do overlap and supervised models are needed to determine the origin. However, the boxplots in Figure 2 have already proven that the differentiation of origin is possible; the eight elements presented here show a visual distinction, and for some countries of origin a marker element can already be identified visually. Most apparent is that walnut samples from China and USA contain higher concentrations of Ba and Sr compared with the other walnut samples. The Chinese and US-American samples can be distinguished by Cu, with it being increased for Chinese samples compared to the US ones. Pakistani samples have a higher Mo content. Walnut samples from Chile contain more Fe, and, like Hungarian walnuts, contain more Al.

3.3.2. Influence of the Harvest Year

The fact that the year of harvest has no significant influence on the elemental pattern is recognized in many isotopolomics studies and is considered an advantage of ICP-MS analysis [12,45]. This also applies for walnuts, as examined in a previous study, where the harvesting year and the climatic conditions showed no significant influence on the element pattern [3]. On the other hand, when analyzing the metabolome/proteome the harvest year affected the fatty acid saturation degree and the protein amount [18]. Even for stable isotope analysis, annual differences in the δ^2H-values occur [17]. However, to verify the potential influence of the harvest year for our own data set, French, German, and Italian samples were examined for their potential influence, since most of the walnut samples originated from these three countries and were evenly distributed for three harvest years (see Figure 1). ANOVA tests were calculated, and the highest *F*-value (2.92) was found for Mn for French walnuts; however, this value was smaller than the critical *F*-value of 3.15 (0.05 significance level). Thus, none of the 17 elements showed a significant influence with regard to the harvest years.

Additionally, the PCA score plot of all ten origins (Figure 3A) was colored by the respective harvest years and is shown in Figure S1A (Supplementary Materials). For a better visual comparison, both score plots are shown in Figure S1B for direct comparison. While the scores tend to cluster according to their origins, as discussed above, no clustering according to the harvest years is noticeable. Consequently, the origin has a higher influence on the data than the harvest year. This might enable this study to be suitable for the prediction of new samples without the necessity of new reference samples in future years. In fact, this is a further advantage of this method, since the origin of walnut samples can be predicted at the beginning of the next harvest season without any need for new reference samples.

3.3.3. Influence of the Cultivar

Besides the harvest year, the cultivar—in other words, the genotype—may have an influence on the walnut's isotopolome and may cause an unwanted bias in this origin authentication study. In previous studies, a dependency of element uptake for different walnut cultivars was found for Cu, K, Fe, Mn, and Zn [3,46]. However, these studies mainly concerned a physiological-nutritional analysis of walnuts, and the number of samples was relatively small, with 24 and 9, respectively. Considering K, its potential influence can be considered as irrelevant anyway, since the signal intensity of K, as the element with the highest concentration in the walnut kernel, was above the calibration range and, therefore, K was not taken into account for statistical evaluation (see Section 3.2).

Another study implies that the genotype solely has only a secondary effect on the isotopolome: Juranović Cindrić et al. analyzed the elemental composition of *Juglans nigra* walnut samples, which are another species compared to *Juglans regia*, as investigated in this study. The authors compared the elemental concentrations to the literature values of *Juglans regia* and found similar results [21]. Thus, when the elemental concentrations of different *Juglans* species are similar, the elemental concentrations of different cultivars, a taxonomic rank below the species, should be similar as well.

It should be emphasized that the potential influence of the cultivar would be problematic for the origin authentication when all samples from one country consisted of a cultivar which would reversely originate solely from this country. This is not the case for the present data set as presented in Table S2. Considering two major cultivars, for example, 47 walnut "Lara" samples originate from Switzerland (2), Germany (2), France (30), and Italy (13). Twenty walnut "Chandler" samples originate from Switzerland (2), Chile (1), China (2), Italy (11), Turkey (1), and the USA (3) (number in brackets indicates the sample size per origin). Consequently, no origin or cultivar was overrepresented. Additionally, PCA plots were calculated for 120 walnut samples, for which the information of the cultivar was available and there were at least three samples of the respective cultivar. The scores plots are shown in Figure S2 (Supplementary Materials), and the scores are colored by origin and cultivar for direct comparison. As seen from Figure S2, the samples do not cluster with respect to the cultivar; thus, the cultivar has a secondary effect on the isotopolome.

3.3.4. Classification of the Geographical Origin

For the origin differentiation, 237 samples from ten origins were considered. Mean concentrations with standard deviations are given in Table S6 (supporting information). For all combinations stated in Table 1, a model was calculated to find the best suited settings. As the response variable, the overall accuracy was investigated, and the overall accuracies are stated in Tables S7 and S8 (Supplementary Materials). As seen from the results, the choice of the classification method has a major impact on the model's performance compared to the data pre-treatment. Especially for the different center and scale approaches, the accuracies only differ in the second decimal. Using stratified nested cross validation, the best accuracy of 72.9% ± 1.6% was found after center (mean) and scale (standard deviation) using SVM. For leave-one-out cross validation, the best accuracy was reached at an improved level of 75.5%, also achieved after center (mean), scale (standard deviation), and SVM. The corresponding confusion matrices are shown in Tables 2 and 3, respectively. To compare these two models, both classification models were evaluated by calculating the sensitivity and the specificity per class to examine the type I and type II errors [47,48]. For the stratified nested cross validation sensitivity, the scores ranged from 20.0% to 84.6%, and the specificity ranged from 25.5% to 87.8%. Meanwhile, for the leave-one-out cross validation, the sensitivity ranged from 16.7% to 86.7%, and the specificity ranged from 20.0% to 100%. The Turkish walnuts are the blind spot for both classification models, with comparably low sensitivity scores of 20.0% (stratified nested cross validation) and 16.7% (leave-one-out cross validation), respectively. In the future, the prediction accuracy of Turkish walnuts may be improved by data fusion—i.e., combining our data set with other omics-disciplines or isotope ratio analysis [9,49]. When comparing the sensitivities per class, the values of the nested cross validation do show less variation or, in other words, the standard deviation is lower. Particularly for the two origins with the fewest number of samples (Chile and Turkey), the accuracies are superior by 3 and 6 percentage points. The stratified approach (see Section 2.5) used for the calculation of the test and training of the nested cross validation set may positively influence the sample's distribution and lead to more evenly distributed accuracies. Therefore, we prefer the classification model validated by stratified nested cross validation, despite the slightly reduced accuracy of 2.7 percentage points. To our knowledge, in the literature authors only seldom comment on their choice of which validation method to apply [15,47]. Although it is mentioned that, via stratified nested cross validation, a generally valid model can be calculated which does not lack overfitting, while leave-one-out cross validation is more prone to this issue [12,38,39,50], we would like to encourage the reader to apply both validation strategies and compare the results.

Table 2. Confusion matrix for the classification of all 237 walnut samples with the SVM model, resulting in a 72.9% ± 1.6% overall accuracy using stratified nested cross validation. The mean values of 20 predictions are given.

		Predicted Origin										
		Switzerland	Chile	China	Germany	France	Hungary	Italy	Pakistan	Turkey	United States	Sensitivity [%]
	Switzerland	22.1	0.0	0.0	6.4	1.6	0.0	0.0	0.0	0.0	1.0	71.3
	Chile	0.0	3.3	0.7	1.0	0.0	0.0	0.0	0.0	0.0	0.0	66.0
	China	0.7	0.0	12.7	1.0	0.4	0.0	0.0	0.0	0.3	0.0	84.3
	Germany	2.8	0.2	0.0	39.5	6.3	0.5	0.1	0.0	0.1	0.6	79.0
Actual	France	1.2	0.0	0.0	6.8	53.3	0.8	0.5	0.0	0.5	0.1	84.6
Origin	Hungary	0.0	0.0	0.0	4.9	0.2	4.8	0.0	1.0	0.1	0.0	43.6
	Italy	0.0	0.9	0.1	2.6	4.7	0.3	22.6	0.7	1.0	0.4	68.5
	Pakistan	0.0	0.0	0.4	2.0	0.0	1.1	0.3	4.4	0.0	0.0	54.4
	Turkey	0.7	0.0	0.0	1.3	1.5	0.4	1.1	0.0	1.2	0.0	20.0
	United States	1.0	0.3	0.7	1.0	0.0	0.1	1.2	0.3	1.6	8.9	59.3
	specificity [%]	77.8	72.5	87.5	59.6	78.6	61.1	87.8	69.0	25.5	81.7	

Table 3. Confusion matrix for the classification of all 237 walnut samples with the SVM model, resulting in a 75.5% overall accuracy using leave-one-out cross validation.

		Predicted Origin										
		Switzerland	Chile	China	Germany	France	Hungary	Italy	Pakistan	Turkey	United States	Sensitivity [%]
	Switzerland	24	0	0	5	1	0	0	0	0	1	77.4
	Chile	0	3	1	1	0	0	0	0	0	0	60.0
	China	1	0	13	1	0	0	0	0	0	0	86.7
	Germany	2	0	0	41	6	0	0	0	0	1	82.0
Actual	France	1	0	0	6	54	1	0	0	1	0	85.7
Origin	Hungary	0	0	0	5	0	5	0	1	0	0	45.5
	Italy	0	0	0	2	4	1	24	1	1	0	72.7
	Pakistan	0	0	0	2	0	1	0	5	0	0	62.5
	Turkey	0	0	0	2	1	0	2	0	1	0	16.7
	United States	1	0	1	1	0	0	1	0	2	9	60.0
	specificity [%]	82.8	100.0	86.7	62.1	81.8	62.5	88.9	71.4	20.0	81.8	

The current sample preparation includes a drying step that is both time and energy consuming and could be optimized for environmental reasons [51]. Without a drying step, this method would be applicable in the food industry where the incoming goods inspection should be carried out as fast as possible. Without drying though, the element contents cannot be expressed in relation to the dry matter, which makes it difficult to evaluate the walnut samples chemometrically. When considering only elemental concentration ratios, however, a comparison is possible. Therefore, the calculation of the classification model (SVM, stratified nested cross validation) was repeated using only the 78 elemental ratios listed in Section 2.5. The prediction accuracy dropped as expected, but only marginally: an overall accuracy of 72.2% ± 1.6% was achieved using stratified nested cross validation (before: 72.9% ± 1.6%). For the sake of completeness, leave-one-out cross validation led to 74.7% (before: 75.5%). The respective loss of accuracy is not significant, and, in this way, fresh walnut samples can also be analyzed in the future without an obligatory drying step

3.3.5. Classification of the Regional Origin of French, German and Italian Walnuts

For France, Germany, and Italy, more than 30 samples could be acquired and, thanks to our project partners, highly authentic samples with detailed and reliable information regarding the origin on a regional level were available (see Table S2). Therefore, the potential of the analytical method to predict the origin even on a regional level was investigated. Mean concentrations with standard deviations for the regions are given in Tables S9–S11 (Supporting Information).

For France, the data set existed of 53 samples from the four regions Auvergne-Rhône-Alpes (ten samples), Nouvelle-Aquitaine (26 samples), Occitanie (11 samples), and Pays de la Loire (six samples). A PCA was calculated for these samples, and clusters can already be recognized for the origin, as seen in the scores and loadings plot shown in Figure S3A,B (Supplementary Materials), respectively. Classification models were calculated for this issue with all combinations stated in Table 1, except leave-one-out cross validation, because of the results in the previous section. The best accuracy of 91.4% ± 2.1% was found after a $\log_{10}$ transformation and an SVM classification model (Table S12, supporting information). The corresponding confusion matrix is shown in Table 4. The separation of samples from Pays de la Loire succeeded almost without error with a sensitivity of 99%. To identify the elements causing the separation, an ANOVA test was calculated. The boxplots of the elements with the highest F-values are shown in Figure S4A (Supplementary Materials). As can be seen here, the Pays de la Loire can be well distinguished from any other French region because of the significant higher concentrations of Co. Ba and Sr are also important for the regional differentiation.

Not yet considered was the Noix de Grenoble, the only walnut with a geographical indication (in French: appellation d'origine protégée (AOP)). These are walnuts originating from certain municipalities in the départements of Isère, Drôme, and Savoie [52]. The three départements are located in Auvergne-Rhône-Alpes; thus, it would be rather challenging to conduct a sub-regional authentication study. However, the promising results obtained so far give the possibility to follow up this classification issue in the future.

For Germany and Italy, the data set was analogously analyzed: PCA scores plots are shown in Figure S3, and the confusion matrices are shown in Tables 5 and 6, respectively.

Table 4. Confusion matrix for the classification of 53 French walnut samples with the SVM model, resulting in a 91.4% ± 2.1% overall accuracy using stratified nested cross validation. The mean values of 20 predictions are given.

		Predicted Regional Origin				
		Pays de la Loire	Nouvelle-Aquitaine	Auvergne-Rhône-Alpes	Occitanie	Sensitivity [%]
Actual Regional Origin	Pays de la Loire	6.0	0.1	0.0	0.0	99.2
	Nouvelle-Aquitaine	0.0	25.2	0.0	0.9	96.7
	Auvergne-Rhône-Alpes	0.0	1.3	8.5	0.2	85.0
	Occitanie	0.3	1.2	0.8	8.9	80.5
	specificity [%]	96.0	91.0	91.9	89.4	

Table 5. Confusion matrix for the classification of 48 German walnut samples with the SVM model, resulting in a 77.4% ± 2.5% overall accuracy using stratified nested cross validation. The mean values of 20 predictions are given.

		Predicted Regional Origin				
		North Rhine-Westphalia	Baden-Württemberg	Hesse	Lower Saxony	Sensitivity [%]
Actual Regional Origin	North Rhine-Westphalia	6.3	0.0	0.0	2.7	70.0
	Baden-Württemberg	0.0	15.1	0.1	0.9	94.1
	Hesse	0.0	2.0	12.0	0.0	85.7
	Lower Saxony	2.2	2.6	0.4	3.8	42.2
	specificity [%]	74.1	76.6	96.4	51.4	

Table 6. Confusion matrix for the classification of 32 Italian walnut samples with the SVM model, resulting in a 94.2% ± 2.8% overall accuracy using stratified nested cross validation. The mean values of 20 predictions are given.

		Predicted Regional Origin				
		Veneto, Padova	Piedmont, Cuneo	Veneto, Rovigo	Campania, Napoli	Sensitivity [%]
Actual Regional Origin	Veneto, Padova	5.0	0.0	0.0	0.0	100.0
	Piedmont, Cuneo	0.0	11.7	1.3	0.1	89.6
	Veneto, Rovigo	0.0	0.4	9.6	0.0	96.0
	Campania, Napoli	0.0	0.0	0.1	3.9	97.5
	specificity [%]	100.0	96.7	87.3	98.7	

The German data set existed of 48 samples from the four German federal states Baden-Württemberg (9 samples), Hesse (16 samples), Lower Saxony (14 samples), and North Rhine-Westphalia (9 samples). The best accuracy of 77.4% ± 2.5% was achieved after center (median) and scale (standard deviation) with an SVM model (Table S13, Supporting Information). Compared to the French samples, the prediction is not as good; here, especially, the samples from Lower Saxony are likely to be misclassified. Again, ANOVA was applied to identify the elements showing significant differences, and the boxplots of the three elements with the highest F-values are shown in Figure S4B. The calculated classification model tends to confuse walnut samples from North Rhine-Westphalia and Lower Saxony, and this is also observable in the boxplots showing similar distributions for Fe and Rb for these regions. Cu shows significant differences between the four regions, but the distributions overlap, making the differentiation difficult.

From Italy, 32 samples could be acquired from Campania, Napoli (four samples); Piedmont, Cuneo (13 samples); Veneto, Padova (five samples); and Veneto, Rovigo (ten samples). An overall accuracy of 94.2% ± 2.8% can be achieved after $\log_{10}$ transformation with an SVM model (Table S14, supporting information). The three Italian regions examined are geographically separated, which may explain the good predictive power, but even within the Veneto region, accurate classification is possible. Most importantly, Fe, Sr, and Zn are relevant marker elements for the differentiation as outlined by an ANOVA test and shown in the corresponding boxplots in Figure S4C.

It should be pointed out that an entirely unknown walnut sample would have to be predicted by the multiclass model first before applying the regional models presented in this section. Mathematically, when applying the classification models one after the other, the prediction accuracies would have to be multiplied (e.g., the chance that a walnut sample will be correctly be predicted as Italian and originating from Napoli would be $0.685 \cdot 0.975 \equiv 66.8\%$). Furthermore, it should be noted that, due to the comparably lower number of samples for the regional classification models, a potential over-fitting is more likely to occur. Thus, more reference samples should be acquired and measured in the future to confirm and enhance the model's reliability.

3.3.6. Further Evaluation of the Classification Models' Performance

Classification models are usually evaluated based on their accuracy, sensitivity, and specificity [47,48,50]. Especially the accuracy is the most important parameter, and of course higher accuracies are desirable. At the same time, the sole expressive power of the accuracy should not be exaggerated. Regarding the classification models of this study, the model for all origins reached 73%, and that for the regional levels in Italy 94%. Both classification issues were validated using nested cross validation; hence, the potential risk of overfitting could be reduced [38,39].

At first glance, 94% seems to be better than 73%, and this appears to be almost paradoxical when considering that the differentiation on a regional level is better than on a worldwide level, although the geographical distances shrunk and, thus, the soil should be more similar. However, it is unreasonable to make a statement about which model is better, because the models are hardly possible to compare with each other—they deal with different issues, and the input data and the classification models' sizes are different.

To our knowledge, there is no additional parameter to evaluate the model's performance and allow us to compare different classification models. Hence, we examined a quotient similar to the signal-to-noise ratio. This parameter is commonly used in analytical chemistry to evaluate a measurement signal. The sole signal's intensity does not allow a meaningful assessment of the signal's quality; instead, the noise has to be considered as well. Then, the higher the signal-to-noise ratio, the more reliable the measurements, and the more robust the results. Regarding the classification models, the signal corresponds to the model's accuracy. The noise is the theoretical accuracy when distributing the samples at random. Mathematically, this value equates to the reciprocal number of classes. Hence, this value correlates to the model's size, and since models with more classes are more challenging to calculate, it accounts for the classification model's difficulty. With the accuracy-to-random ratio, the classification models can be further evaluated:

In Table 7, the characteristics of the classification models discussed in Sections 3.3.4 and 3.3.5 are stated. For comparison, four additional binary classification models were calculated and included in this table: on the one hand, three issues targeting the distinction between Germany and an exporting country (Chile, China, and the USA, respectively); on the other hand, Europe (combining Switzerland, France, Germany, Hungary, and Italy) vs. not-Europe (combining Chile, China, Pakistan, Turkey, and the USA). The very high percentages (>95%) for these binary models should be noted; however, it is maximally possible to be twice as good compared to a random distribution. On the contrary, the classification models on the regional level outperform the random distribution by three times, and the worldwide classification model outperforms the random distribution even by seven, emphasizing the high performance of this model. For the sake of completeness, we also calculated all binary 1-vs.-1 classification models—i.e., all possible two-paired combinations of the ten countries of origins. The results are given in Table S15 (Supplementary Materials), and the accuracies range from 81.7% to 99.0%. Now, also Hungarian and Turkish samples reach fairly good accuracies (>80%), emphasizing again that the calculated accuracies always have to be set in relation to the complexity of the model—i.e., the number of classes. The sole expressive power of the accuracy has limited information value.

Table 7. Evaluation of different classification models for the authentication of walnuts.

Parameter		Accuracy [%]	Number of Classes	Random Distribution [%]	Accuracy-to-Random Ratio
Variables and Equations		a	n	$r = 100\%/n$	a/r
global differentiation	all countries of origin	72.9	10	10	7.29
regional differentiations	FR	91.4	4	25	3.66
	DE	77.4	4	25	3.10
	IT	94.2	4	25	3.77
binary classification models	CN vs. DE	99.5	2	50	1.99
	US vs. DE	96.6	2	50	1.93
	CL vs. DE	98.2	2	50	1.96
	Europe vs. not-Europe	92.4	2	50	1.85

It should therefore be pointed out that the model's accuracy possesses a limited expressive power. However, this is not a generic criticism of the usage of binary classification models; such binary issues often match the authentication problems in practice—e.g., the differentiation of the most expensive white truffle from its cheaper counterfeit [53], or the distinction of a regional product with a protected geographical indication from foreign samples [42].

For walnuts, however, we do not see any options to simplify the multiclass models to binary models, since global trade and/or import to Europe is strongly interlinked, and therefore only the multiclass approach is reasonable for worldwide differentiation.

4. Conclusions

The elemental analysis of walnut with ICP-MS in combination with chemometrics proved to be a powerful technique for geographical origin differentiation on a worldwide and regional level. Although the REE were not considered to be due to too-low concentrations, the worldwide origin was successfully predicted with an overall accuracy of 73%. The most important variables were Al, Ba, Co, Cu, Fe, Mo, Ni, and Sr. No significant loss of accuracy was observed when only the elemental ratios were considered, so fresh walnut samples can be analyzed without the need for a drying step. On a regional level in France, Germany, and Italy, the differentiation of walnut samples was possible, with overall accuracies of 91%, 77%, and 94%, respectively. In the future, we want to broaden walnut authentication with the Noix de Grenoble. Harvest year and cultivar showed no observable influence, which makes this method suitable for predicting new samples without the need for reference samples in future years.

Supplementary Materials: The following figures and tables are available online at http://www.mdpi.com/2304-8158/9/11/1708/s1: Figure S1: PCA plots for the comparison on the influence of harvest year vs. origin; Figure S2: PCA plots for the comparison on the influence of cultivar vs. origin; Figure S3: PCA plots for the regional differentiation of walnut samples within France, Germany and Italy; Figure S4: Boxplots for the significant elements for the walnuts' origins authentication on a regional level in France, Germany and Italy after one-way ANOVA testing. Data expressed as mg/kg in walnut lyophilizate; Table S1: Reagents and materials used in this study; Table S2: Detailed information of all walnut samples analyzed in this study with cultivar, origin and harvest year; Table S3: Microwave digestion procedure; Table S4: Limit of detection (LOD) and limit of quantitation (LOQ) for the measured isotopes for the HR-ICP-MS instrument. Additionally, the respective internal standard element is given; Table S5: Instrumental conditions and measurement parameters for HR-ICP-MS; Table S6: Mean elemental concentrations for the walnut countries of origin in mg/kg; Table S7: Overall accuracy with standard deviation for different data pre-treatment and classification methods for the predictions of all walnut samples using stratified nested cross validation; Table S8: Overall accuracy with standard deviation for different data pre-treatment and classification methods for the predictions of all walnut samples using leave-one-out cross validation; Tables S9–S11: Mean elemental concentrations for the French, German, and Italian walnut regions; Tables S12–S14: Overall accuracy with standard deviation for different data pre-treatment and classification methods for the French, German, and Italian walnut samples using stratified nested cross validation; Table S15: Overall accuracies of binary classification using stratified nested cross-validation of 20 repetitions (classification method: quadratic SVM; data pre-treatment: log10 transformation).

Author Contributions: Conceptualization, T.S.; methodology, T.S.; validation, T.S. and K.v.W.; formal analysis, T.S., A.K. and M.-S.M.; investigation, T.S., K.v.W., and M.-S.M.; resources, M.F.; data curation, T.S. and A.K.; writing—original draft preparation, T.S.; writing—review and editing, T.S., K.v.W., A.K., M.-S.M., and M.F.; visualization, T.S.; supervision, M.F.; project administration, M.F.; funding acquisition, M.F. All authors have read and agreed to the published version of the manuscript.

Funding: This study was performed within the project "Food Profiling—Development of analytical tools for the experimental verification of the origin and identity of food". This project (Funding reference number: 2816500914) is supported by means of the Federal Ministry of Food and Agriculture (BMEL) by a decision of the German Bundestag (parliament). Project support is provided by the Federal Institute for Agriculture and Food (BLE) within the scope of the program for promoting innovation.

Acknowledgments: The authors gratefully thank Johanna Härdter, Christian Marji, Edris Riedel, Caroline Schmitt, and Doreen Teske for their help in sample acquisition and sample preparation. For providing authentic walnut samples, we thank SCA Unicoque, Cancon, France; Matthias Schott, Sasbach-Leiselheim, Germany; Riednuss, Biebesheim am Rhein, Germany; AgroTeamConsulting, Torino, Italy; Nuss—Baumschule Gubler AG, Hörhausen, Switzerland; CRP Food Import Export GmbH, Hamburg, Germany; GoldRiver Orchards, Inc., California, USA. We would like to thank Maike Arndt for her helpful discussion of the manuscript.

Conflicts of Interest: The authors declare no conflict of interest. The funders had no role in the design of the study; in the collection, analysis, or interpretation of data; in the writing of the manuscript; or in the decision to publish the results.

References

1. Martínez, M.L.; Labuckas, D.O.; Lamarque, A.L.; Maestri, D.M. Walnut (*Juglans regia* L.): Genetic resources, chemistry, by-products. *J. Sci. Food Agric.* **2010**, *90*, 1959–1967. [CrossRef] [PubMed]
2. Janick, J.; Paull, R.E. *The Encyclopedia of Fruit and Nuts*; CABI: Wallingford, UK, 2008.
3. Momchilova, S.; Arpadjan, S.; Blagoeva, E. Accumulation of microelements Cd, Cu, Fe, Mn, Pb, Zn in walnuts (*Juglans regia* L.) depending on the cultivar and the harvesting year. *Bulg. Chem. Commun.* **2016**, *48*, 50–54.
4. Creydt, M.; Fischer, M. Omics approaches for food authentication. *Electrophoresis* **2018**, *39*, 1569–1581. [CrossRef] [PubMed]
5. McGrath, T.F.; Shannon, M.; Chevallier, O.P.; Ch, R.; Xu, F.; Kong, F.; Peng, H.; Teye, E.; Akaba, S.; Wu, D. Food Fingerprinting: Using a two-tiered approach to monitor and mitigate food fraud in rice. *J. AOAC Int.* **2020**. [CrossRef]
6. Schelm, S.; Siemt, M.; Pfeiffer, J.; Lang, C.; Tichy, H.-V.; Fischer, M. Food Authentication: Identification and Quantitation of Different Tuber Species via Capillary Gel Electrophoresis and Real-Time PCR. *Foods* **2020**, *9*, 501. [CrossRef]

7. Mannino, G.; Gentile, C.; Maffei, M.E. Chemical partitioning and DNA fingerprinting of some pistachio (*Pistacia vera* L.) varieties of different geographical origin. *Phytochemistry* **2019**, *160*, 40–47. [CrossRef]

8. Grazina, L.; Amaral, J.S.; Mafra, I. Botanical origin authentication of dietary supplements by DNA-based approaches. *Compr. Rev. Food Sci. Food Saf.* **2020**, *19*, 1080–1109. [CrossRef]

9. Kelly, S.; Heaton, K.; Hoogewerff, J. Tracing the geographical origin of food: The application of multi-element and multi-isotope analysis. *Trends Food Sci. Technol.* **2005**, *16*, 555–567. [CrossRef]

10. Aceto, M. The use of ICP-MS in food traceability. In *Advances in Food Traceability Techniques and Technologies*; Elsevier: Amsterdam, The Netherlands, 2016; pp. 137–164.

11. Drivelos, S.A.; Georgiou, C.A. Multi-element and multi-isotope-ratio analysis to determine the geographical origin of foods in the European Union. *TrAC Trends Anal. Chem.* **2012**, *40*, 38–51. [CrossRef]

12. Richter, B.; Gurk, S.; Wagner, D.; Bockmayr, M.; Fischer, M. Food authentication: Multi-elemental analysis of white asparagus for provenance discrimination. *Food Chem.* **2019**, *286*, 475–482. [CrossRef]

13. Segelke, T.; von Wuthenau, K.; Neitzke, G.; Mueller, M.-S.; Fischer, M. Food Authentication: Species and origin determination of truffles (*Tuber* spp.) by inductively coupled plasma mass spectrometry (ICP-MS) and chemometrics. *J. Agric. Food Chem.* **2020**. [CrossRef] [PubMed]

14. Oddone, M.; Aceto, M.; Baldizzone, M.; Musso, D.; Osella, D. Authentication and traceability study of hazelnuts from Piedmont, Italy. *J. Agric. Food Chem.* **2009**, *57*, 3404–3408. [CrossRef] [PubMed]

15. Esteki, M.; Farajmand, B.; Amanifar, S.; Barkhordari, R.; Ahadiyan, Z.; Dashtaki, E.; Mohammadlou, M.; Vander Heyden, Y. Classification and authentication of Iranian walnuts according to their geographical origin based on gas chromatographic fatty acid fingerprint analysis using pattern recognition methods. *Chemom. Intell. Lab. Syst.* **2017**, *171*, 251–258. [CrossRef]

16. Gu, X.; Zhang, L.; Li, L.; Ma, N.; Tu, K.; Song, L.; Pan, L. Multisource fingerprinting for region identification of walnuts in Xinjiang combined with chemometrics. *J. Food Process. Eng.* **2018**, *41*, e12687. [CrossRef]

17. Krauß, S.; Vieweg, A.; Vetter, W. Stable isotope signatures ($\delta 2H$-, $\delta 13C$-, $\delta 15N$-values) of walnuts (*Juglans regia* L.) from different regions in Germany. *J. Sci. Food Agric.* **2020**, *100*, 1625–1634. [CrossRef] [PubMed]

18. Popescu, R.; Ionete, R.E.; Botoran, O.R.; Costinel, D.; Bucura, F.; Geana, E.I.; Alabedallat, Y.F.J.; Botu, M. 1H-NMR profiling and carbon isotope discrimination as tools for the comparative assessment of walnut (*Juglans regia* L.) cultivars with various geographical and genetic origins—A preliminary study. *Molecules* **2019**, *24*, 1378. [CrossRef]

19. Valdés, A.; Beltrán, A.; Mellinas, C.; Jiménez, A.; Garrigós, M.C. Analytical methods combined with multivariate analysis for authentication of animal and vegetable food products with high fat content. *Trends Food Sci. Technol.* **2018**, *77*, 120–130. [CrossRef]

20. Rodushkin, I.; Engström, E.; Sörlin, D.; Baxter, D. Levels of inorganic constituents in raw nuts and seeds on the Swedish market. *Sci. Total Environ.* **2008**, *392*, 290–304. [CrossRef]

21. Juranović Cindrić, I.; Zeiner, M.; Hlebec, D. Mineral composition of elements in walnuts and walnut oils. *Int. J. Environ. Res. Public Health* **2018**, *15*, 2674. [CrossRef]

22. Moodley, R.; Kindness, A.; Jonnalagadda, S.B. Elemental composition and chemical characteristics of five edible nuts (almond, Brazil, pecan, macadamia and walnut) consumed in Southern Africa. *J. Environ. Sci. Heal. Part B* **2007**, *42*, 585–591. [CrossRef]

23. Ozyigit, I.I.; Uras, M.E.; Yalcin, I.E.; Severoglu, Z.; Demir, G.; Borkoev, B.; Salieva, K.; Yucel, S.; Erturk, U.; Solak, A.O. Heavy Metal Levels and Mineral Nutrient Status of Natural Walnut (*Juglans regia* L.) Populations in Kyrgyzstan: Nutritional Values of Kernels. *Biol. Trace Elem. Res.* **2018**, *189*, 277–290. [CrossRef] [PubMed]

24. Food and Agriculture Organization of the United Nations (FAOSTAT). Values of Agricultural Production and Trade of Walnuts in shell. Available online: http://www.fao.org/faostat/en/#home (accessed on 8 September 2020).

25. Wold, S.; Esbensen, K.; Geladi, P. Principal component analysis. *Chemom. Intell. Lab. Syst.* **1987**, *2*, 37–52. [CrossRef]

26. Maaten, L.V.D.; Hinton, G. Visualizing data using t-SNE. *J. Mach. Learn. Res.* **2008**, *9*, 2579–2605.

27. Zhao, H.; Zhang, S.; Zhang, Z. Relationship between multi-element composition in tea leaves and in provenance soils for geographical traceability. *Food Control* **2017**, *76*, 82–87. [CrossRef]

28. Cubero-Leon, E.; Peñalver, R.; Maquet, A. Review on metabolomics for food authentication. *Food Res. Int.* **2014**, *60*, 95–107. [CrossRef]

29. Kim, H.-Y. Analysis of variance (ANOVA) comparing means of more than two groups. *Restor. Dent. Endod.* **2014**, *39*, 74–77. [CrossRef] [PubMed]

30. Bland, J.M.; Altman, D.G. Multiple significance tests: The Bonferroni method. *Bmj* **1995**, *310*, 170. [CrossRef]

31. The Mathworks. Visualize Summary Statistics with Box Plot. Available online: https://de.mathworks.com/help/stats/boxplot.html (accessed on 30 October 2020).

32. Spruyt, V. How to Draw a Covariance Error Ellipse? Available online: https://www.visiondummy.com/2014/04/draw-error-ellipse-representing-covariance-matrix/ (accessed on 9 November 2020).

33. Cortes, C.; Vapnik, V. Support-vector networks. *Mach. Learn.* **1995**, *20*, 273–297. [CrossRef]

34. Bachmann, R.; Klockmann, S.; Haerdter, J.; Fischer, M.; Hackl, T. 1H NMR spectroscopy for determination of the geographical origin of hazelnuts. *J. Agric. Food Chem.* **2018**, *66*, 11873–11879. [CrossRef]

35. The Mathworks. Framework for Ensemble Learning. Available online: https://de.mathworks.com/help/stats/framework-for-ensemble-learning.html (accessed on 1 October 2020).

36. Breiman, L. Random forests. *Mach. Learn.* **2001**, *45*, 5–32. [CrossRef]

37. Gerretzen, J.; Szymańska, E.; Jansen, J.J.; Bart, J.; van Manen, H.-J.; van den Heuvel, E.R.; Buydens, L.M. Simple and effective way for data preprocessing selection based on design of experiments. *Anal. Chem.* **2015**, *87*, 12096–12103. [CrossRef]

38. Krstajic, D.; Buturovic, L.J.; Leahy, D.E.; Thomas, S. Cross-validation pitfalls when selecting and assessing regression and classification models. *J. Chemin.* **2014**, *6*, 10. [CrossRef] [PubMed]

39. Varma, S.; Simon, R. Bias in error estimation when using cross-validation for model selection. *BMC Bioinform.* **2006**, *7*, 91. [CrossRef] [PubMed]

40. Food and Agriculture Organization of the United Nations (FAOSTAT). Import Quantity of shelled and in-shell Walnuts. Available online: http://www.fao.org/faostat/en/#home (accessed on 15 September 2020).

41. Bitter, N.Q.; Fernandez, D.P.; Driscoll, A.W.; Howa, J.D.; Ehleringer, J.R. Distinguishing the region-of-origin of roasted coffee beans with trace element ratios. *Food Chem.* **2020**, *320*, 126602. [CrossRef] [PubMed]

42. Zettl, D.; Bandoniene, D.; Meisel, T.; Wegscheider, W.; Rantitsch, G. Chemometric techniques to protect the traditional Austrian pumpkin seed oil. *Eur. J. Lipid Sci. Technol.* **2017**, *119*, 1600468. [CrossRef]

43. Ballabio, D.; Todeschini, R. Multivariate classification for qualitative analysis. *Infrared Spectrosc. Food Qual. Anal. Control* **2009**, *83*, e102.

44. Mishra, P.; Nordon, A.; Tschannerl, J.; Lian, G.; Redfern, S.; Marshall, S. Near-infrared hyperspectral imaging for non-destructive classification of commercial tea products. *J. Food Eng.* **2018**, *238*, 70–77. [CrossRef]

45. Drivelos, S.A.; Danezis, G.P.; Haroutounian, S.A.; Georgiou, C.A. Rare earth elements minimal harvest year variation facilitates robust geographical origin discrimination: The case of PDO "Fava Santorinis". *Food Chem.* **2016**, *213*, 238–245. [CrossRef]

46. Cosmulescu, S.N.; Baciu, A.; Achim, G.; Mihai, B.; Trandafir, I. Mineral composition of fruits in different walnut (*Juglans regia* L.) cultivars. *Not. Bot. Horti Agrobot. Cluj-Napoca* **2009**, *37*, 156–160.

47. Latorre, C.H.; Crecente, R.P.; Martín, S.G.; García, J.B. A fast chemometric procedure based on NIR data for authentication of honey with protected geographical indication. *Food Chem.* **2013**, *141*, 3559–3565. [CrossRef]

48. da Costa, N.L.; Ximenez, J.P.B.; Rodrigues, J.L.; Barbosa, F.; Barbosa, R. Characterization of Cabernet Sauvignon wines from California: Determination of origin based on ICP-MS analysis and machine learning techniques. *Eur. Food Res. Technol.* **2020**, *246*, 1193–1205. [CrossRef]

49. Borràs, E.; Ferré, J.; Boqué, R.; Mestres, M.; Aceña, L.; Busto, O. Data fusion methodologies for food and beverage authentication and quality assessment—A review. *Anal. Chim. Acta* **2015**, *891*, 1–14. [CrossRef] [PubMed]

50. Arndt, M.; Rurik, M.; Drees, A.; Bigdowski, K.; Kohlbacher, O.; Fischer, M. Comparison of different sample preparation techniques for NIR screening and their influence on the geographical origin determination of almonds (*Prunus dulcis* MILL.). *Food Control* **2020**, *115*, 107302. [CrossRef]

51. Anastas, P.; Eghbali, N. Green chemistry: Principles and practice. *Chem. Soc. Rev.* **2010**, *39*, 301–312. [CrossRef]

52. Comité Interprofessionnel de la Noix de Grenoble. Noix de Grenoble, zone de Production. Available online: https://www.aoc-noixdegrenoble.com/terroir/zone-de-production/ (accessed on 9 September 2020).

53. Segelke, T.; Schelm, S.; Ahlers, C.; Fischer, M. Food Authentication: Truffle (*Tuber* spp.) Species Differentiation by FT-NIR and Chemometrics. *Foods* **2020**, *9*, 922. [CrossRef]

 *foods*

MDPI

Article

Geographic Pattern of Sushi Product Misdescription in Italy—A Crosstalk between Citizen Science and DNA Barcoding

Anna Maria Pappalardo *, Alessandra Raffa, Giada Santa Calogero and Venera Ferrito

Department of Biological, Geological and Environmental Sciences, Section of Animal Biology "M. La Greca", University of Catania, Via Androne 81, 95124 Catania, Italy; alessandra.raffa92@gmail.com (A.R.); giadacalogero@gmail.com (G.S.C.); vferrito@unict.it (V.F.)
* Correspondence: pappalam@unict.it; Tel.: + 39-095-730-6051

Abstract: The food safety of sushi and the health of consumers are currently of high concern for food safety agencies across the world due to the globally widespread consumption of these products. The microbiological and toxicological risks derived from the consumption of raw fish and seafood have been highlighted worldwide, while the practice of species substitution in sushi products has attracted the interest of researchers more than food safety agencies. In this study, samples of sushi were processed for species authentication using the Cytochrome Oxidase I (COI) gene as a DNA barcode. The approach of Citizen Science was used to obtain the sushi samples by involving people from eighteen different Italian cities (Northern, Central and Southern Italy). The results indicate that a considerable rate of species substitution exists with a percentage of misdescription ranging from 31.8% in Northern Italy to 40% in Central Italy. The species most affected by replacement was *Thunnus thynnus* followed by the flying fish roe substituted by eggs of *Mallotus villosus*. These results indicate that a standardization of fish market names should be realized at the international level and that the indication of the scientific names of species should be mandatory for all products of the seafood supply chain.

Keywords: sushi restaurants; COI barcoding; molecular traceability; teleosts

Citation: Pappalardo, A.M.; Raffa, A.; Calogero, G.S.; Ferrito, V. Geographic Pattern of Sushi Product Misdescription in Italy—A Crosstalk between Citizen Science and DNA Barcoding. *Foods* **2021**, *10*, 756. https://doi.org/10.3390/foods10040756

Academic Editors: Margit Cichna-Markl and Isabel Mafra

Received: 16 February 2021
Accepted: 27 March 2021
Published: 2 April 2021

Publisher's Note: MDPI stays neutral with regard to jurisdictional claims in published maps and institutional affiliations.

1. Introduction

In part I of Food Business Regulation (Cap. 132X) of the Government of the Hong Kong Special Administrative Region, the meanings of the terms sushi and sashimi are made explicit. In particular, sushi is described as "food consisting of cooked and pressed rice flavoured with vinegar and garnished with other food ingredients including raw or cooked or vinegared seafood, marine fish or shellfish roe, vegetable, cooked meat or egg on top or in the middle which may or may not be wrapped with seaweed and usually served in pieces", while sashimi is described as "food consisting of fillets of marine fish, molluscs, crustaceans, fish roe or other seafood to be eaten in raw state". Although sushi and sashimi are perceived by consumers as healthy foods, the biological and chemical hazards for human health, derived from the consumption of raw fish and seafood, have been highlighted worldwide, such as the risk of parasitic and/or pathogenic microorganism infection [1–7]; the potential risk arising from a lack of proper control of temperature of these perishable foods [8]; the risk of exposure to toxicants, such as heavy metals and polychlorinated biphenyls (PCBs); polycyclic aromatic hydrocarbons (PAHs) and other contaminants [4,7,9,10]. The food safety of sushi and sashimi and the health of consumers are currently of high concern given that the consumption of these products is now globally widespread [11,12]. As a result, the most important food safety agencies in the world, such as the European Food Safety Authority [13], the Food and Drug Administration [14], the Hong Kong Food and Environmental Hygiene Department's and the World Health Organization, have implemented regulations and guidelines to face all issue related to the consumption of raw fish and seafood. In this context, another important issue that has attracted the interest of researchers is the molecular authentication of fish and seafood species in transformed products, because the processing procedure generally removes the specific diagnostic morphological traits useful to assign the product to a particular species through only morphological inspection. Indeed, the voluntary or involuntary practices of substitution of valuable species with species of less value for economic profit have been detected worldwide in the last decades by using DNA sequencing, which proved to be the most useful method to unveil these frauds. For example, DNA-based surveys carried out in European and non-European countries have highlighted a high rate of food frauds in the fishery sector [15–18]. Among the most used molecular markers, mitochondrial genes, such as Cytochrome b (Cytb), 16S rRNA (16S), Cytochrome Oxidase I (COI) and mtDNA Control Region (CR), have proven to be optimal tools for seafood species authentication. However, while the CR and Cytb have been successfully and widely used to study the genetic population structure [19–26] rather than to authenticate fish species [27–29], COI has become the optimal DNA barcode for the identification of animal species [30–35] and particularly for fish species authentication in seafood products [15,36–41]. Furthermore, researchers have been also encouraged to look for rapid and low-cost molecular strategies to tackle substitution species frauds by large scale screening both using classic and new technologies [42–49]. COI DNA barcoding has been used to unveil the misdescription of sushi products in the United States of America [50], the United Kingdom [11], South-Korea [17], Malaysia [51,52] and Canada [53]. In Italy, the study by Armani et al. [54] performed a molecular-based authentication of the seafood species used in sushi preparations in four provinces of Tuscany. However, when designing a food fraud investigation, the sampling plan is pivotal to ensure that as many products as possible are sampled over a large area. In this context, the contribution of consumers is crucial, and the citizen science (CS) approach, based on involving a large number of people, normally including the local population of a region or a state, with the aim of collecting scientific data, could prove to be of fundamental help. This strategy permits the collection of a vast quantity of data information or samples that cannot be collected by only one researcher or a small research team. The quality of a study is not undermined by the citizen science approach if the work planning includes comprehensible protocols, effective training before starting and accurate oversight during the studying period [55,56]. Based on these premises, in this study, we involved many people from eighteen different cities throughout the Italian territory (North, Central and South Italy) to obtain samples of sushi to be processed for species authentication by using the COI gene as the DNA barcode. To the best of our knowledge, this is the first study on sushi authentication extended to the Italian territory by using the approach of citizen sciences. The aim was to analyze the compliance of the fish names of marketed products with the list of Italian names of fish species of commercial interest included in the Italian Ministerial Decree (MD) n.19105 22 September 2017 of the Italian Ministry of Agricultural, Food and Forestry Policies and then to verify if the information the consumers obtain from the menu meet the transparency requirements established by the European regulations.

2. Materials and Methods

2.1. Sample Collection and Survey

Between January 2018 and January 2019, we collected sushi products sold in restaurants and takeaways in different cities of Northern, Central and Southern Italy (Figure 1). Samples were obtained using a "citizen science" strategy involving people who responded to the invitation to participate in the "sushi survey". People living in various Italian regions were chosen among relatives and friends of our research team and colleagues at the University of Catania. This allowed us to establish a direct contact with them to better program the sampling. Prior to the start of the study, people received a letter from us where we explained our research project and asked them about their willingness to participate. After receiving their consent, we contacted them by phone and also via skype (i) to respond to all queries they would ask us; (ii) to explain how to proceed for sampling sushi products;

and (iii) how to fill in the documents that they would receive by us. In particular, we advised them to focus the sampling on white fish, tuna and eggs. By mail, we provided participants with a step-by-step guide for sampling, including a sample collection table (Figure S1: sushi sampling guide) together with a 1.5 mL eppendorf tube to be used to preserve small pieces of sampled sushi in 95% ethanol. A stamped envelope to be used to send us the samples and the collection table was also included. In the table, participants indicated the sushi venue (restaurant or takeaway) they visited; the name on the menu of the product they consumed; and how many samples among white fish, tuna and/or eggs they collected.

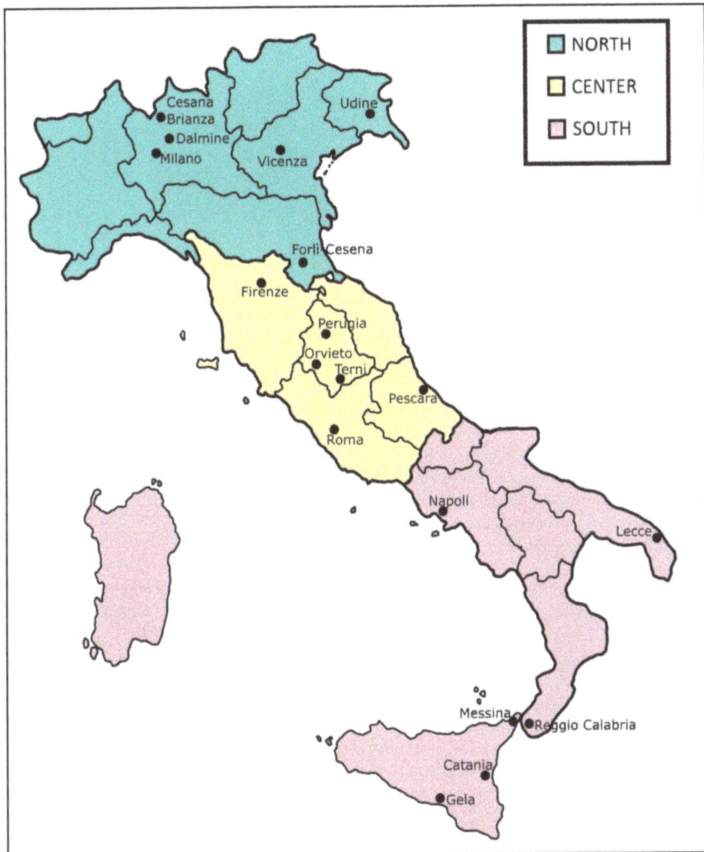

Figure 1. Collection sites of the sushi survey in Northern (green), Central (light yellow) and Southern (pink) Italy.

2.2. DNA Barcoding Analysis

A total of 180 samples were processed for DNA analysis. For each sushi product, 3 DNA extractions were replicated to investigate the presence of multiple fish species in the product. Total genomic DNA was extracted using a DNeasy tissue kit (Qiagen, Hilden, Germany) following the manufacturer's instructions and with some modifications. DNA concentration was measured with a NanoDrop One spectrophotometer (Thermo Scientific, Waltham, MA, USA). A portion of about 650 bases of the COI gene was amplified following the Polymerase Chain Reaction (PCR) conditions reported by [38] in a 50 µL reaction mixture also containing the M13 tailed primers (VF2_t1 and FishR2_t1) described in Ivanova et al. [57] to improve the sequencing quality of the PCR products. Negative controls were included in all PCR runs to check for cross-contamination. Amplicons successfully obtained were verified by electrophoresis on a 0.8% agarose gel and displayed through a Safe Imager TM 2.0 Blue Light Transilluminator (Thermo Fisher, Waltham, MA, USA) using the SYBR® Safe dye (Thermo Fisher, Waltham, MA USA). The QIAquick PCR purification kit (Qiagen, Hilden, Germany) was used to purify all amplicons, which were then bidirectionally sequenced with M13 sequencing primers using an ABI 3730 automated sequencing machine at Genechron Biotech Company (https://www.genechron.com accessed on 30 January 2021).

2.3. Data Analysis

The chromatograms were checked for the quality of peaks and assembled using ChromasPro 2.6.6 software (https://technelysium.com.au/wp/chromaspro/ accessed on 30 January 2021). Barcode multiple-sequence alignment was carried out using the online version of MAFFT v.7 [58]. Sequences were trimmed when the errors occurred near the beginning and again at the end of any sequence. Primer sequences were manually removed by using BioEdit 7.2 (https://bioedit.software.informer.com/versions/ accessed on 30 January 2021). The obtained sequences were carefully checked for the presence of nuclear mitochondrial pseudogenes or nuclear mitochondrial DNA sequences (NUMTs), which could be easily coamplified with orthologous mtDNA sequences [59]. The translation of nucleotide sequences to amino acids was performed by the EMBOSS Transeq tool (https://www.ebi.ac.uk/Tools/st/emboss_transeq accessed on 30 January 2021January) in order to check for premature stop codons and to verify that the open reading frames were maintained in the protein-coding locus. To confirm the identity of the amplified sequences, we conducted Basic Local Alignment Searches (BLAST) (https://blast.ncbi.nlm.nih.gov accessed on 30 January 2021) against GenBank without "Uncultured/environmental sample sequences" with megablast and default parameters (https://www.ncbi.nlm.nih.gov/genbank/ accessed on 30 January 2021) and also used the BOLD database (https://www.boldsystems.org/ accessed on 30 January 2021) to validate our sequences. For species assignment, the highest values of percent identity found between the query sequence and the BLAST matched sequences were selected. If multiple BLAST matches had identical percent identity values, it was confirmed that all matches belonged to the same species. All sequences obtained from the present study were published in the National Center for Biotechnology Information database (NCBI), and their GenBank accession numbers are reported in Tables 1–3.

Table 1. Sushi sampling in Northern Italy. In square brackets, the number of processed samples for each sushi product. In bold, misdescription cases.

Code *	Retail Point	Menu/Label Description	Scientific Nameof Declared Species	Identified Species by DNA Barcoding and BLAST Search	GenBank Acc. Number of Obtained Sequences	Matched GenBank Accession from BLAST °	Matched BOLD ID	% Identity with 100% Coverage
MIL1B [3]	Restaurant	sea bream or common bass	*Sparus aurata/Dicentrarchus labrax*	*Dicentrarchus labrax*	MW714726	KP330301	GBMIN94166-17	99.69
MIL1T [3]	Restaurant	tuna	*Thunnus thynnus*	**Thunnus albacares**	MW714727	MH638785	ANGBF54814-19	99.54
MIL2B [1]	Takeaway	sea bream or common bass	*Sparus aurata/Dicentrarchus labrax*	*Dicentrarchus labrax*	MW714728	KY176457	GBMIN121550-17	99.37
MIL2T [3]	Takeaway	tuna	*Thunnus thynnus*	**Thunnus albacares**	MW714729	MH638777	ANGBF54806-19	98.47
MIL3B [3]	Takeaway	sea bream	*Sparus aurata/Dicentrarchus labrax*	*Dicentrarchus labrax*	MW714730	KP330301	GBMIN94165-17	98.74
MIL4B [3]	Takeaway	common bass	*Sparus aurata/Dicentrarchus labrax*	*Sparus aurata*	MW714731	MF438138	ANGBF45411-19	99.54
CES1B [3]	Takeaway	common bass	*Dicentrarchus labrax*	*Dicentrarchus labrax*	MW714732	KP330300	GBMIN94165-17	99.06
DAL1B [3]	Restaurant	sea bream	*Sparus aurata*	*Sparus aurata*	MW714733	JQ623999	DNATR096-12	99.24
DAL1T [3]	Restaurant	tuna	*Thunnus thynnus*	**Thunnus albacares**	MW714734	MH638785	ANGBF54814-19	99.24
DAL1E [3]	Restaurant	tobiko/flying fish egg	*Hirundichthys affinis* °	*Hirundichthys oxycephalus*	MW714735	KX769042	GBMIN125981-17	99.02
VI1T [3]	Restaurant	tuna	*Thunnus thynnus*	*Thunnus thynnus*	MW714736	KP975912	FCSF387-14	98.92
VI1B [3]	Restaurant	sea bream	*Sparus aurata*	*Sparus aurata*	MW714737	KC501553	DNATR1582-13	98.78
VI1E [3]	Restaurant	tobiko/flying fish egg	*Hirundichthys affinis* °	*Hirundichthys affinis*	MW714738	JQ842898	TOBA086-09	99.52
FC1B [3]	Takeaway	sea bream	*Sparus aurata*	**Seriola lalandi**	MW714739	MH211123	GBMNA18700-19	99.39
FC1T [3]	Takeaway	maguro Yaki (red tuna)	*Thunnus thynnus*	*Thunnus thynnus*	MW714740	KC501694	DNATR1720-13	99.39
FC2B [1]	Takeaway	sea bream	*Sparus aurata*	*Sparus aurata*	MW714741	MF438138	ANGBF45411-19	99.69
UD1B [3]	Restaurant	kajiki roll (swordfish)	*Xiphias gladius*	*Xiphias gladius*	MW714742	MK295657	ANGBF51916-19	99.38
UD1T [2]	Restaurant	tuna	*Thunnus thynnus*	**Thunnus obesus**	MW714743	GU451774	GBGCA1353-13	99.08
UD1E [3]	Restaurant	tobiko/flying fish egg	*Hirundichthys affinis* °	**Mallotus villosus**	MW714744	HM421773	DSFAL635-09	99.39

* MIL = Milano; CES = Cesena Brianza; DAL = Dalmine; VI = Vicenza; FC = Forlì Cesena; UD = Udine. ° These species are not present in the Italian D.M. 2008.

Table 2. Sushi sampling in Central Italy. In square brackets, the number of processed samples for each sushi product. In bold, misdescription cases.

Code *	Retail Point	Menu/Label Description	Scientific Name of Declared Species	Identified Species by DNA Barcoding and BLAST Search	GenBank Acc. Number of Obtained Sequences	Matched GenBank Accession from BLAST °	Matched BOLD ID	% Identity with 100% Coverage
FIR1B [3]	Takeaway	common bass	*Dicentrarchus labrax*	*Dicentrarchus labrax*	MW714657	KP330300	GBMIN94165-17	99.21
FIR1T [3]	Takeaway	tuna	*Thunnus thynnus*	***Thunnus albacares***	MW714658	MH638777	ANGBF54806-19	99.85
FIR1E [3]	Takeaway	tobiko/flying fish egg	*Hirundichthys affinis* °	*Hirundichthys affinis*	MW714659	JQ842898	TOBA9086	99.52
PER1B [3]	Restaurant	sea bream or common bass	*Sparus aurata/Dicentrarchus labrax*	*Sparus aurata*	MW714660	MF438138	ANGBF45411-19	99.54
PER1T [3]	Restaurant	tuna	*Thunnus thynnus*	***Thunnus albacares***	MW714661	MH638785	ANGBF54814-19	99.39
PER1E [1]	Restaurant	tobiko/flying fish egg	*Hirundichthys affinis* °	*Hirundichthys affinis*	MW714662	JQ842898	TOBA9086	99.35
PER2B [3]	Takeaway	sea bream or common bass	*Sparus aurata/Dicentrarchus labrax*	*Dicentrarchus labrax*	MW714663	KP330301	GBMIN94166-17	99.21
ORV1B [3]	Takeaway	common bass	*Dicentrarchus labrax*	*Dicentrarchus labrax*	MW714664	KP330301	GBMIN94166-17	98.58
ORV1T [2]	Takeaway	tuna	*Thunnus thynnus*	***Thunnus albacares***	MW714665	MH638785	ANGBF54814-19	98.17
ORV1E [3]	Takeaway	tobiko/flying fish egg	*Hirundichthys affinis* °	*Hirundichthys affinis*	MW714666	JQ842898	TOBA086-09	99.52
ORV2B [3]	Restaurant	sea bream	*Sparus aurata*	*Sparus aurata*	MW714667	KC501553	DNATR1582-13	98.47
TER1B [3]	Takeaway	sea bream	*Sparus aurata*	*Sparus aurata*	MW714668	KC501557	DNATR1596-13	99.39
TER1T [3]	Takeaway	tuna	*Thunnus thynnus*	***Thunnus orientalis***	MW714669	JN097817	GBGCA1390-13	99.70
TER1E [3]	Takeaway	tobiko/flying fish egg	*Hirundichthys affinis* °	*Hirundichthys affinis*	MW714670	JQ842898	TOBA9086	99.52
PE1B [3]	Restaurant	common bass	*Dicentrarchus labrax*	*Dicentrarchus labrax*	MW714671	KP330301	GBMIN94166-17	98.58
PE1T [3]	Restaurant	tuna	*Thunnus thynnus*	***Thunnus albacares***	MW714672	MH638785	ANGBF54814-19	98.92
PE1E [1]	Restaurant	tobiko/flying fish egg	*Hirundichthys affinis* °	*Hirundichthys coromandelensis*	MW714673	KX379460	ANGBF32076-19	98.73
RO1B [3]	Restaurant	sea bream	*Sparus aurata*	***Seriola lalandi***	MW714674	MF069453	ANGBF17684-19	99.24
RO1T [3]	Restaurant	tuna	*Thunnus thynnus*	***Thunnus albacares***	MW714675	HM007768	ANGBF7098-12	99.39
RO1E [1]	Restaurant	tobiko/flying fish egg	*Hirundichthys affinis* °	***Mallotus villosus***	MW714676	FJ205579	GBGC7486-09	99.23

* FIR = Firenze; PER = Perugia; ORV = Orvieto; TER = Terni; PE = Pescara; RO = Roma. ° These species are not present in the Italian D.M. 2008.

Table 3. Sushi sampling in Southern Italy. In square brackets, the number of processed samples for each sushi product. In bold, misdescription cases.

Code *	Retail Point	Menu/Label Description	Scientific Name of Declared Species	Identified Species by DNA Barcoding and BLAST Search	GenBank Acc. Number of Obtained Sequences	Matched GenBank Accession from BLAST °	Matched BOLD ID	% Identity with 100% Coverage
CAT1B [3]	Restaurant	sea bream or common bass	*Sparus aurata*	*Sparus aurata*	MW714949	KC501553	DNATR1582-13	99.08
CAT1T [3]	Restaurant	Tuna	*Thunnus thynnus*	**Thunnus albacares**	MW714950	MH638785	ANGBF54814-19	99.54
CAT1E [3]	Restaurant	tobiko/flying fish egg	*Hirundichthys affinis* °	**Mallotus villosus**	MW714951	FJ205579	GBGC7486-09	99.39
CAT3B [3]	Restaurant	sea bream or common bass	*Sparus aurata*	**Xiphias gladius**	MW714952	JN049558	ANGBF7251-12	99.54
GE1B [3]	Restaurant	common bass	*Dicentrarchus labrax*	*Dicentrarchus labrax*	MW714953	KP330301	GBMIN94166-17	99.53
GE1T [3]	Restaurant	Tuna	*Thunnus thynnus*	**Thunnus albacares**	MW714954	MH638785	ANGBF54814-19	99.08
GE1E [3]	Restaurant	tobiko/flying fish egg	*Hirundichthys affinis* °	*Hirundichthys affinis*	MW714955	JQ842898	TOBA086-09	99.52
GE2B [3]	Restaurant	common bass	*Dicentrarchus labrax*	*Dicentrarchus labrax*	MW714956	KP330301	GBMIN94166-17	99.53
ME1B [3]	Restaurant	sea bream	*Sparus aurata*	*Sparus aurata*	MW714957	MF438138	ANGBF45411-19	99.85
ME1T [3]	Restaurant	Tuna	*Thunnus thynnus*	**Thunnus albacares**	MW714958	MH638762	ANGBF54791-19	98.93
ME1E [3]	Restaurant	lumpfish roe	*Cyclopterus lumpus*	*Cyclopterus lumpus*	MW714959	MG421634	TZAIC166-05	99.54
ME2B [3]	Restaurant	sea bream	*Sparus aurata*	*Sparus aurata*	MW714960	MF438138	ANGBF45411-19	99.39
ME2E [2]	Restaurant	Ikura	salmon eggs	*Oncorhynchus keta*	MW714961	LC094477	ANGBF41103-19	98.93
RC1B [3]	Takeaway	Anago	*Anguilla sp*	*Anguilla rostrata*	MW714962	KX459333	SERCA165-12	98.31
RC1T [3]	Takeaway	Tuna	*Thunnus thynnus*	**Thunnus albacares**	MW714963	MH638785	ANGBF54814-19	99.39
RC1E [2]	Takeaway	Ikura	salmon eggs	*Oncorhynchus keta*	MW714964	LC094477	ANGBF41103-19	99.54
RC2E [2]	Takeaway	lumpfish roe	*Cyclopterus lumpus*	*Cyclopterus lumpus*	MW714965	MG421634	TZAIC166-05	99.07
LE1B [3]	Restaurant	sea bream	*Sparus aurata*	*Sparus aurata*	MW714966	MF438138	ANGBF45411-19	98.93
LE1E [1]	Restaurant	tobiko/flying fish egg	*Hirundichthys affinis* °	**Mallotus villosus**	MW714967	FJ205579	GBGC7486-09	99.39
NA1B [3]	Takeaway	sea bream or common bass	*Sparus aurata/Dicentrarchus labrax*	*Sparus aurata*	MW714968	KC501554	DNATR1599-13	99.24
NA2B [3]	Takeaway	sea bream or common bass	*Sparus aurata/Dicentrarchus labrax*	**Pomatomus saltatrix**	MW714969	KC501113	DNATR1143-13	99.39

* NA = Napoli; LE = Lecce; RC = Reggio Calabria; ME = Messina; CAT = Catania; GE = Gela. ° These species are not present in the Italian D.M. 2008.

3. Results

3.1. Sampling

A total of 61 sushi samples consisting of 45 fish samples, white fish and tuna, and 16 roe samples were collected from 15 restaurants and 14 takeaways from people living in 18 Italian cities who responded to the invitation to participate in the "sushi survey" (Figure 1). For each sushi venue, participants collected from 1 to 3 samples; in the latter case, "white fish", "tuna" and "fish roe" were sampled. The initial instructions provided by us to the participants in the survey allowed us to obtain a homogeneous, high-quality sampling plan throughout the territory. In Tables 1–3, the names found on the menu/label for each sample were reported, as well the corresponding scientific names of the declared species found in the list of the Italian names of fish species of commercial interest included in the Italian ministerial decree (MD) 21 September 2017. Misdescription was marked up when no match was found among the name on the menu, the scientific name in the list of the MD and the fish species identified by DNA barcoding.

3.2. DNA Barcoding

Three samples of each sushi product for a total of 180 samples were processed; however, DNA extraction was unsuccessful for 17 samples, and a total of 163 COI DNA sequences were obtained. The presence of multiple fish species was not detected after the COI sequencing of three samples for each examined product. The sequence length was between 636 and 655 bp. In these functional mitochondrial COI sequences, no insertions, deletions or stop codons were observed, and NUMTs were not sequenced given that vertebrate NUMTS are generally smaller than 600 bp [59]. A total of 16 fish species were identified in all examined sushi products. The percent identity between the COI query sequences and their top-match sequences ranged from 98.17 to 99.85 with 100% of sequence coverage (Tables 1–3).

3.3. Geographic Pattern of Sushi Product Misdescription

3.3.1. Northern Italy

Red tuna, *Thunnus thynnus*, was substituted by yellowfin tuna, *T. albacares*, in three cases and by bigeye tuna, *T. obesus*, in one case; sea bream, *Sparus aurata*, was substituted in one case by yellowtail amberjack, *Seriola lalandi*. Concerning fish roe, only in one case, under the name tobiko or flying fish roe, the eggs of *Mallotus villosus* were found in place of the eggs of species of the genus *Hirundichthys* (Table 1).

3.3.2. Central Italy

In five cases, *T. thynnus* was substituted by *T. albacares* and in one case by *T. orientalis*, while tobiko or flying fish eggs were substituted by *M. villosus* eggs (Table 2).

3.3.3. Southern Italy

In all cases, red tuna, *T. thynnus*, was substituted by *T. albacares*. Sea bream was substituted in one case by *Xiphias gladius* and in another case by the bluefish, *Pomatomus saltatrix*. Tobiko or flying fish eggs in one case were substituted by the eggs of *M. villosus*.

Based on the names of the products chosen by consumers on the menu, a total of 17 species should have been detected, but we found a total of 29 species (Table 3).

4. Discussion

The results of the survey on the authentication of fish species used for sushi products sold in restaurants and takeaways in Italy indicate that a considerable rate of species substitution exists throughout the territory and that it is focused on certain species. The percentage of misdescription ranges from 31.8% in Northern Italy to 40% in Central Italy. The rate of misdescription affecting takeaways ranges from 25% of cases in Northern Italy to 50% in Southern Italy, while the percentage of misdescription in restaurants ranges from 33.3% in Southern Italy to 50% in Central Italy. The species most affected by replacement was *Thunnus thynnus*, which was substituted in 67% of cases in Northern Italy and 100% of cases in Central and Southern Italy. The so-called "white fish" usually represented by *S. aurata* and *D. labrax* was affected by a low rate of substitution ranging from 11% in Northern Italy to 22% in Southern Italy. Finally, tobiko or flying fish roe was affected by a medium rate of substitution ranging from 20% in Central Italy to 33% in Northern Italy. Before discussing our results, it should be noted that i) we compared them with those obtained from a similar survey carried out in Italy and in European and non-European countries, and ii) the cases of misdescription detected in the present study were based on the incongruence found between the scientific or common names of the species declared on the menu at the retailers (sushi restaurant and takeaway), the specific molecular diagnosis obtained through the COI DNA barcoding and the corresponding denomination in Italian language to be attributed to the detected fish species, as indicated in the decree of the Ministry of Agricultural, Food and Forestry Policies (MD n. 19105 22 September 2017) dealing with the Italian names of fishes of commercial interest. In particular, the MD clearly states that to correctly inform consumers, the name to be used to indicate *T. thynnus* is "tuna" or "red tuna", while the name "yellowfin tuna" must be used to indicate *T. albacares*, and the names, "orientalis or oceanic tuna" and "bigeye tuna", should be used to indicate the species *T. orientalis* and *T. obesus*, respectively. Based on this premise, the high percentage of misdescription found for *T. thynnus* is shown by the fact that only in two cases out of 16, consumers really ate red tuna as declared on the menu, while in 87.5% of cases, they consumed yellowfin tuna (12 cases), orientalis tuna (1 case) and bigeye tuna (1 case) in place of red tuna. The survey carried out in Italy by Armani et al. [54] on misdescription in sushi products sold in Tuscany revealed a generally low rate of misdescription (3.4%), which in any case did not concern tuna-based products. However, the authors identified the products sold as tuna only at the genus level and then as belonging to the genus *Thunnus*, because EU regulations (1379/2013 and 1169/2011) require only the name of the seafood category and not the name of the species at the catering level. Similarly, a moderate level of species substitution (10%) was detected by Vandamme et al. [11] during a screening of seafood labelling accuracy in sushi bars and restaurants across England. The low rate of substitution detected for tuna products was imputed to the United Kingdom labelling regulations allowing the inclusion of all *Thunnus* species under the umbrella term "tuna". Interestingly, high levels of mislabeling (83.3%) for Bluefin tuna, *T. thynnus*, like those detected by us, were detected in French sushi restaurants, compared with the low general substitution rate (3.6%) observed over the whole sampling [15]. An intermediate level of species substitution was detected by Oceana [60] in a survey carried out in sushi restaurants in Brussels, where a 54.5% level of fraud was found, mainly due to the frequent substitution of *T. thynnus* by others cheaper tropical tuna species (*T. albacares* and *T. obesus*). Both in the United States of America and in China, the species of the genus *Thunnus* are sold under the umbrella terms "tuna" according to the Food and Drug Administration and the Food and Drugs (Composition and Labelling) Regulations (Cap. 132W), respectively [50,61]. However, the molecular screening carried out by Lowenstein et al. [50] in the United States of America led to the identification of sushi tuna samples up to the level of species by highlighting the substitution of bluefin tuna by different species in 40% of samples. A case of the substitution of *T. obesus* by *T. thinnus* has also been observed in sushi products in Canada, which could raise suspicion of illegal, unreported and unregulated fishing issue [53]. Instead, the investigation carried out by However et al. [61] in Honk Kong stated that tuna samples, identified only at the genus level, were correctly labeled.

Focusing our attention on the other cases of species substitution observed in our study, three species, *S. lalandi* (Yellowtail amberjack named oceanic amberjack in the Italian list of the species), *X. gladius* (swordfish) and *P. saltatrix* (bluefish), were found in place of *S. aurata* declared on the menu. In this case, there is no doubt that the species substitution was deliberate, although the economic profit may not be the incentive to defraud, but rather the ease of finding the species. The Yellowtail amberjack is an aquaculture species often consumed as sashimi reared in Japan, Australia and New Zealand. In recent decades, the bluefish has undergone a rapid northern range expansion within the Mediterranean from the southern and eastern sectors of the basin. This geographical expansion has been demonstrated to be a result of increasing water temperature [62] and is having an important socio-economic impact due to the voracious behavior of this predator [63]. However, the presence of *X. gladius* in place of *S. aurata* is of major concern, as swordfish is a species of greater economic value than seabream, and in this case, substitution could launder illegally caught swordfish. Another frequent case of species substitution observed by us was the substitution of flying-fish eggs or tobiko by eggs of capelin, *M. villosus*. Flying fish are all included within the family Exocoetidae, and the term tobiko indicates the roe of flying fish of the genera *Cheilopogon* and *Hyrundichthys* generally used in sushi preparation. Tobiko is made of small eggs of 2 mm or less in size, which are crisp and of golden orange color. Due to the small supply of flying fish roe, tobiko are often prepared by using immature roe of capelin or other fish which might be also colored and sold as imitation [64,65]. The Italian MD n. 19105 22 September 2017 includes the names of only two taxa of flying fish: the "oceanic flying-fish", which is an umbrella name for the species of *Cypselurus* spp., and "Indopacific flying-fish", which is used to indicate the species *Cheilopogon atrisignis*. Therefore, we considered only the above cases of substitution concerning *M. villosus* as misdescriptions, which was also reported by Armani et al. [54] in Tuscany and by Wallstrom et al. [66] in sushi bars in Honolulu. The results obtained from the molecular survey carried out in Italy indicate the effectiveness of COI barcoding for fish authentication in sushi products and highlight two main issues: (i) it is evident that a revision of the regulations by making the use of the scientific names of species mandatory for all products of the seafood supply chain is the only way to protect consumers from frauds, to guarantee their health, to protect the threatened species from illegal fishing and to restore the depleted fish stocks; (ii) to achieve these goals, a standardization of fish market names, avoiding using the same trade name to indicate multiple species, should be realized at the international level given that the fish market is now globalized.

Finally, the results of our study were obtained using the approach of Citizen Science, which allowed us to cover a wide portion of the Italian territory for the sushi survey. This relatively new approach was used by Bernard-Capelle et al. [15] to detect the rate of fish mislabeling in France and by Pardo et al. [67] to carry out a survey on seafood mislabeling in restaurants of 23 states across Europe. The most important benefit for researchers engaging citizens to obtain information for scientific investigations is the possibility to collect a high number of samples covering a wide geographical area controlling costs resulting from sampling. On the other hand, citizens, as consumers, will become aware of food safety concerns, which could be difficult to perceive by the end users of the food chain.

Supplementary Materials: The following are available online at https://www.mdpi.com/article/10.3390/foods10040756/s1, Figure S1: sushi sampling guide.

Author Contributions: Conceptualization, A.M.P. and V.F.; methodology and experiments, A.M.P., A.R. and G.S.C.; data analysis, A.M.P. and V.F.; writing—original draft preparation, A.M.P. and V.F.; writing—review and editing, A.M.P. and V.F.; funding acquisition, V.F. All authors have read and agreed to the published version of the manuscript.

Funding: This research was funded by University of Catania, "PIA.CE.RI." grant 2020.

Institutional Review Board Statement: Not applicable.

Informed Consent Statement: Not applicable.

Conflicts of Interest: The authors declare no conflict of interest.

References

1. Kim, N.H.; Yun, A.-R.; Rhee, M.S. Prevalence and classification of toxigenic *Staphylococcus aureus* isolated from refrigerated ready-to-eat foods (sushi, kimbab and California rolls) in Korea. *J. Appl. Microbiol.* **2011**, *111*, 1456–1464. [CrossRef] [PubMed]
2. Muscolino, D.; Giarratana, F.; Beninati, C.; Tornambene, A.; Panebianco, A.; Ziino, G. Hygienic-sanitary evaluation of sushi and sashimi sold in Messina and Catania, Italy. *Ital. J. Food Saf.* **2014**, *3*, 1701. [CrossRef] [PubMed]

3.   Liang, W.-L.; Pan, Y.-L.; Cheng, H.-L.; Li, T.-C.; Yu, P.H.-F.; Chan, S.-W. The microbiological quality of take-away raw salmon finger sushi sold in Hong Kong. *Food Control* **2016**, *69*, 45–50. [CrossRef]

4.   Kulawik, P.; Dordevic, D.; Gambus, F.; Szczurowska, K.; Zajac, M. Heavy metal contamination, microbiological spoilage and biogenic amine content in sushi available on the Polish market. *J. Sci. Food Agric.* **2018**, *98*, 2809–2815. [CrossRef] [PubMed]

5.   Guardone, L.; Armani, A.; Nucera, D.; Costanzo, F.; Mattiucci, S.; Bruschi, F. Human anisakiasis in Italy: A retrospective epidemiological study over two decades. *Parasite* **2018**, *25*, 41. [CrossRef]

6.   Ramires, T.; Iglesias, M.A.; Vitola, H.S.; Nuncio, A.S.P.; Kroning, I.S.; Kleinubing, N.R.; Fiorentini, A.M.; da Silva, W.P. First report of *Escherichia coli* 0157:H7 in ready-to-eat sushi. *J. Appl. Microbiol.* **2019**, *128*, 301–309. [CrossRef]

7.   Lehel, J.; Yaucat-Guendi, R.; Darnay, L.; Palotas, P.; Laczay, P. Possible food safety hazards of ready-to-eat raw fish containing product (sushi, sashimi). *Crit Rev. Food Sci. Nutr.* **2020**. [CrossRef]

8.   Hoel, S.; Mehli, L.; Bruheim, T.; Vadstein, O.; Jakobsen, A.N. Assessment of Microbiological Quality of Retail Fresh Sushi from Selected Sources in Norway. *J. Food Prot.* **2015**, *78*, 977–982. [CrossRef]

9.   Lowenstein, J.H.; Burger, J.; Jeitner, C.W.; Amato, G.; Kolokotronis, S.-O.; Gochfeld, M. DNA barcodes reveal species-specific mercury levels in tuna sushi that pose a health risk to consumers. *Biol. Lett.* **2010**, *6*, 692–695. [CrossRef]

10. Burger, J.; Gochfeld, M.; Jeitner, C.; Donio, M.; Pittfield, T. Sushi consumption rates and mercury levels in sushi: Ethnic and demographic differences in exposure. *J. Risk Res.* **2014**, *17*, 981–997. [CrossRef]

11. Vandamme, S.G.; Griffiths, A.M.; Taylor, S.-A.; Di Muri, C.; Hankard, E.A.; Towne, J.A.; Watson, M.; Mariani, S. Sushi barcoding in the UK: Another kettle of fish. *PeerJ* **2016**, *4*, e1891. [CrossRef]

12. House, J. Sushi in the United States, 1945–1970. *Food Foodways* **2018**, *26*, 40–62. [CrossRef]

13. EFSA (European Food Safety Authority). Scientific and technical assistance on the evaluation of the temperature to be applied to pre-packed fishery products at retail level. *EFSA J.* **2015**, *13*, 4162. [CrossRef]

14. US Food and Drug Administration. Protecting the Food Supply from Intentional Adulteration, Such as Acts of Terrorism. 2017. Available online: https://www.fda.gov/Food/GuidanceRegulation/FSMA/ucm587803.htm (accessed on 24 January 2018).

15. Benard-Capelle, J.; Guillonneau, V.; Nouvian, C.; Fournier, N.; Le Loët, K.; Dettai, A. Fish mislabeling in France: Substitution rates and retail types. *PeerJ* **2015**, *2*, e714. [CrossRef]

16. Christiansen, H.; Dettai, A.; Heindler, F.M.; Collins, M.A.; Duhamel, G.; Hautecoeur, M.; Steinke, D.; Volckaert, A.M.; Van de Putte, A.P. Diversity of Mesopelagic fishes in the Southern Ocean—A phylogeographic perspective using DNA barcoding. *Front. Ecol. Evol.* **2018**, *6*, 120. [CrossRef]

17. Do, T.D.; Choi, T.J.; Kim, J.; An, H.E.; Park, Y.J.; Karagozlu, M.Z.; Kim, C.B. Assessment of marine fish mislabelling in South Korea's markets by DNA barcoding. *Food Control* **2019**, *100*, 53–57. [CrossRef]

18. Garcia-Vazquez, E.; Perez, J.; Martinez, J.L.; Pardiñas, A.F.; Lopez, B.; Karaiskou, N.; Triantafyllidis, A. High level of mislabelling in Spanish and Greek hake markets suggests the fraudulent introduction of African species. *J. Agric. Food Chem.* **2011**, *59*, 475–480. [CrossRef]

19. Rocco, L.; Ferrito, V.; Costagliola, D.; Marsilio, A.; Pappalardo, A.M.; Stingo, V.; Tigano, C. Genetic divergence among and within four Italian populations of *Aphanius fasciatus* (Teleostei, Cyprinodontiformes). *Ital. J. Zool.* **2007**, *74*, 371–379. [CrossRef]

20. Pappalardo, A.M.; Ferrito, V.; Messina, A.; Patarnello, T.; De Pinto, V.; Guarino, F.; Tigano, C. Genetic structure of the killifish *Aphanius fasciatus* Nardo 1827 (Teleostei, Cyprinodontidae), results of mitochondrial DNA analysis. *J. Fish. Biol.* **2008**, *72*, 1154–1173. [CrossRef]

21. Ferrito, V.; Pappalardo, A.M.; Canapa, A.; Barucca, M.; Doadrio, I.; Olmo, E.; Tigano, C. Mitochondrial phylogeography of the killifish *Aphanius fasciatus* (Teleostei, Cyprinodontidae) reveals highly divergent Mediterranean populations. *Mar. Biol.* **2013**, *160*, 3193–3208. [CrossRef]

22. Cuttitta, A.; Patti, B.; Maggio, T.; Quinci, E.M.; Pappalardo, A.M.; Ferrito, V.; De Pinto, V.; Torri, M.; Falco, F.; Nicosia, A.; et al. Larval population structure of *Engraulis encrasicolus* in the Strait of Sicily as revealed by morphometric and genetic analyses. *Fish. Ocean* **2015**, *24*, 135–149. [CrossRef]

23. Pappalardo, A.M.; Federico, C.; Sabella, G.; Saccone, S.; Ferrito, V. A COI nonsynonymous mutation as diagnostic tool for intraspecific discrimination in the European Anchovy *Engraulis encrasicolus* (Linnaeus). *PLoS ONE* **2015**, *10*, e0143297. [CrossRef]

24. Pedrosa-Gerasmio, I.R.; Agmata, A.B.; Santos, M.D. Genetic diversity, population genetic structure, and demographic history of *Auxis thazard* (Perciformes), *Selar crumenophthalmus* (Perciformes), *Rastrelliger kanagurta* (Perciformes) and *Sardinella lemuru* (Clupeiformes) in Sulu-Celebes Sea inferred by mitochondrial DNA sequences. *Fish. Res.* **2015**, *162*, 64–74.

25. Duong, T.; Uy, S.; Chheng, P.; So, N.; Thi Tran, T.; Nguyen, N.T.; Pomeroy, R.; Egna, H. Genetic diversity and structure of striped snakehead (*Channa striata*) in the Lower Mekong Basin: Implications for aquaculture and fisheries management. *Fish. Res.* **2019**, *218*, 166–173. [CrossRef]

26. Perea, S.; Al Amouri, M.; Gonzalez, E.G.; Alcaraz, L.; Yahyaoui, A.; Doadrio, I. Influence of historical and human factors on genetic structure and diversity patterns in peripheral populations: Implications for the conservation of Moroccan trout. *bioRxiv* **2020**. [CrossRef]

27. Quinteiro, J.; Vidal, R.; Izquierdo, M.; Sotelo, C.G.; Chapela, M.J.; Pérez-Martín, R.I.; Rehbein, H.; Hold, G.L.; Russell, V.J.; Pryde, S.E.; et al. Identification of hake species (*Merluccius* genus) using sequencing and PCR-RFLP analysis of mitochondrial DNA control region sequences. *J. Agric. Food Chem.* **2001**, *49*, 5108–5114. [CrossRef] [PubMed]

28. Kumar, G.; Kocour, M.; Kunal, S.P. Mitochondrial DNA variation and phylogenetic relationships among five tune species based on sequencing of D-loop region. *Mitoch. DNA Part A* **2016**, *27*, 1976–1980.

29. Ceruso, M.; Mascolo, C.; De Luca, P.; Venuti, I.; Smaldone, G.; Biffali, E.; Anastasio, A.; Pepe, T.; Sordino, P. A rapid method for the identification of fresh and processed *Pagellus erythrinus* species against frauds. *Foods* **2020**, *9*, 1397. [CrossRef] [PubMed]

30. Hebert, P.D.N.; Ratnasingham, S.; de Waard, J.R. Barcoding animal life: Cytochrome c oxidase subunit 1 divergence, among closely related species. *Proc. R. Soc. Lond. B* **2003**, *270*, S96–S99. [CrossRef]

31. Paquin, R.; Hedin, M. The power and perils of 'molecular taxonomy': A case study of eyeless and endangered *Cicurina* (Araneae: Dictynidae) from Texas caves. *Mol. Ecol.* **2004**, *13*, 3239–3255. [CrossRef]

32. Lefebure, T.; Douady, C.J.; Gouy, M.; Gibert, J. Relationship between morphological taxonomy and molecular divergence within Crustacea: Proposal of a molecular threshold to help species delimitation. *Mol. Phyl. Evol.* **2006**, *40*, 435–447. [CrossRef] [PubMed]

33. Vitale, D.G.M.; Viscuso, R.; D'Urso, V.; Gibilras, S.; Sardella, A.; Marletta, A.; Pappalardo, A.M. Morphostructural analysis of the male reproductive system and DNA barcoding in *Balclutha brevis* Lindberg 1954 (Homoptera, Cicadellidae). *Micron* **2015**, *79*, 36–45. [CrossRef]

34. Conti, E.; Mulder, C.; Pappalardo, A.M.; Ferrito, V.; Costa, G. How soil granulometry, temperature and water predict genetic differentiation in namibian *Ariadna* spiders and explain their behaviour. *Ecol. Evol.* **2019**, *9*, 4382–4391. [CrossRef] [PubMed]

35. Ward, R.D.; Zemlak, T.S.; Innes, B.H.; Last, P.R.; Hebert, P.D.N. DNA barcoding Australia's fish species. *Philos. Trans. R. Soc. B* **2005**, *360*, 1847–1857. [CrossRef] [PubMed]

36. Cutarelli, A.; Amoroso, M.G.; De Roma, A.; Girardi, S.; Galiero, G.; Guarino, A.; Corrado, F. Italian market fish species identification and commercial frauds revealing by DNA sequencing. *Food Control* **2014**, *37*, 46–50. [CrossRef]

37. Pappalardo, A.M.; Ferrito, V. DNA barcoding species identification unveils mislabeling of processed flatfish products in southern Italy markets. *Fish. Res.* **2015**, *164*, 153–158. [CrossRef]

38. Pappalardo, A.M.; Cuttitta, A.; Sardella, A.; Musco, M.; Maggio, T.; Patti, B.; Mazzola, S.; Ferrito, V. DNA barcoding and COI sequence variation in Mediterranean lanternfishes larvae. *Hydrobiologia* **2015**, *745*, 155–167. [CrossRef]

39. Pappalardo, A.M.; Copat, C.; Ferrito, V.; Grasso, A.; Ferrante, M. Heavy metal content and molecular species identification in canned tuna: Insights into human food safety. *Mol. Med. Rep.* **2017**, *15*, 3430–3437. [CrossRef]

40. Pappalardo, A.M.; Copat, C.; Raffa, A.; Rossitto, L.; Grasso, A.; Fiore, M.; Ferrante, M.; Ferrito, V. Fish-based baby food concern—From species authentication to exposure risk assessment. *Molecules* **2020**, *25*, 3961. [CrossRef]

41. Acutis, P.L.; Cambiotti, V.; Riina, M.V.; Meistro, S.; Maurella, C.; Massaro, M.; Stacchini, P.; Gili, S.; Malandra, R.; Pezzolato, M.; et al. Detection of fish species substitution frauds in Italy: A targeted national Monitoring plan. *Food Control* **2019**, *101*, 151–155. [CrossRef]

42. Pappalardo, A.M.; Ferrito, V. A COIBar-RFLP strategy for the rapid detection of *Engraulis encrasicolus* in processed anchovy products. *Food Control* **2015**, *57*, 385–392. [CrossRef]

43. Ferrito, V.; Bertolino, V.; Pappalardo, A.M. White fish authentication by COIBar-RFLP: Toward a common strategy for the rapid identification of species in convenience seafood. *Food Control* **2016**, *70*, 130–137. [CrossRef]

44. Pappalardo, A.M.; Federico, C.; Saccone, S.; Ferrito, V. Differential flatfish species detection by COIBar-RFLP in processed seafood products. *Eur. Food Res. Technol.* **2018**, *244*, 2191–2201. [CrossRef]

45. Pappalardo, A.M.; Petraccioli, A.; Capriglione, T.; Ferrito, V. From fish eggs to fish name: Caviar species discrimination by COIBar-RFLP, an efficient molecular approach to detect fraud in the caviar trade. *Molecules* **2019**, *24*, 2468. [CrossRef] [PubMed]

46. Ferrito, V.; Raffa, A.; Rossitto, L.; Federico, C.; Saccone, S.; Pappalardo, A.M. Swordfish or shark slice? A rapid response by COIBar–RFLP. *Foods* **2019**, *8*, 537. [CrossRef]

47. Yao, L.; Lu, J.; Qu, M.; Jiang, Y.; Li, F.; Guo, Y.; Wang, L.; Zhai, Y. Methodology and application of PCR-RFLP for species identification in tuna sashimi. *Food Sci. Nutr.* **2020**, *8*, 3138–3146. [CrossRef] [PubMed]

48. Xiong, X.; Yuan, F.; Huang, M.; Lu, L.; Xiong, X.; Wen, J. DNA Barcoding revealed mislabeling and potential health concerns with roasted fish products sold across China. *J. Food Prot.* **2019**, *82*, 1200–1209. [CrossRef]

49. Xiong, X.; Huang, M.; Xu, W.; Li, Y.; Cao, M.; Xiong, X. Using real time fluorescence loop-mediated isothermal amplification for rapid species authentication of Atlantic salmon (*Salmo salar*). *J. Food Compos. Anal.* **2021**, *95*, 103659. [CrossRef]

50. Lowenstein, J.H.; Amato, G.; Kolokotronis, S.O. The real maccoyii: Identifying tuna sushi with DNA barcodes—Contrasting characteristic attributes and genetic distances. *PLoS ONE* **2009**, *4*, e7866. [CrossRef]

51. Chin Chin, T.; Adibah, A.B.; Danial Hariz, Z.A.; Siti Azizah, M.N. Detection of mislabelled seafood products in Malaysia by DNA barcoding: Improving transparency in food market. *Food Control* **2016**, *64*, 247–256. [CrossRef]

52. Adibah, A.B.; Syazwan, S.; Haniza Hanim, M.Z.; Badrul Munir, M.Z.; Intan Faraha, A.G.; Siti Azizah, M.N. Evaluation of DNA barcoding to facilitate the authentication of processed fish products in the seafood industry. *LWT-Food Sci. Technol.* **2020**, *129*, 109585. [CrossRef]

53. Hu, Y.; Huang, S.Y.; Hanner, R.; Levin, J.; Lu, X. Study of fish products in Metro Vancouver using DNA barcoding methods reveals fraudulent labeling. *Food Control* **2018**, *94*, 38–47. [CrossRef]

54. Armani, A.; Tinacci, L.; Lorenzetti, R.; Benvenuti, A.; Susini, F.; Gasperetti, L.; Ricci, E.; Guarducci, M.; Guidi, A. Is raw better? A multiple DNA barcoding approach (full and mini) based on mitochondrial and nuclear markers reveals low rates of misdescription in sushi products sold on the Italian market. *Food Control* **2017**, *79*, 126–133. [CrossRef]

55. Bonney, R.; Shirk, J.L.; Phillips, T.B.; Wiggins, A.; Ballard, H.L.; Miller-Rushing, A.J.; Parrish, J.K. Next Steps for Citizen Science. *Science* **2014**, *343*, 1436–1437. [CrossRef]

56. Kosmala, M.; Wiggins, A.; Swanson, A.; Simmons, B. Assessing Data Quality in Citizen Science. *Front. Ecol. Environ.* **2016**, *14*, 551–560. [CrossRef]

57. Ivanova, N.V.; Zemlak, T.S.; Hanner, R.H.; Hebert, P.D.N. Universal primer cocktails for fish DNA barcoding. *Mol. Ecol. Notes* **2007**, *7*, 544–548. [CrossRef]

58. Katoh, K.; Rozewicki, J.; Yamada, K.D. MAFFT online service: Multiple sequence alignment interactive sequence choice and visualization. *Brief. Bioinform.* **2019**, *20*, 1160–1166. [CrossRef] [PubMed]

59. Zhang, D.X.; Hewitt, G.M. Nuclear integrations: Challenges for mitochondrial DNA markers. *Trends Ecol. Evol.* **1996**, *11*, 247–251. [CrossRef]

60. Oceana Europe 2015. Too Cheap to Be True, Seafood Fraud in Brussel. Available online: https://eu.oceana.org/sites/default/files/421/oceana_factsheet_seafood_fraud_brussels_eng.pdf (accessed on 30 January 2021).

61. But, G.W.-C.; Wu, H.-Y.; Shaw, P.-C. Identification of fish species of sushi products in Hong Kong. *Food Control* **2019**, *98*, 164–173. [CrossRef]

62. Sabatés, A.; Martín, P.; Raya, V. Changes in life-history traits in relation to climate change: Bluefish (*Pomatomus saltatrix*) in the north-western Mediterranean. *ICES J. Mar. Sci.* **2012**, *69*, 1000–1009. [CrossRef]

63. Azzurro, E.; Cerri, J. The bluefish *Pomatomus saltatrix* (Pisces: Pomatomidae) in the Adriatic and Tyrrhenian Seas, can we call it climate invader? *OSF Prepr.* **2020**. [CrossRef]

64. Bledsoe, G.E.; Bledsoe, C.D.; Rasco, B. Caviars and fish roe products. *Crit. Rev. Food Sci. Nutr.* **2003**, *43*, 317–356. [CrossRef] [PubMed]

65. Kokina, A.V.; Syromyatnikov, M.Y.; Savinkova, O.V.; Popov, V.N. The Use of DNA Barcoding and Metabarcoding for Food and Environment Quality Control. In *Green Technologies and Infrastructure to Enhance Urban Ecosystem Services*; Vasenev, V., Dovletyarova, E., Cheng, Z., Valentini, R., Calfapietra, C., Eds.; SSC 2018; Springer Geography Springer: Cham, Switzerland, 2020. [CrossRef]

66. Wallstrom, M.A.; Morris, K.A.; Carlson, L.V.; Marko, P.B. Seafood mislabeling in Honolulu, Hawai'i. *Forensic Sci. Int. Rep.* **2020**, *2*, 100154. [CrossRef]

67. Pardo, M.A.; Jimenez, E.; Viðarsson, J.R.; Olafsson, K.; Olafsdottir, G.; Daníelsdottir, A.K.; Perez-Villareal, B. DNA barcoding revealing mislabeling of seafood in European mass caterings. *Food Control* **2018**, *92*, 7–16. [CrossRef]

 *foods*

Article

Classification of Botrytized Wines Based on Producing Technology Using Flow-Modulated Comprehensive Two-Dimensional Gas Chromatography

Olga Vyviurska, Nemanja Koljančić, Ha Anh Thai, Roman Gorovenko and Ivan Špánik *

Faculty of Chemical and Food Technology, Institute of Analytical Chemistry, Slovak University of Technology in Bratislava, 81237 Bratislava, Slovakia; olga.vyviurska@stuba.sk (O.V.); nemanja.koljancic@stuba.sk (N.K.); ha.thai@stuba.sk (H.A.T.); romagorovenko@gmail.com (R.G.)
* Correspondence: ivan.spanik@stuba.sk

Abstract: The enantiomeric ratio of chiral compounds is known as a useful tool to estimate wine quality as well as observe an influence of wine-producing technology. The incorporation of flow-modulated comprehensive two-dimensional gas chromatography in this type of analysis provides a possibility to improve the quality of results due to the enhancement of separation capacity and resolution. In this study, flow-modulated comprehensive two-dimensional gas chromatography was incorporated in enantioselective analysis to determine the influence of winemaking technology on specific features of botrytized wines. The samples included Tokaj essences (high-sugar wines), Tokaj botrytized wines and varietal wines (Furmint, Muscat Lunel, Lipovina) and wines matured on grape peels. The obtained data was processed with hierarchic cluster analysis to reveal variations in composition and assess classification ability for botrytized wines. A significant difference between the samples was observed for the enantiomeric distribution of ethyl lactate and presence of monoterpene alcohols. The varietal wines were successfully separated from the other types, which showed more similar results and could be divided with additional parameters. We observed a correlation between the botrytized wines and the varietal wines fermented with grape skins. As to the essences produced from juice of botrytized grapes, the results were quite similar to those of the botrytized wines, even though monoterpenes were not detected in the extracts.

Keywords: enantioselective analysis; flow-modulated comprehensive two-dimensional gas chromatography; botrytized wines; Tokaj wine region

Citation: Vyviurska, O.; Koljančić, N.; Thai, H.A.; Gorovenko, R.; Špánik, I. Classification of Botrytized Wines Based on Producing Technology Using Flow-Modulated Comprehensive Two-Dimensional Gas Chromatography. *Foods* **2021**, *10*, 876. https://doi.org/10.3390/foods10040876

Academic Editors: Margit Cichna-Markl and Isabel Mafra

Received: 1 March 2021
Accepted: 15 April 2021
Published: 16 April 2021

1. Introduction

The pleasant honey-like taste and unique fruit flavor of botrytized (noble rot) wines are the result of a specific winemaking technology, which includes overripe grapes infected by *Botrytis cinerea*. The fungus induces increased content of sugar and fatty acid aroma precursors, and the formation of new compounds in grapes [1]. According to Schmitt-Kopplin et al. [2], Botrytis infection initiates fermentation retardation of the yeast metabolomics activity during alcoholic fermentation of wine. The distinctive climate condition of Tokaj wine region (soil slopes of volcanic origin and surrounding wetlands) supports growth of *Botrytis cinerea* on grapes [3]. Furmint, Muscat Lunel and Lipovina are the main grape varieties in the region for production of wine specialities and dry white wines. For example, Tokaj essence is made as juice of botrytized berries obtained by gravitation during harvest season [4]. High sugar content (65 to 752 g/L) supports long term-fermentation resulting in 5–7% alcohol [5]. At the same time, aging of the essence increases the concentration of polyphenols and its antioxidant properties [6]. In the case of the botrytized wines, infected berries are picked up and macerated in grape must for one or two days. A ratio of noble-rotten grape to grape (tub, "putňa" in Slovak, "puttonyos" in Hungarian) determines the sweet, smooth taste, and pleasant aroma of the wine. One "putňa" includes one barrel of noble-rotten grape berries (20–25 kg) per 136–140 L of one-year-old young wine. Tokaj botrytized wines are commonly aged from 3 to 5 years in oak barrels. A more detailed description of winemaking technologies and classification of botrytized wines from Tokaj wine region is reported by [7]. As for the volatile organic compounds, significant difference in volatile ethyl esters, fatty acids and sherry lactones, was revealed for botrytized Amarone wine in comparison to the same wine produced from healthy grapes [8]. Furthermore, an increase in 3-sulfanylhexan-1-ol was reported for in the botrytized wines produced from Sauvignon blanc and Semillon grapes [9]. A dominance of S-3-sulfanylhexan-1-ol was typical for these botrytized wines, whereas a racemic ratio was observed for dry white Sauvignon blanc and Semillon wines [10].

Chirality is one of the most important properties of organic compounds included in the composition of food products [11]. The importance of determining enantiomers and their ratio lies on the fact that enantiomers volatile molecules have different aroma characteristics and odor detection threshold [12]. Changes in enantiomeric ratio could yield information about product quality, technical processing, different biological activity, and contamination [13]. In some cases, discrimination of samples can be problematic due to an additional racemization of chiral compounds through fermentation or distillation processes. Thus, a complex approach with a number of enantiomeric data and chemometric modelling has been demonstrated in the recent studies. For example, the enantiomeric concentrations of terpenes and (1R, 2R)-methyl jasmonate were exploited in statistics discrimination of 22 tea cultivars according to their geographical origin [14]. Influence of different storage conditions and manufacturing process on green tea was shown with a total distribution of catechins and methylxanthines [15]. The PCA analysis was included in the metabolomics profiling of chiral amino acids for classification of cheese depending ripening period (6, 18, 26 months) [16]. Castells et al. [17] proposed to discriminate honey origin with the enantiomeric ratio of dintitrophenyl amino acids. The authors [18] developed a chemometric method based on the composition of triacylglycerols and volatiles for varietal classification of extra virgin olive oils. Comprehensive two-dimensional gas chromatography (GC×GC) shows clear advantages for the analysis of complex samples, such as improved separation capacity, increased number of identified compounds, structured chromatograms and significant signal enhancement [19]. Enantioselective GC×GC analysis was successfully used for the evaluation of essential oils [20,21] and herbal products [22]. Flow-modulated comprehensive two-dimensional gas chromatography proposed more available equipment as well gives a possibility to adjust amount of sample directed to the second column. In our study, we tried to exploit enantioselective GC×GC analysis for the evaluation of botrytized wines in comparison to the corresponding varietal grape wines, selected essences, and the varietal wines fermented with grape skins.

2. Materials and Methods

2.1. Chemicals

n-Hexane and standards of enantiomers were supplied from Sigma Aldrich (St. Louis, MO, USA), and sodium chloride was obtained from Chemapol (Prague, Czech Republic). A mixture of n-alkanes (C7-C30) used for the calculation of retention indices was purchased from Supelco (Belleforte, PA, USA).

2.2. Samples

The samples included Tokaj essences (ES), wine maturated on grape peel (GP30 for 30 and GP90 for 90 days), botrytized wines (2, 3, 4, 5 and 6 putňove, 2P1-6P8) and varietal wines (Furmint, Muscat Lunel and Lipovina variety, F1-L5). More detailed information about the used samples is presented in Table 1.

Table 1. The samples used for investigation.

Code	Year	Producer	Code	Year	Producer
ES1	1999	TOKAJ & CO.	5P7	2000	Zlatý Strapec
ES2	2000	J & J OSTROŽOVIČ	5P8	2003	J & J OSTROŽOVIČ
GP30	2016	J & J OSTROŽOVIČ	5P9	2003	J & J OSTROŽOVIČ
GP90	2016	J & J OSTROŽOVIČ	5P10	2004	J & J OSTROŽOVIČ
2P1	1989	TOKAJ & CO.	6P1	1977	Zlatý Strapec
2P2	1990	TOKAJ & CO.	6P2	1983	Zlatý Strapec
3P1	1988	Zlatý Strapec	6P3	1989	J & J OSTROŽOVIČ
3P2	1990	TOKAJ & CO.	6P4	1989	TOKAJ & CO.
3P3	1995	J & J OSTROŽOVIČ	6P5	1999	J & J OSTROŽOVIČ
3P4	1995	Zlatý Strapec	6P6	2002	J & J OSTROŽOVIČ
3P5	1999	J & J OSTROŽOVIČ	6P7	2003	J & J OSTROŽOVIČ
3P6	2000	Zlatý Strapec	6P8	2006	TOKAJ & CO.
3P7	2009	TOKAJ & CO.	F1	2014	J & J OSTROŽOVIČ
4P1	1993	Zlatý Strapec	F2	2015	TOKAJ & CO.
4P2	1995	TOKAJ & CO.	F3	2015	J & J OSTROŽOVIČ
4P3	2000	Zlatý Strapec	M1	2016	J & J OSTROŽOVIČ
4P4	2002	J & J OSTROŽOVIČ	M2	2016	J & J OSTROŽOVIČ
4P5	2004	J & J OSTROŽOVIČ	M4	2015	J & J OSTROŽOVIČ
4P6	2009	TOKAJ & CO.	M5	2015	TOKAJ & CO.
5P1	1959	Zlatý Strapec	L1	2015	J & J OSTROŽOVIČ
5P2	1972	Zlatý Strapec	L2	2015	TOKAJ & CO.
5P3	1989	J & J OSTROŽOVIČ	L3	2015	TOKAJ & CO.
5P4	1990	TOKAJ & CO.	L4	2015	J & J OSTROŽOVIČ
5P5	1993	J & J OSTROŽOVIČ	L5	2015	J & J OSTROŽOVIČ
5P6	1993	Zlatý Strapec	-	-	-

2P, 3P, 4P, 5P and 6P are shortcut for "puttony" wines with the corresponding number of tubs was added.

2.3. Sample Preparation

The studied samples were prepared by liquid-liquid extraction procedure using n-hexane. First, 2.0 g of sodium chloride was added into a 20 mL aliquot of wine. Then, the mixture was transferred to a separatory funnel in order to facilitate extraction. Next, 5 mL of n-hexane was added to the funnel, the mixture was shaken by hand for 5 min and extraction was repeated two more times under the same conditions. The combined extracts were centrifuged at 3600 rpm for 10 min. The resulting organic extract was evaporated to 1 mL under nitrogen flow in a 55 °C water bath. The development of the method is described in detail in [23].

2.4. Instrumentation

Agilent 7890A gas chromatograph (Wilmington, DE, USA) coupled with a reverse fill/flush (RFF) flow modulator (Agilent G3486A CFT Modulator, Folsom, CA, USA), flame-ionization detector (FID) and quadrupole mass spectrometer (qMS) were used to determine an enantiomer ratio in the wine samples. The GC column setup contains 30 m × 0.25 mm × 0.25 μm Rt-ßDEXse (Restek, Bellefonte, PA, USA) in the first dimension and 5 m × 0.25 mm × 0.15 μm INNOWax (Agilent Technologies, Folsom, CA, USA) in the second dimension. A supplementary restrictor (5 m × 250 μm ID) facilitates a switch of the carrier gas direction in the modulator between loading and injection mode. The modulation period was set to 6 s and included a 0.11 s sampling time. The second column effluent is directed to a splitter connected to qMS and FID detectors. Such approach is connected to compatibility of the detectors with elevated second flow of carrier gas and high acquisition frequency required for GC×GC analysis. A 0.5 m × 100 μm ID restrictor and a 1.2 m × 250 μm ID restrictor were installed to qMS and FID, respectively.

An initial temperature of the oven program was 40 °C and kept for 10 min. Further, temperature was increased with a rate 2 °C/min to 220 °C and maintained for 25 min. A total analysis time was 125 min. 1 µL of the sample extract was injected in splitless mode into 250 °C heated inlet. Helium (99.999% purity) in constant flow mode was used as the carrier gas. 0.7 mL/min flow rate was set in the first dimension and 23 mL/min for the second dimension. Flow rates to FID and MS detectors were determined with parameters of the restrictors and set as 23.3 mL/min and 2.1 mL/min, respectively. The flame-ionization detector was operated at 250 °C with a hydrogen flow rate of 30 mL/min, an air flow rate of 450 mL/min, and a makeup flow rate of 25 mL/min. A data acquisition rate of 100 Hz was used for FID detector. A transfer line to MS detector was kept at 250 °C for whole run time. Ion source temperature and quadrupole temperature were maintained at 180 °C and 300 °C, respectively. The MS signal acquisition rate was 21.43 spectra/s (40–400 m/z range). The primary processing of the obtained chromatograms was performed using GC Image software version v. 2.1. (Zoex Corporation, Houston, TX, USA), and MSD ChemStation software (version F.01.01.2317, Agilent Technologies, Santa Clara, CA, USA) with NIST14, FFNSC2, MPW2007 and W9N11 databases. GC×GC-MS identification of compounds was also supported with injection of standard compounds.

3. Results

A chiral column with the stationary phase based on 2,3-di-O-ethyl-6-O-*tert*-butyl dimethylsilyl-β-cyclodextrin, was used in the first dimension for GC×GC analysis. A polar INNOWax column was selected for the second dimension to separate analytes according to polarity. GC×GC-MS data was used to determine chiral volatile compounds presented in the samples. Due to higher acquisition rate of flame-ionization detector and narrower peaks, GC×GC-FID data was preferred for calculation of enantiomeric ratio. GC×GC-MS and GC×GC-FID chromatograms of Furmint varietal wine (2015) are represented in Figure 1. More detailed information about volatile organic compounds composition of Tokaj varietal wines and Tokaj selection wines can be found in the previous studies [23,24]. The target chiral compounds include ethyl lactate, linalool, α-terpineol, γ-nonalactone and whiskey lactone. Retention times and retention indices of stereoisomers are shown in Table 2. Enantiomeric ratio of the chiral compounds was estimated according to:

$$E_R = \frac{A_R}{A_R + A_S} \times 100$$

where A_R is obtained peak area of R enantiomer and A_S is obtained peak area of the following S configuration [25]. RSD values of enantiomeric ratios based on GC×GC-FID were less than 10%.

Table 2. Target chiral compounds.

Compounds	RT1, min	RT2, s	RI	Resolution
R-(+)-ethyl lactate	44.288	1.491	972	-
S-(-)-ethyl lactate	44.988	1.491	979	3.29
R-(-)-linalool	60.388	1.177	1204	-
S-(+)-linalool	60.688	1.177	1210	1.56
R-(+)-α-terpineol	68.288	1.491	1322	-
S-(-)-α-terpineol	68.588	1.491	1328	2.59
cis-whiskey lactone	78.788	1.648	1506	-
trans-whiskey lactone	81.488	1.805	1543	11.00
R-γ-nonalactone	84.388	1.833	1592	-
S-γ-nonalactone	84.588	1.833	1595	8.20

RT1 corresponds to retention time of compounds eluted from first dimension and RT2 corresponds to retention time of compounds eluted from second dimension, RI—retention index for the 1st column.

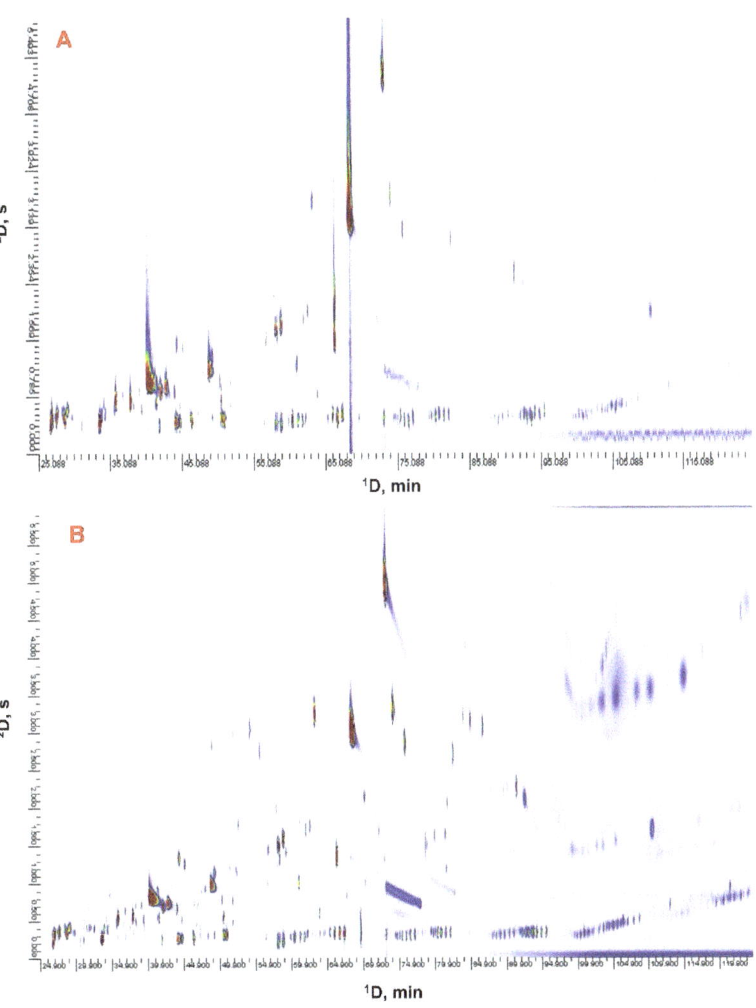

Figure 1. GC×GC-MS (**A**) and GC×GC-FID (**B**) chromatogram of Furmint varietal wine (2015).

3.1. Ethyl Lactate

Ethyl lactate is an important aroma compound which contributes to the "broader", "fuller" taste of wine and could be used for the determination of the microbiological infection of wine [26]. Enantiomers of ethyl lactate are obtained through different fermentation processes that are typical for winemaking. R-(+)-ethyl lactate is a product of sugar fermentation by yeast, whereas presence of S-(-)-ethyl lactate is caused by activity of lactic acid bacteria during malolactic fermentation [27]. Lactic bacteria (Lactobacillus, Pediococcus, and Oenococcus) support conversion of L-malic acid to L-lactic acid and additional biosynthesis of aroma compounds [28]. Table 3 shows that both of R- and S-enantiomers were detected in 46 from 49 samples. The exception was observed for varietal wine produced from Muscat Lunel (2016) and Lipovina (2015). Overall, R-(+)-ethyl lactate was dominant for Tokaj varietal wines in comparison to the other types. The highest value of R-(+)-ethyl lactate (91%) was detected in Furmint (2015). The excess of S-(-)-ethyl lactate over R-(+)-ethyl lactate varied from 2 times (4P-1993, 4P-2009) to 6 times (2P-1990, 3P-1999) in the botrytized wines. A few samples like 3P-2009, and 6P-2006 showed a reverse tendency with the R:S ratio as 70:30, 81:19, respectively. It is worthy to remark that higher excess of S-enantiomer (nearly 8 times) was recorded in the wine fermented with grape skins for 30 days. A content of R-enantiomer increased almost double in after 90 days of fermentation with grape skins. Mills et al. [29] showed a presence of atypical lactic acid bacteria community in botrytized wines (Leuconostoc and Lactococcus), which could be connected to Botrytis colonization on the grape berry. Increased content of S-enantiomer was shown for the other types of wine [27,30] after malolactic fermentation. For example, Freitas et al. [31] claimed a decrease of R-(+)-ethyl lactate by 50–68%, and an increase of 85–75% for S-(-)-ethyl lactate as a result of activity of lactic acid bacteria.

3.2. Terpenes

A majority of terpenes are bonded to sugar molecules and occurs in grapes in non-volatile form. Their concentration increases during grape ripening and wine ageing [32] and become important components of wine flavor and aroma [33]. A racemic mixture of terpenes is commonly observed in raw fruits or as a product of fermentation process. In our case, linalool was mostly presented in the varietal Tokaj wines (ten out of twelve samples). For the botrytized wines, linalool was detected only in two samples, whereas other types of samples did not contain linalool at all. In particularly, only 3 and 4 "puttony" wines (both from 2009) were reported to contain linalool enantiomers. It is worthy to note that a racemic mixture was obtained in almost half of total samples, e.g., Furmint–2015 (48:52), Muscat Lunel–2015 (49:51). A slight dominance of S-stereoisomer (59–63%) was observed for some Muscat Lunel and Lipovina samples. In botrytized wines, linalool generally metabolizes to (E)-2,6-dimethyl-2,7-octadiene-1,6-diol (>95%) (Figure 2) [34]. The other biotransformation by-products include (Z)-2,6-dimethyl-2,7-octadiene-1,6-diol, 3,9-epoxy-p-meth-1-ene, furanoid (Z)- and (E)-linalool oxides, pyranoid (Z)- and (E)-linalool oxides, and 2-vinyl-2-methyl-tetrahydrofuran-5-one. Overall, linalool level tends to reduce with prolonged wine aging due to conversion to α-terpineol and furan linalool oxides [35]. For example, concentration of linalool decreased by 3.3 and 71.6 times for Alvarinho and Loureiro wines after 20 months of maturation [36].

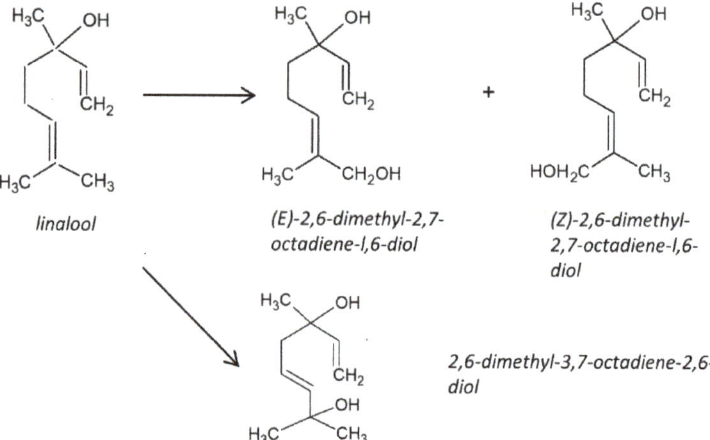

Figure 2. Transformation of linalool (modified [34]).

Table 3. Results of GC×GC-FID analysis.

Code	Enantiomer Ratio [%]									
	R-(+)-Ethyl Lactate	S-(-)-Ethyl Lactate	R-(-)-Linalool	S-(+)-Linalool	R-(+)-α-Terpineol	S-(-)-α-Terpineol	R-γ-Nonalactone	S-γ-Nonalactone	*trans*-Whiskey Lactone	*cis*-Whiskey Lactone
Essences										
ES1	24	76	-	-	-	-	61	39	38	62
ES2	30	70	-	-	-	-	74	26	38	62
Wines fermented with grape skins										
GP30	11	89	-	-	-	-	77	23	24	76
GP90	28	72	-	-	-	-	71	29	34	66
Botrytized wines										
2P1	21	79	-	-	-	-	70	30	43	57
2P2	14	86	-	-	-	-	70	30	-	-
3P1	27	73	-	-	-	-	69	31	46	54
3P2	24	76	-	-	-	-	66	34	48	52
3P3	15	85	-	-	-	-	71	29	39	61
3P4	22	78	-	-	45	55	61	39	36	64
3P5	14	86	-	-	-	-	77	23	36	64
3P6	18	82	-	-	-	-	75	25	44	56
3P7	70	30	43	57	-	-	-	-	48	52
4P1	35	65	-	-	-	-	-	-	40	60
4P2	24	76	-	-	-	-	-	-	39	61
4P3	20	80	-	-	-	-	61	39	43	57
4P4	32	68	-	-	-	-	-	-	39	61
4P5	20	80	-	-	-	-	68	32	42	58
4P6	36	64	39	61	51	49	-	-	46	54
5P1	38	62	-	-	-	-	-	-	43	57
5P2	24	76	-	-	-	-	-	-	27	73
5P3	26	74	-	-	-	-	-	-	46	54
5P4	21	79	-	-	-	-	65	35	37	63
5P5	20	80	-	-	-	-	58	42	46	54
5P6	25	75	-	-	-	-	64	36	42	58
5P7	22	78	-	-	-	-	70	30	40	60
5P8	43	57	-	-	52	48	-	-	33	67
5P9	42	58	-	-	51	49	-	-	32	68
5P10	41	59	-	-	-	-	74	26	33	67
6P1	29	71	-	-	48	52	71	29	39	61
6P2	25	75	-	-	-	-	-	-	40	60
6P3	54	46	-	-	-	-	-	-	58	42
6P4	15	85	-	-	-	-	-	-	37	63
6P5	23	77	-	-	-	-	75	25	38	62
6P6	27	73	-	-	36	64	71	29	38	62
6P7	30	70	-	-	44	56	80	20	39	61
6P8	81	19	-	-	54	46	58	42	42	58
Varietal wines										
F1	77	23	-	-	-	-	73	27	-	-
F2	79	21	-	-	55	45	-	-	42	58
F3	91	9	52	48	58	42	72	28	-	-
M1	-	-	56	44	67	33	-	-	-	-
M2	-	-	48	52	63	37	63	37	-	-
M4	83	17	51	49	59	41	68	32	-	-
M5	73	27	40	60	54	46	-	-	-	-
L1	-	-	38	62	59	41	74	26	-	-
L2	88	12	37	63	-	-	64	36	-	-
L3	50	50	47	53	54	46	-	-	53	47
L4	91	9	41	59	56	44	-	-	-	-
L5	90	10	45	55	56	44	71	29	-	-

2P, 3P, 4P, 5P and 6P are shortcut for "puttony" wines with the corresponding number of tubs was added.

α-Terpineol is another monoterpene alcohol which concentration correlates with ageing period of wine. Ferreira et al. [35] showed a significant increase in α-terpineol content after one week of accelerated aging, and the further decrease at the end of aging period (nine weeks). A positive effect of *Botrytis cinerea* on α-terpineol concentration in wine was found in comparison with sweet Chardonnay wines [37]. In contrast to linalool, α-terpineol was more typical for botrytized wine samples. However, the GC×GC-MS analysis did not confirm the presence of α-terpineol stereoisomers in Tokaj essences and fermented with grape skins. Interestingly, α-terpineol mostly occurs in the samples with higher "putňa" number, e.g., four samples for 6P wines vs. one sample for 3P wines. Almost racemic ratio was found in those samples and slightly higher enantiomeric ratio (64%) was obtained for S-enantiomer in 6P wine (2002). As to the varietal wines, the majority of samples contained α-terpineol as racemate. Two samples of Muscat wine showed small dominance (>60%) of R-(+)-α-terpineol. Unfortunately, the results of α-terpineol enantiomers cannot be used for the differentiation of wines according to grape variety or "putňa" number.

3.3. Lactones

γ-Lactones are commonly identified in wine, where they play an important role as aroma active compounds. Organoleptic properties of lactones are mainly determined by carbon chain attached to a carbon ring [38]. In winemaking, formation of lactone could occur in grapes, through fermentation or during aging processes by their extraction from oak wood [39]. Azpilicueta et al. [39] reported that accumulation of γ-nonalactone from American oak barrels did not dependent on wine type (Merlot or Cabernet Sauvignon), and constant concentration could be achieved after two months of aging. Aroma characteristics of γ-nonalactone enantiomers are slightly different, e.g., soft coconut and sweet taste is typical for R-stereoisomer, whilst weak coconut is observed for S-stereoisomer [40]. As can be seen from Table 3, both R and S configurations could be found in all wine categories. However, γ-nonalactone was observed in less samples than ethyl lactate. In case of 3P wines, γ-nonalactone was detected in 6 out of 7 samples. Whereas for the other botrytized wines, this compound was presented in a half of the samples. The similar results were observed for the varietal wines (7/12). Overall, R-γ-nonalactone is dominant (58–80%) for all the samples. The similar findings were also reported for Australian botrytized white wines [40] and Bordeaux dessert wines [41]. As can be seen from the results for the wines fermented with grape skins, fermentation period did not significantly affect stereoisomeric distribution.

Originally, whiskey (oak) lactone was identified as *cis-* and *trans-*5-n-butyl-4-methyl-4,5-dihydro-2(3H)-furanone in burbon whiskey [42]. Whiskey lactone molecule has four stereoisomers, but only *trans-*(3S,4R)-and *cis-*(3S,4S)-whiskey lactones are naturally occurring [43]. Although *cis-*stereoisomer has a lower odor threshold, both *cis-* and *trans-*whiskey lactones contribute to fresh wood and coconut aroma of wines [44]. A number of factors influences a degree of whiskey lactone extraction from oak barrels, e.g., composition of wood, toasting processes, and wine ageing period in barrels [45]. At higher concentration, whiskey lactones can become a predominant flavor compound in wines. This problem can occur in new winery where new wood barrels are used for wine ageing or if wine maker is not careful and large amount of whiskey lactones is extracted from wooden barrels [46]. The two main oak species (French oak and American oak) are traditionally used for wine aging. It was shown that American oak releases particularly *cis-*whiskey lactone, especially with new oak barrels [47]. A *cis/trans* ratio of whiskey lactone has been suggested as a parameter to distinguish wines aged in American or French oak barrels. For example, Alamo-Sanza et al. [48] found that the content of *cis-*whiskey lactone is 5-fold greater higher than the content of *trans-*enantiomer for wines aged in American oak barrels, whereas for French oak barrels this value was only doubled. Nearly 10% of *trans-*stereoisomer was detected in wine after aging in American oak [49], and almost racemic mixture was measured in the case of French oak. In our case, almost all botrytized samples contain whiskey lactone, except of one sample produced 1990. The reversed situation was observed for the varietal wines, where whiskey lactone was detected in two samples from Furmint and Lipovina (2015). Overall, *cis*-whiskey lactone dominated in the samples (52–73% range), and a *cis/trans* ratio varies from 1.1 to 2.7. Interestingly, this correlation between stereoisomers was higher (3.2) for the wine fermented with grape skins 30 days, and it decreased to 1.9 value after 90 days. The similar enantiomeric distribution (62% *cis*-whisky lactone) was shown for essences, which undergo long-term aging.

3.4. Hierarchical Clustering Analysis of the Wine Samples

Clustering analysis is commonly used to determine data structure and look for similarities between multiple objects [50]. It has been successfully incorporated in food analysis to emphasize bioactive components and functional properties of products [51]. In order to check the significance of variations in the target compounds composition, hierarchical cluster analysis was selected as a classification tool. The ratios for R-(+)-ethyl lactate, R-(-)-linalool, R-(+)-α-terpineol, R-γ-nonalactone, *cis*-whiskey lactone were included in the calculations. The data obtained for the botrytized wines were averaged to simplify a dendrogram. Distances between the samples were estimated with Euclidean distances by the following formula:

$$d = \sqrt{(x_1 - y_1)^2 + (x_2 - y_2)^2 + \ldots + (x_n - y_n)^2},$$

where $x_1, x_2 \ldots x_n$ and $y_1, y_2 \ldots y_n$ represents co-ordinates of two points in n-dimensional space. A distance between two clusters was determined by the distance of the furthest neighbours in two clusters. This approach called complete linkage is recommended to decrease number of undistinguished clusters.

The dendrogram in Figure 3 illustrates the stages of linkages and reveals successful separation of the samples on the varietal wines and the others. Unfortunately, it was not possible to classify the varietal wines accordingly to grape variety, and three groups are clustered based on the analysed variables. From other side, with a few exceptions (L2 and L4), some dependence from wine producer could be supposed. The first group (F1, L3, L4, M5) contains the samples supplied from Tokaj & Co (Malá Tŕňa, Slovakia, whereas the varietal wines from Ostrožovič (Veľká Tŕňa, Slovakia) mainly belong to the second (F2, L5, M4, F3, L2) and the third (L1, M1, M2) groups. This assumption requires a larger number of samples or the incorporation of additional variables to be confirmed.

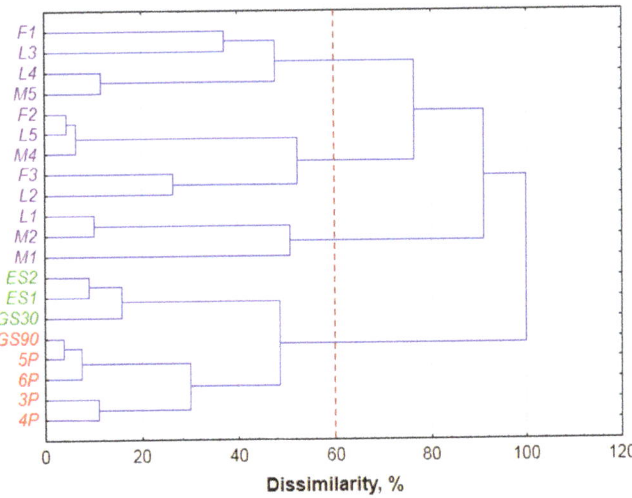

Figure 3. HCA dendrogram of the obtained data. Dissimilarity is calculated as a ratio of d_{link} to d_{max}.

As for the other samples, this group is divided on two subgroups with essences and botrytized wines. It is worthwhile noting, that influence of period of fermentation period with grape skins was confirmed with cluster analysis. The sample obtained after 30 days of fermentation (GP30) was sorted to essences, and the sample with prolonged fermentation (90 days) was more similar to the botrytized wines.

4. Conclusions

The results obtained through chiral analysis with flow-modulated comprehensive two-dimensional gas chromatography, show a significant difference between wine categories. Particularly, it can be seen for the varietal wines and the botrytized wines, where dissimilarity is also confirmed with hierarchic cluster analysis. In this case, the variations in data are mainly related to ethyl lactate, linalool and whiskey lactone. The dominance of S-(-)-ethyl lactate in the botrytized wines is supposed to be a result of malolactic fermentation supported by *Botrytis cinerea* colonization on the grape berry. Another finding which could correlates with the influence of the fungus on winemaking technology is the low presence of monoterpene alcohols (especially linalool) in comparison with the varietal wines. R-γ-nonalactone prevails in all the samples, whereas a content of whiskey lactone is directly connected to wine aging condition. The essence samples belong to a special high-sugar wine category and obtained from juice of from botrytized berries, show similar results to the botrytized wines. Moreover, monoterpenes were not observed at all in the extracts. Interestingly, that the enantiomeric distribution of the target compounds changes with simultaneous fermentation of the varietal wines with grape skins. According to cluster analysis the sample is classified to an essence subgroup after 30 days of fermentation. In the case of 90 days of fermentation, the results were more comparable with the botrytized wines. Increased enantiomeric ratio of R-ethyl lactate and a reduction in a *cis/trans* ratio of whiskey lactone are observed with the extension of the fermentation period.

Author Contributions: Conceptualization and methodology, I.Š. and O.V.; GC×GC measurements, R.G.; data analysis, H.A.T., O.V.; investigation, O.V.; writing—original draft preparation, O.V., N.K. and H.A.T.; writing—review and editing, I.Š.; visualization, O.V., H.A.T., N.K.; funding acquisition, I.Š. All authors have read and agreed to the published version of the manuscript.

Funding: The authors would like to thank for financial support from Slovak Research and Development Agency for contract number APVV-15-0333 and Grant Agency of Ministry of Education of Slovak Republic for contract VEGA 1/0521/19.

Acknowledgments: Many thanks to Tokaj wineries Ostrožovič spol. s.r.o., Anna Nagyová–Zlatý Strapec and Tokaj & Co., s.r.o. for their cooperation and granting the samples.

Conflicts of Interest: The authors declare no conflict of interest.

Abbreviations

GC×GC	comprehensive two-dimensional gas chromatography
RT	retention time
RI	retention index

References

1. Magyar, I.; Soós, J. Botrytized wines–current perspectives. *Int. J. Wine Res.* **2016**, *8*, 29–39. [CrossRef]
2. Hong, Y.S.; Cilindre, C.; Liger-Belair, G.; Jeandet, P.; Hertkorn, N.; Schmitt-Kopplin, P. Metabolic Influence of *Botrytis cinerea* Infection in Champagne Base Wine. *J. Agric. Food Chem.* **2011**, *59*, 7237–7245. [CrossRef] [PubMed]
3. Ribéreau-Gayon, J.; Riberau-Gayon, P.; Seguin, G. *Botrytis cinerea* in enology. In *Biology of Botrytis*; Coley-Smith, K.J.R., Verhoeff, W.R., Jarvis, Eds.; Academic Press: London, UK, 1980; pp. 251–274.
4. Sipiczki, M. Yeasts in Botrytized Wine Making. *Yeasts Prod. Wine* **2019**. [CrossRef]
5. Csoma, H.; Kállai, Z.; Antunovics, Z.; Czentye, K.; Sipiczki, M. Vinification without Saccharomyces: Interacting Osmotolerant and "Spoilage" Yeast Communities in Fermenting and Ageing Botrytised High-Sugar Wines (Tokaj Essence). *Microorganisms* **2021**, *9*, 19. [CrossRef]
6. Eftimova, Z.; Eftimová, J.; Balážová, Ľ. Antioxidant activity of tokaj essence. *Potravinarstvo* **2018**, *12*, 323–329. [CrossRef]

7.   Furdikova, K.; Machynakova, A.; Drtilova, T.; Spanik, I. Comparison of Different Categories of Slovak Tokaj Wines in Terms of Profiles of Volatile Organic Compounds. *Molecules* **2020**, *25*, 669. [CrossRef] [PubMed]

8.   Tosi, E.; Fedrizzi, B.; Azzolini, M.; Finato, F.; Simonato, B.; Zapparoli, G. Effects of noble rot on must composition and aroma profile of Amarone wine produced by the traditional grape withering protocol. *Food Chem.* **2012**, *130*, 370–375. [CrossRef]

9.   Thibon, C.; Shinkaruk, S.; Jourdes, M.; Bennetau, B.; Dubourdieu, D.; Tominaga, T. Aromatic potential of botrytized white wine grapes: Identification and quantification of new cysteine-S-conjugate flavor precursors. *Anal. Chim. Acta* **2010**, *660*, 190–196. [CrossRef]

10.   Tominaga, T.; Niclass, Y.; Frerot, E.; Dubourdieu, D. Stereoisomeric distribution of 3-mercaptohexan-1-ol and 3-mercaptohexyl acetate in dry and sweet white wines made from Vitis vinifera (var. Sauvignon blanc and Semillon). *J. Agric. Food Chem.* **2006**, *54*, 7251–7255. [CrossRef]

11.   Ribeiro, C.; Gonçalves, R.; Tiritan, M. Separation of Enantiomers Using Gas Chromatography: Application in Forensic Toxicology, Food and Environmental Analysis. *Crit. Rev. Anal. Chem.* **2020**, 1–25. [CrossRef]

12.   Zawirska-Wojtasiak, R. Chirality and the Nature of Food Authenticity of Aroma. *Acta Sci. Polon. Technol.* **2006**, *5*, 21–36.

13.   Alvarez-Rivera, G.; Bueno, M.; Ballesteros-Vivas, D.; Cifuentes, A. Chiral analysis in food science. *Trac-Trend Anal. Chem.* **2020**, *123*, 151–171. [CrossRef]

14.   Mu, B.; Zhu, Y.; Lv, H.P.; Yan, H.; Peng, Q.H.; Lin, Z. The enantiomeric distributions of volatile constituents in different tea cultivars. *Food Chem.* **2018**, *265*, 329–336. [CrossRef] [PubMed]

15.   Pasquini, B.; Orlandini, S.; Goodarzi, M.; Caprini, C.; Gotti, R.; Furlanetto, S. Chiral cyclodextrin-modified micellar electrokinetic chromatography and chemometric techniques for green tea samples origin discrimination. *Talanta* **2016**, *150*, 7–13. [CrossRef] [PubMed]

16.   Nakano, Y.; Taniguchi, M.; Fukusaki, E. High-sensitive liquid chromatography-tandem mass spectrometry-based chiral metabolic profiling focusing on amino acids and related metabolites. *J. Biosci. Bioeng.* **2019**, *127*, 520–527. [CrossRef] [PubMed]

17.   Acquaviva, A.; Siano, G.; Quintas, P.; Filgueira, M.R.; Castells, C.B. Chiral x achiral multidimensional liquid chromatography. Application to the enantioseparation of dintitrophenyl amino acids in honey samples and their fingerprint classification. *J. Chromatogr. A* **2020**, *1614*, 460729. [CrossRef] [PubMed]

18.   Blasi, F.; Pollini, L.; Cossignani, L. Varietal Authentication of Extra Virgin Olive Oils by Triacylglycerols and Volatiles Analysis. *Foods* **2019**, *8*, 58. [CrossRef]

19.   Cordero, C.; Schmarr, H.G.; Reichenbach, S.E.; Bicchi, C. Current Developments in Analyzing Food Volatiles by Multidimensional Gas Chromatographic Techniques. *J. Agric. Food Chem.* **2018**, *66*, 2226–2236. [CrossRef]

20.   Krupcik, J.; Gorovenko, R.; Spanik, I.; Armstrong, D.W.; Sandra, P. Enantioselective comprehensive two-dimensional gas chromatography of lavender essential oil. *J. Sep. Sci.* **2016**, *39*, 4765–4772. [CrossRef]

21.   Shellie, R.; Marriott, P.; Cornwell, C. Application of comprehensive two-dimensional gas chromatography (GC× GC) to the enantioselective analysis of essential oils. *J. Sep. Sci.* **2001**, *24*, 823–830. [CrossRef]

22.   Wang, M.; Marriott, P.J.; Chan, W.H.; Lee, A.W.M.; Huie, C.W. Enantiomeric separation and quantification of ephedrine-type alkaloids in herbal materials by comprehensive two-dimensional gas chromatography. *J. Chromatogr. A* **2006**, *1112*, 361–368. [CrossRef]

23.   Machynakova, A.; Khvalbota, L.; Furdikova, K.; Drtilova, T.; Spanik, I. Characterization of volatile organic compounds in Slovak Tokaj wines. *J. Food Nutr. Res. Slov.* **2019**, *58*, 307–318.

24.   Vyviurska, O.; Spanik, I. Assessment of Tokaj varietal wines with comprehensive two-dimensional gas chromatography coupled to high resolution mass spectrometry. *Microchem. J.* **2020**, *152*, 104385. [CrossRef]

25.   Pazitna, A.; Dzurova, J.; Spanik, I. Enantiomer Distribution of Major Chiral Volatile Organic Compounds in Selected Types of Herbal Honeys. *Chirality* **2014**, *26*, 670–674. [CrossRef] [PubMed]

26.   Kaunzinger, A.; Wüst, M.; Gröbmiller, H.; Burow, S.; Hemmrich, U.; Dietrich, A.; Beck, T.; Hener, U.; Mosandl, A.; Rapp, A. Enantiomer distribution of ethyl lactate—A new criterion for quality assurance of wine. *Z. Für Lebensm. Unters. Forsch.* **1996**, *203*, 499–500. [CrossRef]

27.   Lloret, A.; Boido, E.; Lorenzo, D.; Medina, K.; Carrau, F.; Dellacassa, E.; Versini, G. Aroma variation in tannat wines: Effect of malolactic fermentation on ethyl lactate level and its enantiomeric distribution. *Ital. J. Food Sci.* **2002**, *14*, 175–180.

28.   Pozo-Bayón, M.; G-Alegría, E.; Polo, M.; Tenorio, C.; Martín-Álvarez, P.; Calvo De La Banda, M.; Ruiz-Larrea, F.; Moreno-Arribas, M. Wine volatile and amino acid composition after malolactic fermentation: Effect of Oenococcus oeni and Lactobacillus plantarum starter cultures. *J. Agric. Food Chem.* **2005**, *53*, 8729–8735. [CrossRef]

29.   Bokulich, N.A.; Joseph, C.M.L.; Allen, G.; Benson, A.K.; Mills, D.A. Next-Generation Sequencing Reveals Significant Bacterial Diversity of Botrytized Wine. *PLoS ONE* **2012**, *7*, e36357. [CrossRef] [PubMed]

30.   Lasik-Kurdyś, M.; Majcher, M.; Nowak, J. Effects of different techniques of malolactic fermentation induction on diacetyl metabolism and biosynthesis of selected aromatic esters in cool-climate grape wines. *Molecules* **2018**, *23*, 2549. [CrossRef]

31.   Fernandes, L.; Relva, A.M.; da Silva, M.D.R.G.; Freitas, A.M.C. Different multidimensional chromatographic approaches applied to the study of wine malolactic fermentation. *J. Chromatogr. A* **2003**, *995*, 161–169. [CrossRef]

32.   Baron, M.; Prusova, B.; Tomaskova, L.; Kumsta, M.; Sochor, J. Terpene content of wine from the aromatic grape variety 'Irsai Oliver'(*Vitis vinifera* L.) depends on maceration time. *Open Life Sci.* **2017**, *12*, 42–50. [CrossRef]

33. Roussis, I.G.; Lambropoulos, I.; Papadopoulou, D. Inhibition of the decline of volatile esters and terpenols during oxidative storage of Muscat-white and Xinomavro-red wine by caffeic acid and N-acetyl-cysteine. *Food Chem.* **2005**, *93*, 485–492. [CrossRef]

34. Bock, G.; Benda, I.; Schreier, P. Biotransformation of Linalool by Botrytis-Cinerea. *J. Food Sci.* **1986**, *51*, 659–662. [CrossRef]

35. Loscos, N.; Hernandez-Orte, P.; Cacho, J.; Ferreira, V. Evolution of the aroma composition of wines supplemented with grape flavour precursors from different varietals during accelerated wine ageing. *Food Chem.* **2010**, *120*, 205–216. [CrossRef]

36. Oliveira, J.M.; Oliveira, P.; Baumes, R.L.; Maia, O. Changes in aromatic characteristics of Loureiro and Alvarinho wines during maturation. *J. Food Compos. Anal.* **2008**, *21*, 695–707. [CrossRef]

37. Wang, X.J.; Tao, Y.S.; Wu, Y.; An, R.Y.; Yue, Z.Y. Aroma compounds and characteristics of noble-rot wines of Chardonnay grapes artificially botrytized in the vineyard. *Food Chem.* **2017**, *226*, 41–50. [CrossRef] [PubMed]

38. Jackson, R.S. *Wine Science: Principles and Applications*; Academic Press: London, UK, 2008.

39. Cerdan, T.G.; Goni, D.T.; Azpilicueta, C.A. Accumulation of volatile compounds during ageing of two red wines with different composition. *J. Food Eng.* **2004**, *65*, 349–356. [CrossRef]

40. Cooke, R.C.; Van Leeuwen, K.A.; Capone, D.L.; Gawel, R.; Elsey, G.M.; Sefton, M.A. Odor detection thresholds and enantiomeric distributions of several 4-alkyl substituted γ-lactones in Australian red wine. *J. Agric. Food Chem.* **2009**, *57*, 2462–2467. [CrossRef]

41. Staniatopoulos, P.; Brohan, E.; Prevost, C.; Siebert, T.E.; Herderich, M.; Darriet, P. Influence of Chirality of Lactones on the Perception of Some Typical Fruity Notes through Perceptual Interaction Phenomena in Bordeaux Dessert Wines. *J. Agric. Food Chem.* **2016**, *64*, 8160–8167. [CrossRef]

42. Suomalainen, H.; Nykänen, L. Composition of whisky flavour. *Process Biochem.* **1970**, *5*, 13–18.

43. Shimotori, Y.; Hoshi, M.; Okabe, H.; Miyakoshi, T.; Kanamoto, T.; Nakashima, H. Synthesis, odour characteristics and antibacterial activities of the stereoisomeric forms of whisky lactone and its thiono analogues. *Flavour Frag. J.* **2017**, *32*, 29–35. [CrossRef]

44. Arapitsas, P.; Antonopoulos, A.; Stefanou, E.; Dourtoglou, V. Artificial aging of wines using oak chips. *Food Chem.* **2004**, *86*, 563–570. [CrossRef]

45. Spillman, P.J.; Sefton, M.A.; Gawel, R. The effect of oak wood source, location of seasoning and coopering on the composition of volatile compounds in oak-matured wines. *Aust. J. Grape Wine Res.* **2004**, *10*, 216–226. [CrossRef]

46. Moreno, N.J.; Azpilicueta, C.A. Binding of oak volatile compounds by wine lees during simulation of wine ageing. *LWT-Food Sci. Technol.* **2007**, *40*, 619–624. [CrossRef]

47. Brown, R.C.; Sefton, M.A.; Taylor, D.K.; Elsey, G.M. An odour detection threshold determination of all four possible stereoisomers of oak lactone in a white and a red wine. *Aust. J. Grape Wine Res.* **2006**, *12*, 115–118. [CrossRef]

48. del Alamo-Sanza, M.; Nevares, I.; Martinez-Gil, A.; Rubio-Breton, P.; Garde-Cerdan, T. Impact of long bottle aging (10 years) on volatile composition of red wines micro-oxygenated with oak alternatives. *LWT-Food Sci. Technol.* **2019**, *101*, 395–403. [CrossRef]

49. Perez-Coello, M.S.; Sanz, J.; Cabezudo, M.D. Determination of volatile compounds in hydroalcoholic extracts of French and American oak weed. *Am. J. Enol. Viticult.* **1999**, *50*, 162–165.

50. Liu, N.; Koot, A.; Hettinga, K.; de Jong, J.; van Ruth, S.M. Portraying and tracing the impact of different production systems on the volatile organic compound composition of milk by PTR-(Quad)MS and PTR-(ToF)MS. *Food Chem.* **2018**, *239*, 201–207. [CrossRef]

51. Granato, D.; Santos, J.S.; Escher, G.B.; Ferreira, B.L.; Maggio, R.M. Use of principal component analysis (PCA) and hierarchical cluster analysis (HCA) for multivariate association between bioactive compounds and functional properties in foods: A critical perspective. *Trends Food Sci. Technol.* **2018**, *72*, 83–90. [CrossRef]

 *foods*

Article

Chia Oil Adulteration Detection Based on Spectroscopic Measurements

Monica Mburu [1], **Clement Komu** [1], **Olivier Paquet-Durand** [2], **Bernd Hitzmann** [2] and **Viktoria Zettel** [2,*]

[1] Institute of Food Bioresources Technology, Dedan Kimathi University of Technology, Private Bag, Dedan Kimathi, Nyeri 10143, Kenya; monica.mburu@dkut.ac.ke (M.M.); clementmumo@gmail.com (C.K.)

[2] Department of Process Analytics and Cereal Science, Institute of Food Science and Biotechnology, University of Hohenheim, Garbenstr. 23, 70599 Stuttgart, Germany; O.Paquet-Durand@uni-hohenheim.de (O.P.-D.); Bernd.Hitzmann@uni-hohenheim.de (B.H.)

* Correspondence: Viktoria.Zettel@uni-hohenheim.de; Tel.: +49-711-459-24460

Abstract: Chia oil is a valuable source of omega-3-fatty acids and other nutritional components. However, it is expensive to produce and can therefore be easily adulterated with cheaper oils to improve the profit margins. Spectroscopic methods are becoming more and more common in food fraud detection. The aim of this study was to answer following questions: Is it possible to detect chia oil adulteration by spectroscopic analysis of the oils? Is it possible to identify the adulteration oil? Is it possible to determine the amount of adulteration? Two chia oils from local markets were adulterated with three common food oils, including sunflower, rapeseed and corn oil. Subsequently, six chia oils obtained from different sites in Kenya were adulterated with sunflower oil to check the results. Raman, NIR and fluorescence spectroscopy were applied for the analysis. It was possible to detect the amount of adulterated oils by spectroscopic analysis, with a minimum R^2 of 0.95 for the used partial least square regression with a maximum $RMSEP_{range}$ of 10%. The adulterations of chia oils by rapeseed, sunflower and corn oil were identified by classification with a median true positive rate of 90%. The training accuracies, sensitivity and specificity of the classifications were over 90%. Chia oil B was easier to detect. The adulterated samples were identified with a precision of 97%. All of the classification methods show good results, however SVM were the best. The identification of the adulteration oil was possible; less than 5% of the adulteration oils were difficult to detect. In summary, spectroscopic analysis of chia oils might be a useful tool to identify adulterations.

Keywords: chia oil; adulteration; spectroscopy; NIR; Raman; fluorescence

Citation: Mburu, M.; Komu, C.; Paquet-Durand, O.; Hitzmann, B.; Zettel, V. Chia Oil Adulteration Detection Based on Spectroscopic Measurements. *Foods* **2021**, *10*, 1798. https://doi.org/10.3390/foods10081798

Academic Editors: Isabel Mafra and Margit Cichna-Markl

Received: 30 June 2021
Accepted: 3 August 2021
Published: 4 August 2021

Publisher's Note: MDPI stays neutral with regard to jurisdictional claims in published maps and institutional affiliations.

1. Introduction

Chia, *Salvia hispanica* L., a member of the Labiatae family, is cultivated in environments ranging from tropical to subtropical conditions and used as a food ingredient. Native from southern Mexico and northern Guatemala, chia has been cultivated on a commercial basis in Australia, Colombia, Argentina, Peru, Ecuador, Bolivia and Paraguay [1]. Research has proved that chia seeds are a good source of oil, protein, dietary fiber, minerals and polyphenolic compounds [2]. Quantitatively, chia seeds contain 91–93 g/100 g dry matter, 26–41 g/100 g carbohydrates, 32–39 g/100 g oil, 22–24 g/100 g protein, 18–30 g/100 g dietary fiber, and 4–6 g/100 g ash, vitamins, antioxidants, minerals contents [3].

Chia oil is known to lower the risks of cardiovascular disease, inflammation, hepatoprotective effect and also to prevent the likelihood of obesity-related disorders [4]. According to research carried out by Gazem et al. [5], investigating in vitro the cancer cytotoxic properties of chia seeds oil and its blends, chia seed oil was found to significantly inhibit anti-lipoxygenase activity, and demonstrated potent and differential anticancer activity. The team concluded that supplementation of a modern diet with chia seeds oil may delay or prevent the incidence of degenerative disorders. Additionally, according to research carried out by Albert et al. [6], it was observed that supplementation of a diet with long-chain omega-3 polyunsaturated fatty acids can prevent cardiovascular and inflammatory diseases. Current research has not shown any adverse effects of chia seed consumption, but toxicological data on controlled human trials on the safety and efficacy of chia seed oils are still limited. With the emerging concepts around the combination of chemotherapy and nutritional therapy, there is need to increase data on fatty acid composition in various foods that can be applied in chemotherapeutic subjects. Chia seed oil is becoming an appealing and preferred choice for healthy food and cosmetic applications due to its lower content of saturated fatty acids (palmitic and stearic acids) and adequate concentration of linolenic fatty acids (55–60%) and linoleic acids (18–20%) [3]. Both chia seeds and chia seed oil have been safely applied in animal feeds to decrease the cholesterol levels and increase the polyunsaturated fatty acids and in egg and meat products [7]

Extraction of chia oils apply different methods with diverse oil yields including cold-pressing followed by centrifugation to remove physical matter, hot-pressing, solvent extraction and supercritical fluid. Chia oil yield and quality in terms of fatty acids composition are affected by several factors including agroecological zones of growth, seed variety, seed storage conditions, pre-treatment method, size reduction practices and the aforementioned extraction procedures [8]. Due to the high value of chia oil, some unscrupulous sellers may adulterate with cheaper oils in order to increase profit. This adulteration will also make the long-chain polyunsaturated fatty acids highly susceptible to lipid hydrolysis and oxidation, thus loosing shelf-life, consumer acceptability, nutritional value, functionality and safety.

Vegetable oils are valuable component of human nutrition. Adulteration of valuable expensive oils with cheaper oils is very common practice. Applying spectroscopic methods provides an opportunity quickly detect these adulterations. There are several works available on olive oil adulteration detection by fluorescence spectroscopy [9–11]. Sikorska et al. [12] were able to distinguish between different edible oils using fluorescence spectroscopy. Near Infrared spectroscopy (NIR) is also well established for food analysis [13]. With data obtained from NIR, UV-Vis and GC, the ComDim chemometrics method was able to distinguish 32 vegetable oil samples by their characteristics and compositions [14]. Rodríguez et al. [15] showed that it is possible to detect adulteration of sesame and chia oils by Fourier transform infrared spectroscopy with prediction errors between 1% and 5%. Studies on oil adulteration detection with spectroscopic methods have been published by several authors. For example, La Mata et al. [16] used ATR-FTIR spectroscopy and were able to differentiate between blends with olive oil content higher than 50% (*w/w*) and those below 50% (*w/w*). More examples for the application of FTIR on olive oil adulteration can be found in literature [17–20]. (FT- or M-) IR spectroscopy was also successfully used for sesame oil adulteration [21–25]. Extra virgin olive oil adulteration with hazelnut oil was evaluated using mid-infrared and Raman spectroscopic data [26]. The application of Raman spectroscopy on olive oil adulteration [27] or the combination of Raman and NIR spectroscopy [28] is another way of combining the spectroscopic methods. Adulteration detection by FT-Raman and NIR spectroscopy, combined with data fusion and Soft Independent Modelling of Class Analogy, was performed on a case study to determine the adulteration of hazelnut paste with almonds or chickpeas [29]. Other examples of combinations of NIR and fluorescence were given by Hu et al. [30], who worked on the fraud detection of Chinese tea oil or by Li et al. [31], who applied these spectroscopic methods to detect adulteration and authenticity of walnut oil.

This study focuses on the adulteration of chia oils with cheaper oils that are available in European and African markets. The more expensive chia oils are currently paid a great deal of attention in African countries, and therefore it is necessary to prevent the valuable oil from adulteration. Adulteration detection is mostly dependent on discriminant analysis, where the spectrum of the test sample is compared to a reference library. The establishment of the reference library usually takes a long time due to the amount of data that has to be covered, e.g., known adulterated samples. Important questions must be answered throughout the process, such as whether a test sample belongs to the native samples or the adulterated samples and whether the adulteration can actually be identified. The last but most difficult question is to which amount the test sample has been adulterated.

2. Materials and Methods

2.1. Sample Preparation

Two different samples of chia oil were purchased, A: Bio Chia Öl (Ölmühle Fandler GmbH, Pöllau, Austria with best before dates of 21 January 2020 and 28 February 2020, origin: Mexico) and B: Chiaöl (Ölmühle Solling GmbH, Boffzen, Germany with best before dates of 7 September 2019 and 26 December 2019, origin: Mexico). For adulteration, common food preparation oils were purchased at the local markets: rapeseed oil (R): Reines Rapsöl, raffiniert (Bökelmann + Co. Ölmühle GmbH & Co. KG, Hamm, Germany, with best before date 24 April 2020), sunflower oil (S): Reines Sonnenblumenöl, raffiniert (Walter Rau Lebensmittelwerke GmbH, Hilter, Germany, with best before date 17 May 2020), and corn oil (C): Mazola, reines Maiskeimöl (Peter Kölln GmbH & Co. KG, Elmsholm, Germany, with best before date 27 May 2020). The nutritional values of the oil samples are presented in Table 1.

Table 1. T Nutritional values of the oil samples, for A and B (chia oils) per 100 g, for R, S and C per 100 mL.

Sample Name	Energy [kJ]	Energy [kcal]	Fat [g]	Saturated Fatty Acids [g]	Single Unsat. Fatty Acids [g]	Multiple Unsat. Fatty Acids [g]	Vitamin E [mg]
A (chia oil)	3700	900	100	10.1	7.8	82.1	-
B (chia oil)	3700	900	100	10.6	7.2	82.2	-
R (rapeseed)	3404	828	92	6.5	60	25.5	46
S (sunflower)	3404	828	92	10	28	54	30
C (corn)	3404	828	92	13	28	51	37

In Table 2, the sample preparation and its labelling for the Mexican chia oils is presented. Every sample was prepared three times, and 114 samples were collected. The sample volume remained constant at 3.5 mL.

Table 2. Sample preparation and labelling for the spectroscopic analysis. A and B are the two Mexican chia oils, S is sunflower oil, R is rapeseed oil and C is corn oil. All values are mass percentages.

Samples		Materials				
		A	B	S	R	C
native oils	A100	100%				
	B100		100%			
	S100			100%		
	R100				100%	
	C100					100%
samples with A	AS90	90%		10%		
	AS95	95%		5%		
	AS98	98%		2%		
	AS99	99%		1%		
	AR90	90%			10%	
	AR95	95%			5%	
	AR98	98%			2%	
	AR99	99%			1%	
	AC90	90%				10%
	AC95	95%				5%
	AC98	98%				2%
	AC99	99%				1%

Table 2. *Cont.*

Samples		Materials				
		A	B	S	R	C
samples with B	BS90		90%	10%		
	BS95		95%	5%		
	BS98		98%	2%		
	BS99		99%	1%		
	BR90		90%		10%	
	BR95		95%		5%	
	BR98		98%		2%	
	BR99		99%		1%	
	BC90		90%			10%
	BC95		95%			5%
	BC98		98%			2%
	BC99		99%			1%
additional combinations	RS50			50%	50%	
	RC50				50%	50%
	SC50			50%		50%
	AS50	50%		50%		
	AR50	50%			50%	
	AC50	50%				50%
	BS50		50%	50%		
	BR50		50%		50%	
	BC50		50%			50%

For Kenyan chia oil samples, named oil U, V, W, X, Y, Z (from chia seeds obtained from different growth sites in Kenya) a smaller sample volume (2 mL) was chosen because of the small number of samples available. Its samples were prepared, according to Table 3, two times with exceptions (indicated with *), which were prepared once. Therefore, 28 different samples were obtained from Kenyan chia oil. All samples were directly prepared in a quartz glass cuvette and mixed by gently shaking. Then the cuvettes were placed in the respective spectrometer.

Table 3. Sample preparation for the additional Kenyan chia oil samples (U–Z) that were adulterated with sunflower oil (S). Samples indicated with * were prepared only once, the others were prepared two times.

Samples	Materials						
	U	V	W	X	Y	Z	S
U100 *	100%						
V100		100%					
W100			100%				
X100 *				100%			
Y100					100%		
Z100						100%	
US90 *	90%						10%
US50 *	50%						50%
VS90		90%					10%
VS50		50%					50%
WS50 *			50%				50%
XS90 *				90%			10%
XS50				50%			50%
YS90					90%		10%
YS50					50%		50%
ZS90						90%	10%
ZS50						50%	50%

* All together 142 samples are used for spectroscopic measurement.

2.2. Spectroscopic Measurements

Three spectrometers were used to obtain near infrared (NIR), Raman and fluorescence spectra of the oil samples. NIR spectroscopy measurements were performed in the Multi-Purpose NIR Analyzer (Bruker Optik GmbH, Ettlingen, Germany), varying wavelengths from 800 nm to 2800 nm, in absorbance, with a resolution of 15 nm and 8 scans per measurement.

Raman spectroscopy was performed with a FT-Raman785 spectrometer (Inno-spec GmbH, Model 11-0130005-119, Nürnberg, Germany), equipped with a 784.98 nm Laser applying a measurement range from 350 cm^{-1} to 3200 cm^{-1}. The integration time was 1 s and 3 scans were performed for each measurement. The background was measured with an empty cuvette.

3D-fluorescence spectra were obtained with FluoroMax4 Spectrofluorometer (HORIBA JOBIN YVON Technology, Edison, NY, USA). Spectra were analysed in a range between 300 nm and 550 nm of excitation and 350 nm and 700 nm emission with 10 nm distance steps and a slit width of 1 nm. In total, the resulting spectra contained 936 measured intensities of wavenumber and wavelength combinations.

Every prepared sample was measured 5 times. In total, 142 samples were measured. Every single spectrum was used for the analysis, in total 710 spectra were obtained for each spectroscopic method. The resulting combined spectra contained 2751 points.

2.3. Spectra Evaluation: Preprocessing

The evaluation of the spectra was performed with Matlab R2020a (version 9.8). The spectra were pre-processed with different methods to extract the desired information. A baseline correction and a standard normal variate (SNV) transformation was applied to Raman and NIR spectra. For the baseline correction, the following Matlab code, presented in Equation (1), was applied in a loop using the intensity values of all wavenumbers k in a spectrum.

$$I_{BC}(k) = I(k) - cumsum[smooth(diff\{I(k)\}, 20)] \tag{1}$$

$I_{BC}(k)$ is the baseline corrected intensity value, $I(k)$ the raw intensity, cumsum, smooth and diff are Matlab functions. To harmonize the spectra further, a standard normal variate transformation, presented in Equation (2), was applied as follows

$$I_{SNV}(k) = \frac{I_{BC}(k) - \bar{I}_{BC}}{SD_{BC}} \tag{2}$$

$I_{SNV}(k)$ is the transformed intensity, $\bar{I}_{BC}$ and SD_{BC} are the mean value and standard deviation of the base line corrected spectrum. For the fluorescence spectra, no pre-processing was applied. The spectra were then evaluated separately for each spectrometer typ. For further evaluations NIR, fluorescence and Raman spectra were combined. The intensities of the fluorescence spectra were therefore scaled down with a SNV transformation, subsequently the NIR and Raman spectra were appended to the fluorescence spectra to produce combined spectra.

2.4. Spectra Evaluation: Classification

The classification was performed by using the Classification Learner App, which is implemented in Matlab. The following classification algorithms were tested: decision tree (DT), linear discriminant analysis (LD), k nearest neighbour classification (KNN), support vector machine linear (SVMl) and cubic (SVMc). The classification was performed with 5 classes: 1. A, 2. Adult A, 3. B, 4. Adult B, 5. Adult. The classification was performed to check if A and B samples of the native oils could be distinguished and if an adulteration was present. For Adult A and Adult B, the 12 samples with A and B were complemented by 3 corresponding samples of the additional combinations presented in Table 2. Therefore, 225 spectra were in both classes. To obtain equal number of spectra in every class some simulation spectra were calculated, so that every class was enlarged to 225 spectra.

The number of pure oil samples of class A and B resulted just in 15 spectra each, therefore new spectra were simulated out of them. First, the means m and the standard deviations SD of intensity values for all wavenumbers (Raman and NIR) or wavelength combination (fluorescence) for both classes were individually calculated. 150 spectra for each pure oil sample were simulated by adding to each value in the mean (75) or the original (75) spectrum the corresponding standard deviation times a standard normal distributed random number, which has a cero mean and a standard deviation of one, as shown in Equation (3).

$$\widetilde{I}(k) = I(k) + SD(k) \times ran(k) \tag{3}$$

Here $\widetilde{I}(k)$ is the simulated intensity value, k is either the wavenumber (for Raman and NIR) or an index for the wavelength combinations (fluorescence), $I(k)$ is the corresponding mean or original value and $SD(k)$ the corresponding standard deviation, $ran(k)$ is a standard normal distributed random number, which is calculated for each k. To complete the class A and B data sets to 225 spectra, the original spectra were used five times.

For the "Adult" class, the 100% pure samples of S, R and C as well as the corresponding additional combinations (RS50, RC50, SC50), which were 90 spectra together, were complemented by 90 simulated spectra obtained in the same manner as discussed before (Equation (3)) from the samples S, R and C. To complete the data set of class "Adult", 45 replication spectra from S, R and C samples were added. In total 1125 spectra were obtained, where each class consisted of 225 spectra.

The quality of the classification is assessed with the amount of correct detected samples, which is calculated as % of samples in the validation dataset and is presented as True Positive Rate (TPR). The sensitivity (Equation (4)), specificity (Equation (5)), accuracy (Equation (6)) and precision (Equation (7)) are calculated with the values of true positive (TP), true negative (TN), false negative (FN) and false positive (FP) identified samples [32].

$$\text{Sensitivity (\%)} = \text{TP}/(\text{TP+FN}) \cdot 100\%, \tag{4}$$

$$\text{Specificity (\%)} = \text{TN}/(\text{TN+FP}) \cdot 100\%, \tag{5}$$

$$\text{Accuracy (\%)} = (\text{TP+TN})/(\text{TP+TN+FP+FN}) \cdot 100\%, \tag{6}$$

$$\text{Precision (\%)} = \text{TP}/(\text{TP+FP}) \, 100\%, \tag{7}$$

2.5. Spectra Evaluation: Partial Least Squares Regression

Partial Least Squares Regression (PLSR) models are calculated for each oil to predict the adulteration levels. For the Mexican chia oil samples, A and B, 1 up to 32 principal components (3–10 for Kenyan samples, depending on the number of measured samples) are tested for the PLSR model. A leave-one-out-cross-validation (CV) is performed for each dataset. The coefficient of determination R^2 and the root mean square error of prediction $RMSEP_{range}$ are calculated.

The detection limit dl for the PLSR was calculated from the blank sample (100% pure chia oil) with Equation (8), where m is the mean and SD is the standard deviation.

$$dl = m_{100\% \text{ chia oil}} + 3\,SD_{100\% \text{ chia oil}} \tag{8}$$

3. Results and Discussion

The native oils could easily be distinguished by their fluorescence spectra (Figure 1). All of the oils differ in intensities and slight intensity regions. It was assumed that the best results would be obtained through fluorescence spectra evaluation. The visible peaks can be assigned to pigments of groups belonging to NADH, tocopherols, riboflavin (emission 524 nm), oxidation products of oil ingredients e.g., vitamin E derivates at 525 nm emission and chlorophyll at excitation 405 nm and emission 670 nm [10–12,33–35]. However, the oils were not prepared in a special way or measured in a solvent; therefore, the ranges might have shifted and/or the intensities might be lower. Since we work with raw materials that are subject to natural variations, it is definitely possible that the spectra of two oils are not one in the same. The fluorescence spectra of the chia oils show the same intensity regions. Overall, all of the oils examined show higher intensities in the regions of carotenoids, tocopherols, polyphenols and chlorophylls. Lower intensities in the regions of 350 nm excitation and 400 nm to 450 nm emission indicate the presence of oxidation products formed during oil ageing. Observing Figure 1 in-depth, it is obvious that the intensities of the oils used for adulteration (sunflower, rape seed and corn) have higher intensities in the respective regions.

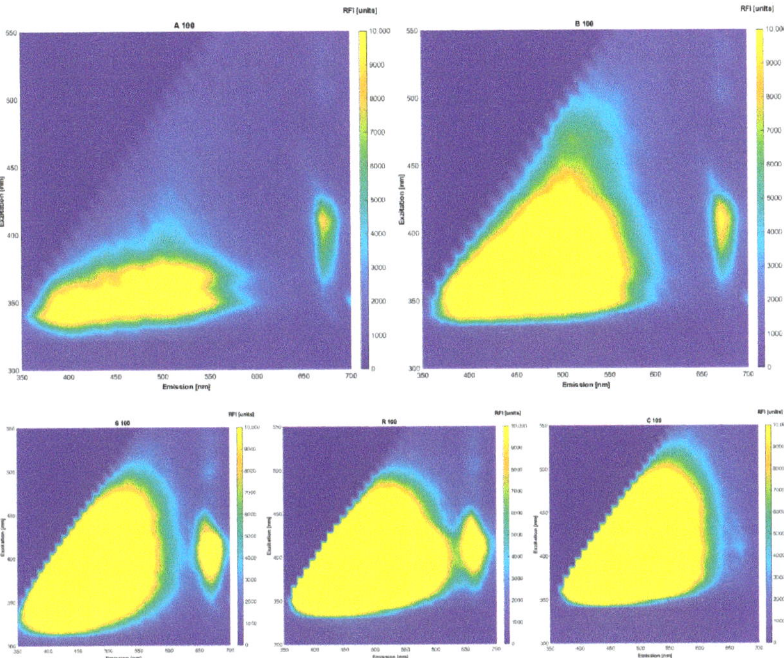

Figure 1. Fluorescence spectra of native oils, A: chia, B: chia, S: sunflower, R: rape seed, C: corn.

For NIR and Raman spectra, the native oil spectra are presented in Figures 2 and 3. The left side shows the raw spectra, whereas the right side shows the pre-processed spectra. For the combined evaluation, the fluorescence spectra were also pre-processed by SNV, and therefore the intensities are comparable. For NIR spectra, no big differences between the samples are obvious, but in the Raman spectra different intensities for the samples are visible. In Figure 4, the combined spectra of all native oils are presented. The spectra of A and B show differences compared to the other oils.

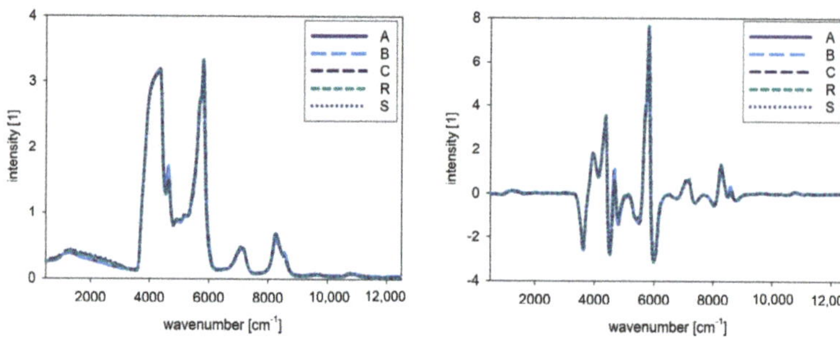

Figure 2. NIR spectra of native oils, A: chia, B: chia, R: rape seed, S: sunflower, C: corn, raw spectra (**left**), pre-processed spectra (**right**).

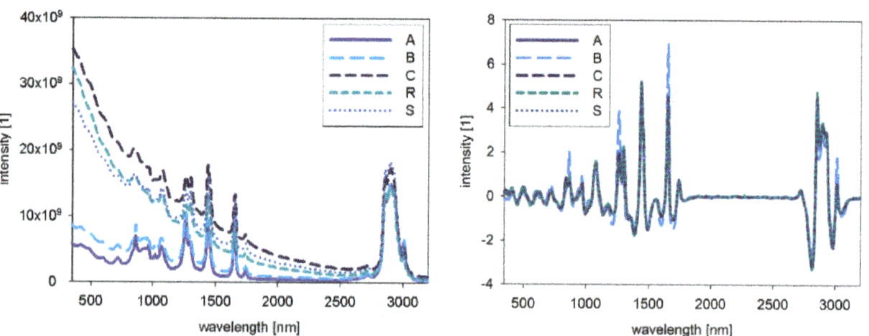

Figure 3. Raman spectra of native oils, A: chia, B: chia, R: rape seed, S: sunflower, C: corn, raw spectra (**left**), pre-processed spectra (**right**).

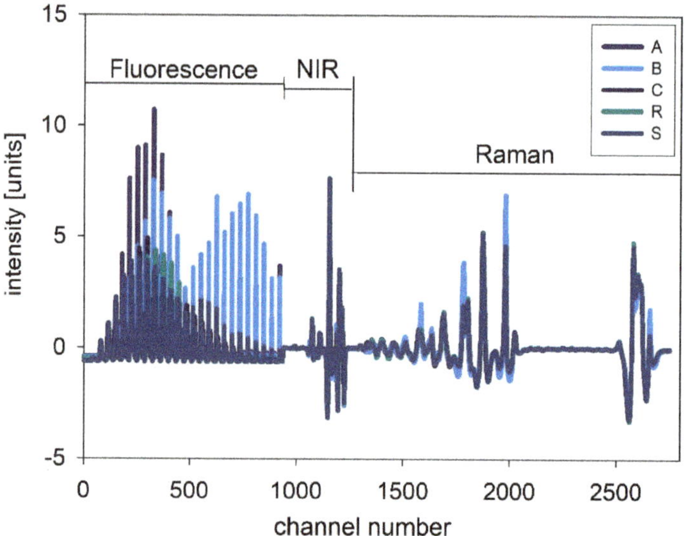

Figure 4. Combination of all spectra for evaluation.

Mean values and standard deviations for the ten classification runs can be found in Tables 4 and 5. The best results for the classification were obtained with a TPR of 99.7% for the classification with SVMc and the combination of all of the spectra together (Table 5). The combination of fluorescence and NIR spectra were classified with a TPR of 99.5% with SVMc, and SVMc is also the best classification method for all single spectra. The medians for the TPR, sensitivity, specificity and accuracy of the classification are presented in Figure 5. The median TPR is over 90% for most of the calculations. As usual, the training accuracies are, with exceptions, all over 90%, higher than the validation accuracies which were between 71% and 79.9%. The sensitivity as well as the specificity were over 90% for all of the samples. However, B was better detected. The precision was around 100% for pure B samples whereas for A, the precision was poor with 54.2 ± 3% for the Raman spectra classification by KNN. The precision for adulterated samples was over 90%. It is obvious that KNN results in the poorest classification results for A and B as well as for all measured spectra and their combinations. Adulterations for A were incorrectly classified.

Table 4. Results of the classification of samples with single spectra; Means and standard deviations of 10 classification runs.

Fluorescence	1 Tree	2 LD	3 KNN	4 SVMl	5 SVMc
TPR	94.9 ± 1.6	94.9 ± 1.6	93.2 ± 1.4	92.8 ± 1	98.1 ± 2
Accuracy training	95 ± 0.7	97.1 ± 0.5	92.9 ± 0.4	92.9 ± 0.4	97.5 ± 0.7
Accuracy validation	78 ± 0.6	78 ± 0.6	77.3 ± 0.6	77.1 ± 0.4	79.3 ± 0.8
Sensitivity A	94.6 ± 2.9	94.6 ± 2.9	100 ± 0	100 ± 0	100 ± 0
Sensitivity B	100 ± 0	100 ± 0	100 ± 0	100 ± 0	100 ± 0
Specificity A	97.9 ± 1.1	97.9 ± 1.1	93.3 ± 1.8	99.6 ± 0.5	99.8 ± 0.3
Specificity B	100 ± 0	100 ± 0	100 ± 0	100 ± 0	100 ± 0
Precision A	91.2 ± 4.9	91.2 ± 4.9	78.1 ± 6.3	98.1 ± 2.7	99.3 ± 1.2
Precision B	100 ± 0	100 ± 0	100 ± 0	100 ± 0	100 ± 0
NIR	**6 tree**	**7 LD**	**8 KNN**	**9 SVMl**	**10 SVMc**
TPR	95.8 ± 2.2	95.8 ± 2.2	90.2 ± 1.8	97.2 ± 1.3	98.4 ± 0.7
Accuracy training	95.3 ± 0.5	95.5 ± 0.8	90.7 ± 0.3	97.6 ± 0.4	98.5 ± 0.3
Accuracy validation	78.3 ± 0.9	78.3 ± 0.9	76.1 ± 0.7	78.9 ± 0.5	79.4 ± 0.3
Sensitivity A	95.5 ± 2.9	95.5 ± 2.9	100 ± 0	100 ± 0	100 ± 0
Sensitivity B	99.5 ± 1.1	99.5 ± 1.1	100 ± 0	100 ± 0	100 ± 0
Specificity A	98.7 ± 0.9	98.7 ± 0.9	90.2 ± 1.7	98.6 ± 0.9	99 ± 0.6
Specificity B	100 ± 0	100 ± 0	100 ± 0	100 ± 0	100 ± 0
Precision A	94.7 ± 3.1	94.7 ± 3.1	71 ± 5.4	94.4 ± 3.2	96.1 ± 2.2
Precision B	100 ± 0	100 ± 0	100 ± 0	100 ± 0	100 ± 0
Raman	**11 tree**	**12 LD**	**13 KNN**	**14 SVMl**	**15 SVMc**
TPR	92.2 ± 2.5	92.2 ± 2.5	79.6 ± 2.6	96.4 ± 0.7	98 ± 0.4
Accuracy training	92.3 ± 1.2	97 ± 0.3	79 ± 0.6	97.3 ± 0.3	98.4 ± 0.2
Accuracy validation	76.9 ± 1	76.9 ± 1	71.8 ± 1	78.6 ± 0.3	79.2 ± 0.2
Sensitivity A	96 ± 3.6	96 ± 3.6	100 ± 0	100 ± 0	100 ± 0
Sensitivity B	98.5 ± 2	98.5 ± 2	100 ± 0	100 ± 0	100 ± 0
Specificity A	96.5 ± 2.1	96.5 ± 2.1	81.5 ± 2.9	99.5 ± 0.5	99.7 ± 0.3
Specificity B	99.9 ± 0.2	99.9 ± 0.2	100 ± 0	100 ± 0	100 ± 0
Precision A	87.1 ± 7.1	87.1 ± 7.1	56.7 ± 5.8	97.9 ± 2.4	98.8 ± 1.3
Precision B	99.8 ± 0.8	99.8 ± 0.8	100 ± 0	100 ± 0	100 ± 0

Table 5. Results of the classification of samples with combinations of spectra; means and standard deviations of 10 classification runs.

Fluo + NIR + Raman	16 Tree	17 LD	18 KNN	19 SVMl	20 SVMc
TPR	98.1 ± 0.6	98.1 ± 0.6	91.2 ± 2	99.2 ± 0.4	99.7 ± 0.3
Accuracy training	97.5 ± 0.6	99.4 ± 0.1	90.7 ± 0.3	99.2 ± 0.2	99.6 ± 0.2
Accuracy validation	79.2 ± 0.3	79.2 ± 0.3	76.5 ± 0.8	79.7 ± 0.2	79.9 ± 0.1
Sensitivity A	98.3 ± 2	98.3 ± 2	100 ± 0	100 ± 0	100 ± 0
Sensitivity B	99.7 ± 0.9	99.7 ± 0.9	100 ± 0	100 ± 0	100 ± 0
Specificity A	98.8 ± 0.5	98.8 ± 0.5	89.7 ± 1.9	99.5 ± 0.6	99.8 ± 0.3
Specificity B	99.9 ± 0.4	99.9 ± 0.4	100 ± 0	100 ± 0	100 ± 0
Precision A	95.2 ± 2.2	95.2 ± 2.2	70 ± 5.4	97.9 ± 2.9	99.3 ± 1.2
Precision B	99.7 ± 1	99.7 ± 1	100 ± 0	100 ± 0	100 ± 0
Fluo + NIR	**21 tree**	**22 LD**	**23 KNN**	**24 SVMl**	**25 SVMc**
TPR	97.1 ± 1	97.1 ± 1	94.5 ± 1.1	98.4 ± 1.1	99.5 ± 0.5
Accuracy training	97.1 ± 0.7	99.5 ± 0.2	94.5 ± 0.3	98 ± 0.3	99.5 ± 0.1
Accuracy validation	78.8 ± 0.4	78.8 ± 0.4	77.8 ± 0.4	79.3 ± 0.4	79.8 ± 0.2
Sensitivity A	94.6 ± 3.8	94.6 ± 3.8	100 ± 0	100 ± 0	100 ± 0
Sensitivity B	99.7 ± 0.9	99.7 ± 0.9	100 ± 0	100 ± 0	100 ± 0
Specificity A	99 ± 0.7	99 ± 0.7	93.6 ± 1.4	99 ± 0.8	99.3 ± 0.5
Specificity B	100 ± 0	100 ± 0	100 ± 0	100 ± 0	100 ± 0
Precision A	95.7 ± 3.2	95.7 ± 3.2	78.8 ± 5.1	95.7 ± 3.5	97.2 ± 2.6
Precision B	100 ± 0	100 ± 0	100 ± 0	100 ± 0	100 ± 0

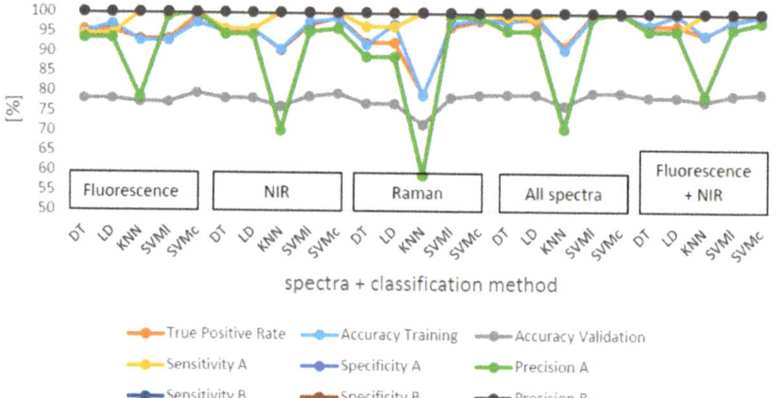

Figure 5. Median of 10 runs for TPR, accuracy of training and validation, sensitivity, specificity and precision of the classification for all evaluated spectra and combined variations as well as different classification methods; DT: decision tree, LD: linear discriminant analysis, KNN: nearest neighbour classification, SVMl and SVMc: support vector machines linear and cubic.

For Raman spectra evaluation, KNN resulted in a false classification of 64.3% for Adult A and a false classification of 33.3% for Adult B for one out of ten classifications (Figure 6). The same classification method leads to the combined evaluation of fluorescence and NIR spectra (Figure 7) to only a false classification of 35.7% of Adult A, which indicated that somehow the adulteration samples of chia oil A are more difficult to detect in general. The best results were obtained for the combined evaluation of fluorescence and NIR spectra, the confusion matrix of one classification run is presented in Figure 8. The wrong classifications are more or less equally distributed over all samples and remain below 10%. A successful classification is hence possible for 5 classes. KNN does not seem to be sufficient for these classification processes.

The presented method was capable of identifying most of the samples in the validation trial. It is a fast method which is easy to use after a calibration. The quantification of other compounds in the oil might also be possible with this method but this was not the focus of this study. The time-saving after the calibration of a spectroscopic method is around 2 to 3 times faster [36]. This underlines the necessity of the validation, which was successfully performed in this study.

The best results of the PLSR are presented in Table 6. The coefficients of determination are above 0.95 for all samples. Given the fact that the extreme points (the native oils) could be distinguished quite easily, this is not surprising.

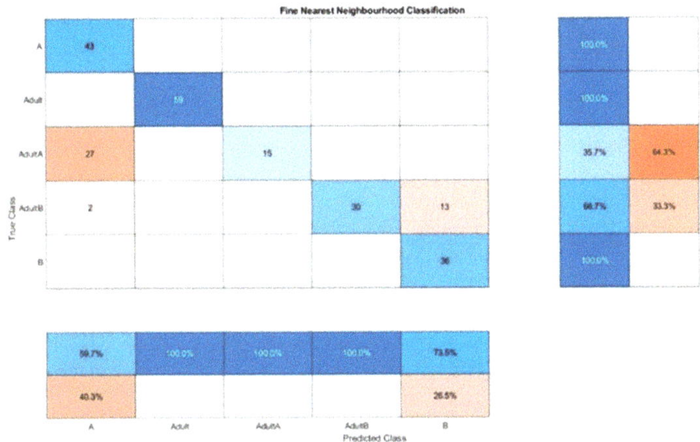

Figure 6. Confusion matrix for the oil classification by KNN out of the Raman spectra.

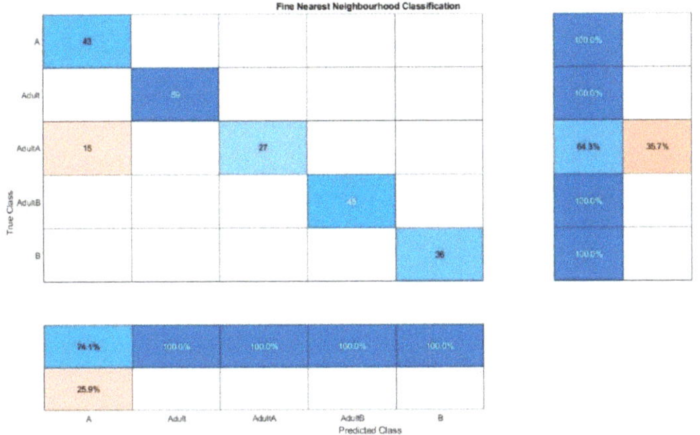

Figure 7. Confusion matrix for the oil classification by KNN out of the combined evaluation of the fluorescence and NIR spectra.

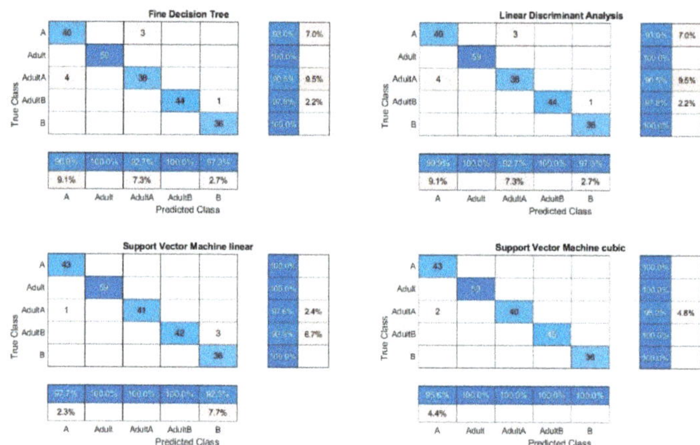

Figure 8. Confusion matrix for the oil classification by DT, LDA, SVMl & SVMc out of the combined evaluation of the fluorescence and NIR spectra.

Table 6. Results of the best PLSR predictions for the single oils A and B with single evaluations of the adulteration oils R, S, and C and the combination of all adulterations with the samples all separately for all methods and the combination of all methods.

	Oil	R^2	RMSEPrange	Detection Limit [%]
Fluorescence with Preprocessing	AC	0.994	2.7	8.3
	AR	0.988	3.8	8.6
	AS	0.993	2.9	7.2
	A all	0.991	3.4	8.7
	BC	0.992	3.2	6.1
	BR	0.992	3.1	17.9
	BS	0.983	4.7	18.6
	B all	0.990	3.6	15.4
Fluorescence without Preprocessing	AC	0.994	2.7	10.0
	AR	0.988	3.9	9.4
	AS	0.993	2.9	7.4
	A all	0.991	3.4	7.8
	BC	0.991	3.3	7.6
	BR	0.992	3.1	17.9
	BS	0.981	4.8	23.5
	B all	0.991	3.5	10.5
NIR	AC	0.997	2.1	6.0
	AR	0.995	2.4	7.4
	AS	0.997	2.1	4.5
	A all	0.995	2.5	6.3
	BC	0.997	2.0	6.6
	BR	0.997	1.9	5.8
	BS	0.997	2.0	4.4
	B all	0.996	2.3	6.2

Table 6. *Cont.*

	Oil	R^2	RMSEPrange	Detection Limit [%]
Raman	AC	0.992	3.2	5.5
	AR	0.993	3.0	9.0
	AS	0.994	2.7	5.3
	A all	0.984	4.6	11.2
	BC	0.920	10.0	18.1
	BR	0.986	4.1	11.6
	BS	0.932	9.2	7.0
	B all	0.938	8.9	12.0
Combined	AC	0.998	1.7	3.9
	AR	0.996	2.3	7.4
	AS	0.999	1.3	3.0
	A all	0.997	2.1	5.8
	BC	0.997	1.9	6.9
	BR	0.998	1.6	6.8
	BS	0.998	1.5	4.1
	B all	0.997	1.8	4.3

The $RMSEP_{range}$ values are more interesting; they were, with one exception, all below 5%. For the regression of samples with chia oil A, the best results were obtained with NIR spectra. For B, the best results were obtained with combined spectra. The highest error, corrected to the range of the considered samples (A, B, R, S, C), was $RMSEP_{range} = 10\%$ for the evaluated Raman spectra alone, the lowest 1.3% for the combined evaluation of the spectra. The determination of the detection limit was not suitable for fluorescence spectra, as the smallest is 6.1%. However, for NIR (4.4%) and Raman, lower detection limits were obtained. It was found to be best with 3% of the spectra obtained with chia oil A adulterated with sunflower oil S for the combined spectra evaluation. The best result for combined spectra evaluation for chia oil B was also obtained with S as adulteration oil with a detection limit of 4.1%.

As can be seen in Table 7, for the Kenyan chia oils the $RMSEP_{range}$ was between 0.6% and 16.7%. The detection limit varied according to the adulteration oil and it was better for the combined evaluations of the spectra. The measurements are regarded as unrepresentative because only a limited amount of sample was present. The detection limits were low (0.7/0.8 for Raman of U and Y), but the models had high $RMSEP_{ranges}$, so the reliability of these results is questionable.

For this study, two Mexican chia oils and six Kenyan chia oils were evaluated. Therefore, the range within this study is higher than in the study presented by Rodríguez et al. [15]. The comparison is difficult as the methods and the study designs were different and it is not clear how they calculated their RMSEP. Here, six different classification methods were evaluated and a PLSR regression was performed to get an idea of the amount of adulteration and, furthermore the $RMSEP_{ranges}$ were quite low in this study. The combination of all of the spectra was beneficial for the $RMSEP_{range}$ and the PLSR as the range is here between 1.3% and 2.3%. This is better as presented by Rodríguez et al. [15] for the FT-IR analysis by SIMCA and OC-P-PLS. The comparison of the RMSEPs for the adulterated samples with A and B shows that, with one exception, the presented PLSR method is better than the other method, because the range of the $RMSEP_{range}$ was between 1.3% and 4.8%. The classification sensitivity and specificity depended on the classification method which was sometimes lower, but mostly higher or at the same level. Oil B was easier to detect. However, it is difficult to compare the methods point by point, as the calculation of the RMSEP might be different as our RMSEP is standardized to the measurement range. For the Kenyan samples, the sample size was limited and the results might therefore be unrepresentative, but it proves the method working.

Table 7. Results of the best PLSR predictions for the single chia oils U, V, W, X, Y and Z with single evaluations of the adulteration oil S separately for all methods and the combination of all methods.

	Oil	R^2	RMSEPrange	Detection Limit [%]
Fluorescence with Preprocessing	U	0.999	1.5	3.8
	V	0.995	3.0	5.3
	W	0.999	1.5	3.1
	X	0.994	3.1	2.5
	Y	0.993	3.5	4.7
	Z	0.997	2.3	6.7
Fluorescence without Preprocessing	U	0.999	1.3	3.5
	V	0.995	3.1	5.0
	W	0.999	1.6	2.9
	X	0.994	3.1	2.6
	Y	0.993	3.5	4.6
	Z	0.997	2.1	7.8
NIR	U	1.000	0.6	1.6
	V	0.995	3.1	11.3
	W	0.998	1.9	5.6
	X	0.999	1.0	2.6
	Y	0.999	1.3	2.2
	Z	0.999	1.1	3.0
Raman	U	0.910	12.8	0.7
	V	0.838	16.7	52.0
	W	0.872	16.1	57.8
	X	0.889	13.2	2.2
	Y	0.865	14.9	0.8
	Z	0.838	16.7	52.0
Combined	U	1.000	0.6	1.6
	V	0.996	2.7	8.0
	W	0.999	1.4	2.7
	X	1.000	0.6	1.7
	Y	0.999	1.3	3.3
	Z	0.999	1.1	3.8

4. Conclusions

The aim of the study was to answer following questions. Is it possible to detect chia oil adulteration by spectroscopic analysis of the oils? Is it possible to identify the adulteration oil? Is it possible to determine the amount of adulteration? The presented results suggest that it is possible to distinguish between different oils by fluorescence, NIR and Raman spectroscopy. It is possible to detect adulterations of chia oils and to distinguish between different adulterations. Here, adulterations of chia oils by rapeseed, sunflower and corn oil were identified with a median of 90% for the TPR. The training accuracies were over 90%, the sensitivity and specificity of the classifications were over 90% too. B was easier to detect, so the precision was around 100% and the adulterated samples were identified with a precision of 97%. All classification methods show good results, however SVM were the best. However, the classification by KNN is not suitable for this situation. The PLSR of A + B showed R^2 over 0.95 for all models. The best RMSEP$_{range}$ of chia oil A was obtained by NIR spectra evaluation whereas it was best for oil B by combined evaluation of all spectra. The worst RMSEP$_{range}$ was obtained for Raman prediction of BC (10%), the best for combined spectra predicting AS (1.3%). For the Kenyan chia oils, the RMSEP$_{range}$ was between 0.6% and 16.7%. However, only a small number of samples were measured. Detection limits varied according to the adulteration oil and were better for the combined evaluations of the spectra. It is also possible to identify the amount of adulteration, though less than 5% adulteration is difficult to identify. Further evaluations might lead to even better results, as there was not enough sample provided from the Kenyan oils. In conclusion, it is possible to identify adulterations from native samples by spectral analysis of the oils, depending on the adulteration oil. It is also better to combine all methods because a lower RMSEP$_{range}$ can be obtained. The best results might be obtained with a classification by SVM, to identify if an adulteration took place, with a following PLSR of all combined spectra to quantify it.

Author Contributions: Conceptualization, C.K., M.M. and V.Z.; methodology, V.Z.; software, O.P.-D. and V.Z.; validation, M.M. and V.Z.; formal analysis, B.H.; investigation, C.K., M.M. and V.Z.; resources, B.H.; data curation, O.P.-D. and V.Z.; writing—original draft preparation, V.Z. and M.M.; writing—review and editing, V.Z. and M.M.; project administration, V.Z.; funding acquisition, M.M. and B.H. All authors have read and agreed to the published version of the manuscript.

Funding: This study was carried out at the Department of Process Analytics and Cereal Science at the Institute of Food Science and Biotechnology of the University of Hohenheim, supported by the DAAD German Academic Exchange Service under the Funding programme/-ID: Research Stays for University Academics and Scientists, 2019 (57440915).

Conflicts of Interest: The authors declare no conflict of interest.

References

1. Busilacchi, H.; Quiroga, M.; Bueno, M.; Di Sapio, O.; Flores, V.; Severin, C. Evaluación de *Salvia hispanica* L. Cultivada en el Sur de Santa fe (República Argentina). *Cultiv. Trop.* **2013**, *34*, 55–59. Available online: http://www.redalyc.org/articulo.oa?id=193228546009 (accessed on 11 September 2019).
2. Ayerza, R.; Coates, W. Composition of chia (*Salvia hispanica*) grown in six tropical and subtropical ecosystems of South America. *Trop. Sci.* **2004**, *44*, 131–135. [CrossRef]
3. Ixtaina, V.Y.; Martinez, M.L.; Spotorno, V.; Mateo, C.M.; Maestri, D.M.; Diehl, B.W.; Nolasco, S.M.; Tomás, M.C. Characterization of chia seed oils obtained by pressing and solvent extraction. *J. Food Compos. Anal.* **2011**, *24*, 166–174. [CrossRef]
4. Wu, H.; Sung, A.; Burns-Whitmore, B.; Jo, E.; Wien, M. Effect of Chia Seed (*Salvia hispanica*, L) Supplementation on Body Composition, Weight, Post-prandial Glucose and Satiety. *FASEB J.* **2016**, *30* (Suppl. 1), lb221. [CrossRef]
5. Gazem, R.A.A.; Puneeth, H.R.; Madhu, C.S.; Sharada, A.C. In Vitro Anticancer and Anti-Lipoxygenase Activities of Chia Seed Oil and Its Blends with Selected Vegetable Oils. *Asian J. Pharm. Clin. Res.* **2017**, *10*, 124. [CrossRef]
6. Albert, C.M.; Oh, K.; Whang, W.; Manson, J.E.; Chae, C.U.; Stampfer, M.J.; Willett, W.C.; Hu, F.B. Dietary α-Linolenic Acid Intake and Risk of Sudden Cardiac Death and Coronary Heart Disease. *Circulation* **2005**, *112*, 3232–3238. [CrossRef] [PubMed]
7. Antruejo, A.; Azcona, J.O.; Garcia, P.T.; Gallinger, C.; Rosmini, M.; Ayerza, R.; Coates, W.; Perez, C.D. Omega-3 enriched egg production: The effect of α-linolenic ω-3 fatty acid sources on laying hen performance and yolk lipid content and fatty acid composition. *Br. Poult. Sci.* **2012**, *52*, 750–760. [CrossRef]
8. Dąbrowski, G.; Konopka, I.; Czaplicki, S. Variation in oil quality and content of low molecular lipophilic compounds in chia seed oils. *Int. J. Food Prop.* **2018**, *21*, 2016–2029. [CrossRef]
9. Öztürk, B.; Ankan, A.; Özdemir, D. Olive Oil Adulteration with Sunflower and Corn Oil Using Molecular Fluorescence Spectroscopy. In *Victor R. Preedy (Hg.): Olives and Olive Oil in Health and Disease Prevention*, 1st ed.; Academic Press: Cambridge, MA, USA, 2010; pp. 451–461.
10. Kongbonga, Y.G.M.; Ghalila, H.; Onana, M.B.; Majdi, Y.; Ben Lakhdar, Z.; Mezlini, H.; Sevestre-Ghalila, S. Characterization of Vegetable Oils by Fluorescence Spectroscopy. *Food Nutr. Sci.* **2011**, *2*, 692–699. [CrossRef]
11. Zandomeneghi, M.; Carbonaro, A.L.; Caffarata, C. Fluorescence of Vegetable Oils: Olive Oils. *J. Agric. Food Chem.* **2005**, *53*, 759–766. [CrossRef]
12. Sikorska, E.; Górecki, T.; Khmelinskii, I.V.; Sikorski, M.; Kozioł, J. Classification of edible oils using synchronous scanning fluorescence spectroscopy. *Food Chem.* **2005**, *89*, 217–225. [CrossRef]
13. Osborne, B.G. Near-Infrared Spectroscopy in Food Analysis. In *Robert A. Meyers (Hg.): Encyclopedia of Analytical Chemistry, Bd. 91*; Wiley: Hoboken, NJ, USA, 2006; p. 49.
14. Rosa, L.N.; de Figueiredo, L.C.; Bonafe, E.; Coqueiro, A.; Visentainer, J.V.; Março, P.H.; Rutledge, D.N.; Valderrama, P. Multi-block data analysis using ComDim for the evaluation of complex samples: Characterization of edible oils. *Anal. Chim. Acta* **2017**, *961*, 42–48. [CrossRef] [PubMed]
15. Rodríguez, S.D.; Gagneten, M.; Farroni, A.E.; Percibaldi, N.M.; Buera, M.P. FT-IR and untargeted chemometric analysis for adulterant detection in chia and sesame oils. *Food Control.* **2019**, *105*, 78–85. [CrossRef]
16. de la Mata, P.; Dominguez-Vidal, A.; Bosque-Sendra, J.M.; Ruiz-Medina, A.; Rodríguez, L.C.; Ayora-Cañada, M.J. Olive oil assessment in edible oil blends by means of ATR-FTIR and chemometrics. *Food Control.* **2012**, *23*, 449–455. [CrossRef]
17. Maggio, R.M.; Cerretani, L.; Chiavaro, E.; Kaufman, T.S.; Bendini, A. A novel chemometric strategy for the estimation of extra virgin olive oil adulteration with edible oils. *Food Control.* **2010**, *21*, 890–895. [CrossRef]
18. Rohman, A.; Man, Y.C. Fourier transform infrared (FTIR) spectroscopy for analysis of extra virgin olive oil adulterated with palm oil. *Food Res. Int.* **2010**, *43*, 886–892. [CrossRef]

19. Tay, A.; Singh, R.; Krishnan, S.; Gore, J. Authentication of Olive Oil Adulterated with Vegetable Oils Using Fourier Transform Infrared Spectroscopy. *LWT* **2002**, *35*, 99–103. [CrossRef]
20. Jović, O.; Smolić, T.; Primožič, I.; Hrenar, T. Spectroscopic and Chemometric Analysis of Binary and Ternary Edible Oil Mixtures: Qualitative and Quantitative Study. *Anal. Chem.* **2016**, *88*, 4516–4524. [CrossRef]
21. Fadzlillah, N.A.; Man, Y.C.; Rohman, A. FTIR Spectroscopy Combined with Chemometric for Analysis of Sesame Oil Adulterated with Corn Oil. *Int. J. Food Prop.* **2014**, *17*, 1275–1282. [CrossRef]
22. Ozulku, G.; Yildirim, R.M.; Toker, O.S.; Karasu, S.; Durak, M.Z. Rapid detection of adulteration of cold pressed sesame oil adultered with hazelnut, canola, and sunflower oils using ATR-FTIR spectroscopy combined with chemometric. *Food Control.* **2017**, *82*, 212–216. [CrossRef]
23. Rohman, A.; Man, Y.B.C. Palm oil analysis in adulterated sesame oil using chromatography and FTIR spectroscopy. *Eur. J. Lipid Sci. Technol.* **2011**, *113*, 522–527. [CrossRef]
24. Rohman, A.; Man, Y.B.C. Authentication of Extra Virgin Olive Oil from Sesame Oil Using FTIR Spectroscopy and Gas Chromatography. *Int. J. Food Prop.* **2012**, *15*, 1309–1318. [CrossRef]
25. Zhao, X.; Dong, D.; Zheng, W.; Jiao, L.; Lang, Y. Discrimination of Adulterated Sesame Oil Using Midinfrared Spectroscopy and Chemometrics. *Food Anal. Methods* **2015**, *8*, 2308–2314. [CrossRef]
26. Georgouli, K.; Del Rincon, J.M.; Koidis, A. Continuous statistical modelling for rapid detection of adulteration of extra virgin olive oil using mid infrared and Raman spectroscopic data. *Food Chem.* **2017**, *217*, 735–742. [CrossRef] [PubMed]
27. Zou, M.-Q.; Zhang, X.-F.; Qi, X.-H.; Ma, H.-L.; Dong, Y.; Liu, C.-W.; Guo, X.; Wang, H. Rapid Authentication of Olive Oil Adulteration by Raman Spectrometry. *J. Agric. Food Chem.* **2009**, *57*, 6001–6006. [CrossRef] [PubMed]
28. Nunes, C.A. Vibrational spectroscopy and chemometrics to assess authenticity, adulteration and intrinsic quality parameters of edible oils and fats. *Food Res. Int.* **2014**, *60*, 255–261. [CrossRef]
29. Márquez, C.; López, M.I.; Ruisánchez, I.; Callao, M. FT-Raman and NIR spectroscopy data fusion strategy for multivariate qualitative analysis of food fraud. *Talanta* **2016**, *161*, 80–86. [CrossRef]
30. Hu, O.; Chen, J.; Gao, P.; Li, G.; Du, S.; Fu, H.; Shi, Q.; Xu, L. Fusion of near-infrared and fluorescence spectroscopy for untargeted fraud detection of Chinese tea seed oil using chemometric methods. *J. Sci. Food Agric.* **2018**, *99*, 2285–2291. [CrossRef]
31. Li, B.; Wang, H.; Zhao, Q.; Ouyang, J.; Wu, Y. Rapid detection of authenticity and adulteration of walnut oil by FTIR and fluorescence spectroscopy: A comparative study. *Food Chem.* **2015**, *181*, 25–30. [CrossRef] [PubMed]
32. Cheong, A. Confusionmat Stats (Group, Grouphat). MATLAB Central File Exchange. 2021. Available online: https://www.mathworks.com/matlabcentral/fileexchange/46035-confusionmatstats-group-grouphat (accessed on 13 January 2021).
33. Wolfbeis, O.S.; Leiner, M. Charakterisierung von Speiselen mit Hilfe der Fluoreszenztopographie. *Microchim. Acta* **1984**, *82*, 221–233. [CrossRef]
34. Sikorska, E.; Gliszczyńska-Świgło, A.; Khmelinskii, I.; Sikorski, M. Synchronous Fluorescence Spectroscopy of Edible Vegetable Oils. Quantification of Tocopherols. *J. Agric. Food Chem.* **2005**, *53*, 6988–6994. [CrossRef] [PubMed]
35. Kyriakidis, N.B.; Skarkalis, P. Fluorescence Spectra Measurement of Olive Oil and Other Vegetable Oils. *J. AOAC Int.* **2000**, *83*, 1435–1439. [CrossRef] [PubMed]
36. Papoti, V.T.; Tsimidou, M.Z. Looking through the qualities of a fluorimetric assay for the total phenol content estimation in virgin olive oil, olive fruit or leaf polar extract. *Food Chem.* **2009**, *112*, 246–252. [CrossRef]

Article

ATR-FTIR Spectroscopy Combined with Multivariate Analysis Successfully Discriminates Raw Doughs and Baked 3D-Printed Snacks Enriched with Edible Insect Powder

Nerea García-Gutiérrez [1], Jorge Mellado-Carretero [1], Christophe Bengoa [1], Ana Salvador [2], Teresa Sanz [2], Junjing Wang [1], Montse Ferrando [1], Carme Güell [1] and Sílvia de Lamo-Castellví [1,*]

1   Departament d'Enginyeria Química (DEQ), Campus Sescelades, Universitat Rovira i Virgili, Av. Països Catalans, 26, 43007 Tarragona, Spain; nerea.garcia@urv.cat (N.G.-G.); jorge.mellado@urv.cat (J.M.-C.); christophe.bengoa@urv.cat (C.B.); junjing.wang@urv.cat (J.W.); montse.ferrando@urv.cat (M.F.); carme.guell@urv.cat (C.G.)

2   Instituto de Agroquímica y Tecnología de Alimentos (IATA-CSIC), C/Catedràtic Agustín Escardino Benlloch, 7, 46980 Paterna, Spain; asalvador@iata.csic.es (A.S.); tesanz@iata.csic.es (T.S.)

*   Correspondence: silvia.delamo@urv.cat

Abstract: In a preliminary study, commercial insect powders were successfully identified using infrared spectroscopy combined with multivariate analysis. Nonetheless, it is necessary to check if this technology is capable of discriminating, predicting, and quantifying insect species once they are used as an ingredient in food products. The objective of this research was to study the potential of using attenuated total reflection Fourier transform mid-infrared spectroscopy (ATR-FTMIR) combined with multivariate analysis to discriminate doughs and 3D-printed baked snacks, enriched with *Alphitobius diaperinus* and *Locusta migratoria* powders. Several doughs were made with a variable amount of insect powder (0–13.9%) replacing the same amount of chickpea flour (46–32%). The spectral data were analyzed using soft independent modeling of class analogy (SIMCA) and partial least squares regression (PLSR) algorithms. SIMCA models successfully discriminated the insect species used to prepare the doughs and snacks. Discrimination was mainly associated with lipids, proteins, and chitin. PLSR models predicted the percentage of insect powder added to the dough and the snacks, with determination coefficients of 0.972, 0.979, and 0.994 and a standard error of prediction of 1.24, 1.08, and 1.90%, respectively. ATR-FTMIR combined with multivariate analysis has a high potential as a new tool in insect product authentication.

Keywords: insect powder; authentication; 3D food printer; mid-infrared spectroscopy; multivariate analysis

Citation: García-Gutiérrez, N.; Mellado-Carretero, J.; Bengoa, C.; Salvador, A.; Sanz, T.; Wang, J.; Ferrando, M.; Güell, C.; Lamo-Castellví, S.d. ATR-FTIR Spectroscopy Combined with Multivariate Analysis Successfully Discriminates Raw Doughs and Baked 3D-Printed Snacks Enriched with Edible Insect Powder. *Foods* **2021**, *10*, 1806. https://doi.org/10.3390/foods10081806

Academic Editors: Margit Cichna-Markl and Isabel Mafra

Received: 30 June 2021
Accepted: 2 August 2021
Published: 5 August 2021

1. Introduction

The world population is increasing dramatically mainly as a result of the quality-of-life improvement in developing countries and will reach over 9.8 billion by 2050 [1]. An increasing demand for protein-rich sources will be a threat to world food and feed availability [2]. Nowadays, animal protein sources come mostly from livestock, such as poultry, swine, and cattle. Stockbreeding requires large spaces and large quantities of natural resources and also produces significant greenhouse emissions among other contaminants [3].

Insects have been proposed as a suitable alternative to conventional livestock [4]. In fact, insect breeding needs fewer resources compared to conventional livestock: water consumption is lower; less space occupancy is required; and insects can be fed using food waste products, such as potato peels, rotten fruits, and bakery by-products [5].

Moreover, when comparing the protein content of both insects and livestock, the first one shows similar or even higher amounts of protein. For instance, the amount of protein in 100 g of fried grasshoppers is almost triple compared to the protein content in 100 g of grilled beef [6].

Besides the high protein content, most insects have high amounts of polyunsaturated fatty acids, vitamins, and micronutrients (e.g., calcium, iron, and phosphorus) essential for human life [7]. However, it has been observed that the consumption of insects can be risky due to the presence of possible pathogens, antinutrients, and allergenic substances, assessed in several cross-reactivity studies [8,9].

Before being able to open the market to the commercialization of insects, the European Commission needed to confirm that their consumption was totally safe for human health. Unfortunately, in 2019, there was not enough information available about the potential risk associated with insect consumption [10]. Since then, several studies have been performed to evaluate the composition of common edible insect species. In early 2021, the European Food Safety Authority (EFSA) revealed its opinion, positioning itself in favor of the consumption of a very well-studied insect species, *Tenebrio molitor* [11]. A recent study in the European insect market, conducted by the International Platform of Insects for Food and Feed (IPIFF), forecasts that the following insect species to be approved will be the lesser mealworm (*Alphitobius diaperinus*) and the migratory locust (*Locusta migratoria*) among others [12]. The final approval by the European Commission of the use of *T. molitor* as a new ingredient in food products is smoothing the path to the introduction of insects in Western diets [13].

Another challenge to overcome is insect consumption in Western countries. One strategy that has been followed to introduce insects in Western diets is the use of edible insect powders (i.e., dehydrated insects that have been ground to obtain a fine powder) as ingredients for innovative food product design. However, the use of insect powders as ingredients can lead to adulteration and fraud.

There are several techniques used by the food industry to verify the authenticity of their products. Molecular techniques (e.g., genomics, proteomics, and polymerase chain reaction with denaturing gradient gel electrophoresis (PCR-DGGE)) have been widely used for origin authentication. Despite their high sensitivity, further characterization of markers is required for real unknown samples [14]. In the case of complex food matrices, several fingerprinting techniques, such as chromatographic techniques (e.g., high-performance liquid chromatography (HPLC) and mass spectrometry (MS)), have been selected since they are rapid and easy to execute [15]. As an example, matrix-assisted laser desorption/ionization coupled with a time-of-flight mass spectrometer (MALDI-TOF) has been used as a method for the authentication and classification of several commercial edible insect powders [16]. Although MALDI-TOF mass spectrometry has been proven to be a useful technique for insect powder authentication, it has several drawbacks, such as time of analysis and cost, that could hinder its implementation in the food industry.

Other fingerprinting techniques applied in food authentication are based on rotational–vibrational spectroscopy (infrared and Raman spectroscopy). Fourier transform infrared (FTIR) spectroscopy has been applied in different food matrices, such as juices, meat, and extra virgin olive oil authentication, due to the ease of obtaining powerful results quickly [17,18]. Nevertheless, fingerprinting techniques produce huge volumes of data that need to be processed. For this reason, multivariate analysis is essential to process the data collected based on fingerprint techniques [19].

The objective of the present work was to study the potential of using attenuated total reflectance Fourier transform mid-infrared spectroscopy (ATR-FT-MIR) combined with multivariate analysis to rapidly discriminate and predict the concentration of *A. diaperinus* and *L. migratoria* powder added into a raw dough and 3D-printed baked snacks.

2. Materials and Methods

2.1. Materials

Lesser mealworm (*Alphitobius diaperinus*) and migratory locust (*Locusta migratoria*) powder from three different batches (100 g each) were supplied by Kreca Ento-Food BV (Ermelo, The Netherlands). The chickpea flour was purchased from La Finestra Sul Cielo S.A. (Madrid, Spain) and the hot madras curry powder from Westmill Foods (London, UK). The extra virgin olive oil was obtained from Hacienda Ortigosa S.L. (Navarra, Spain) and the salt from Sal Costa, S.L.U (Barcelona, Spain).

2.2. Dough Preparation

The ingredients selected to create a blank dough (B) were chickpea flour (46.3 wt%), water (39.4 wt%), extra virgin olive oil (11.6 wt%), curry powder (1.8 wt%), and salt (0.9 wt%). This dough formulation was made to obtain gluten- and dairy-free tasty dough with good printability. Based on this formulation, different quantities of *A. diaperinus* or *L. migratoria* powder were added, replacing 10% (A_{d1}, L_{d1}), 20% (A_{d2}, L_{d2}), or 30% (A_{d3}, L_{d3}) of the total amount of chickpea flour used (Table 1). Both insect powders were previously milled using an electric coffee grinder (TM-CG-03, Vilapur, Larkhall, UK) for 1 min. The milling process reduced the particle size of *A. diaperinus* powder from $D_{[4,3]}$ values of 677 µm (span of 2.0) to 310 µm (span of 2.2). For *L. migratoria* commercial powder, almost 30% of the particles were bigger than 2000 µm, and, after the milling process, a $D_{[4,3]}$ of 967 µm (span of 1.1) was obtained (Figures S1 and S2 are available as Supplementary Files). A total of 100 g was prepared for each formulation, mixing all the ingredients with a hand blender (HB-10C.019A, HAEGER SPAIN, Barcelona, Spain) for 2 min.

Table 1. Insect powder percentage and measured weight of each ingredient for different formulations.

Ingredients (g)	Dough Formulation [a]						
	B	A_{d1}	A_{d2}	A_{d3}	L_{d1}	L_{d2}	L_{d3}
Chickpea flour	100.03 ± 0.01	90.05 ± 0.02	80.02 ± 0.01	70.03 ± 0.01	90.03 ± 0.05	80.03 ± 0.02	70.04 ± 0.05
A. diaperinus powder	-	10.02 ± 0.05	20.03 ± 0.04	30.02 ± 0.06	-	-	-
L. migratoria powder	-	-	-	-	10.07 ± 0.08	20.08 ± 0.09	30.06 ± 0.08
Water	85.01 ± 0.01	85.01 ± 0.01	85.00 ± 0.01	85.00 ± 0.01	85.01 ± 0.01	85.01 ± 0.01	85.01 ± 0.01
Olive oil	25.00 ± 0.01	25.01 ± 0.01	25.00 ± 0.01	25.01 ± 0.01	25.01 ± 0.01	25.00 ± 0.01	25.00 ± 0.01
Curry powder	4.00 ± 0.01	4.01 ± 0.01	4.00 ± 0.01	4.00 ± 0.01	4.01 ± 0.01	4.00 ± 0.01	4.00 ± 0.01
Salt	2.05 ± 0.03	2.04 ± 0.03	2.06 ± 0.02	2.02 ± 0.05	2.05 ± 0.04	2.07 ± 0.02	2.03 ± 0.03
Insect powder (%)	0.00 ± 0.00	4.64 ± 0.02	9.27 ± 0.02	13.89 ± 0.03	4.65 ± 0.02	9.28 ± 0.02	13.90 ± 0.02

Abbreviations used: B, blank dough with 0% insect powder; A_{d1}, dough with 4.6% of *A. diaperinus* powder; A_{d2}, dough with 9.3% of *A. diaperinus* powder; A_{d3}, dough with 13.9% of *A. diaperinus* powder; L_{d1}, dough with 4.6% of *L. migratoria* powder; L_{d2}, dough with 9.3% of *L. migratoria* powder; L_{d3}, dough with 13.9% of *L. migratoria* powder. [a] Values are means of three different batches ± standard deviation (SD).

2.3. 3D Printing and Post-Processing Analysis

A portable 3D food printer (Focus, ByFlow Company, Eindhoven, The Netherlands) was used to print the blank dough (B) and several doughs enriched with 4.6% (A_{d1}), 9.3% (A_{d2}), and 13.9% (A_{d3}) of *A. diaperinus* powder. In the case of the dough enriched with *L. migratoria*, the printing process could not be carried out because the average particle size obtained in the ground *L. migratoria* powder was larger than the maximum size of the nozzle of the 3D printer used. The printing was performed at room temperature using a Voronoi circle model of $100 \times 100 \times 10$ mm (Figure 1a). A total of 32.0 g of each dough was printed using a 1.6 mm nozzle at a speed of 30 mms^{-1}. Then, the dough was baked at 180 °C for 12 min using the ventilation mode of a vapor oven (3HV469X/02 model, Balay, Pamplona, Spain). Each formulation was made in triplicate.

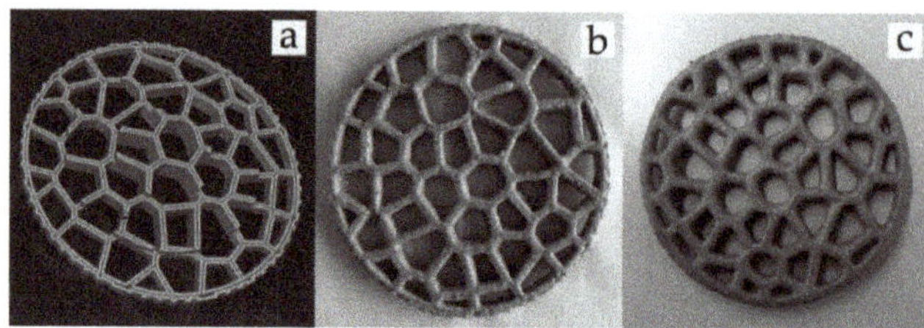

Figure 1. The following photos show (**a**) a 3D visualization created by Webgcode of the Voronoi circle model, (**b**) 3D-printed raw snack, and (**c**) 3D-printed baked snack.

In order to find out the amount of water lost during the baking process, the total weight of the 3D-printed snack was measured before and after baking it (Table 2) (Figure 1b,c). The concentration of the insect powder was recalculated based on the percentage of water loss of the 3D-printed baked snacks.

Table 2. Water loss of the different 3D-printed baked snacks and the concentration of insect powder after the baking process. Abbreviations used: B, blank snack with 0% insect powder; As_1, snack with 7.2% of *A. diaperinus* powder; As_2, snack with 14.4% of *A. diaperinus* powder; As_3, snack with 21.9% of *A. diaperinus* powder.

Snack Formulation	B	As_1	As_2	As_3
Water loss (%) [a]	36.2 ± 0.2	36.0 ± 0.5	35.5 ± 1.5	36.5 ± 1.2
A. diaperinus powder (wt%) [a]	0.0 ± 0.0	7.2 ± 0.1	14.4 ± 0.4	21.9 ± 0.4

[a] Values are means of three different batches $\pm$ standard deviation.

2.4. Spectral Data Acquisition by FTIR

Spectral profiles were collected in the mid-infrared (MIR) region (4000–800 cm^{-1}) with 8 cm^{-1} resolution using portable spectrometer Cary 630 (Agilent Technologies Spain SL, Madrid, Spain) equipped with a single bounce ATR diamond crystal accessory and a deuterated triglycine sulfate (DTGS) detector. Spectral information acquisition was carried out using MicroLab PC software (Agilent Technologies SL, Madrid, Spain). The final spectra were obtained from the average of 128 scans to improve the signal-to-noise ratio, and a background scan was taken before every sample scan to avoid noise in the spectral data from the environment.

For the dough spectral acquisition, 4 mg of dough was placed on the sample stage using a rolling pin (Excel Blades Corp, Paterson NJ, USA). Then, the sample was dried by vacuum to remove the water contribution in the final spectra.

The baked snacks were milled using an electric grinder (TM-CG-03, Vilapur, Larkhall, UK). Then, 4 mg of ground snack was placed in the sample stage. To enhance the contact between the diamond crystal and the sample, a press clamp was used, standardizing the layers' width of the sample over the detector.

Spectral data from three different batches (prepared at three different days) of each blank dough, dough enriched with 4.6, 9.3, and 13.9% of *A. diaperinus* and *L. migratoria*, and 3D-printed baked snack with 7.2, 14.4, and 21.9% of *A. diaperinus* were obtained, collecting 10 spectra per day and per sample. A total number of 30 spectra were obtained for each type of sample at room temperature.

2.5. Multivariate Analysis

Multivariate analysis and data preprocessing were performed using a chemometric software (Pirouette, version 4.5. Infometrix Inc., Bothell, WA, USA). Spectral data from the doughs were mean centered, vector length normalized and transformed using second derivative polynomial-fit Savitzky–Golay function (13 points). The spectral data from the ground snacks were mean centered, transformed using multiplicative scatter correction

(MSC) and second derivative polynomial-fit Savitzky–Golay function (13 points). A statistical Principal Component Analysis (PCA)-based supervised algorithm, soft independent modelling of class analogy (SIMCA), was used to create a discrimination and classification model with the spectral data obtained [20]. In order to detect potential outliers, sample residuals and the Mahalanobis distances were taken into account [21]. Three different outputs were used to interpret SIMCA models created: class projections (i.e., 3-dimensional PCA score plots), interclass distances, and discriminating power. Total misclassifications were analyzed and interpreted for the input data. Models were validated using a predicted set of samples, excluding 5 spectra per sample from the initial input data and creating new models with the remaining data.

The pre-treated spectra were also analyzed by partial least squares regression (PLSR) that was cross-validated (leave-one-out, internal validation approach) to generate calibration models. The same transformations used for the creation of the SIMCA models were applied to the spectral data used for the PLSR analysis: mean centered, vector length normalized, and transformed by second derivative polynomial-fit Savitzky–Golay function (13 points) in the case of the dough and MSC and second derivative polynomial-fit Savitzky–Golay function (13 points) for the 3D-printed baked snacks. Thus, the x variable was the absorbance per each wavenumber, and the y variable (reference data) was the percentage of edible insect powder added to the dough, or, in the case of the 3D-printed snacks, the percentage of insect powder present in the final product. Furthermore, a predicted set of samples was used in order to validate the model, temporarily leaving out 5 spectra per sample from the training set. All the models were evaluated in terms of regression vector, standard error of cross-validation (SECV), standard error of calibration (SEC), determination coefficient (R^2), and outlier diagnostics [22].

3. Results and Discussion

3.1. Spectral Information of Ingredients and Doughs

Previous studies have shown the ability of MIR spectroscopy to provide information about the chemical composition of complex samples, such as food matrices, correlating different IR bands with specific functional groups [23]. Moreover, it is important to analyze all the ingredients alone to help elucidate the origin of the IR bands associated with certain components in these mixtures (Figure 2).

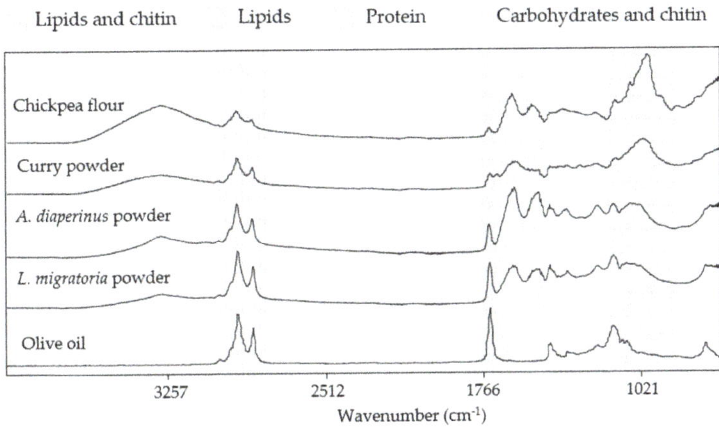

Figure 2. Raw mid-infrared spectra (4000 to 800 cm^{-1}) from the ingredients used to prepare the doughs and the snacks.

Comparing the spectral information from the ingredients used in the dough and snack formulation (Figure 2), it can be observed that chickpea, curry, *A. diaperinus*, and *L. migratoria* powder showed a broad IR band from 3000 to 3500 cm^{-1}, which originated from the stretching of O-H bonds most likely coming from carbohydrates, fiber, and water [24]. Moreover, all these spectra exhibited two narrower IR bands at around 2900 and 2850 cm^{-1}, characteristics of asymmetrical and symmetrical stretching vibrations from C-H bonds from methyl groups presumably caused by lipids or, in the case of insect powders, lipids and chitin. Another common IR band between all the ingredients was observed around 1740 cm^{-1}. This IR band can be associated with the stretching of C=O bonds from ester groups related to lipids [25,26]. Chickpea flour and insect powders presented IR bands in the region of 1650 and 1500 cm^{-1}, and these were related to C-N stretching, C=O vibrations of N-acetyl groups, and N-H bending from amide II groups, most likely from the presence of proteins. In the case of insect powders, these IR bands can also be associated with chitin presence [27]. Finally, a distinctive IR band at 1100–900 cm^{-1} from C-O stretching vibrations, most probably corresponding to the presence of non-structural carbohydrates, such as starch and sugars, was also detected in the chickpea and curry powder spectra [28].

The raw MIR spectra of the blank dough (B) and the doughs with different concentrations of *A. diaperinus* (A$_{d1}$, A$_{d2}$, and A$_{d3}$) and *L. migratoria* (L$_{d1}$, L$_{d2}$, and L$_{d3}$) with their respective transformed spectra are shown in Figure 3. Differences between the dough samples are not easy to detect from the raw spectra, since all of them exhibit the same IR bands related to the presence of lipid and chitin (2900 and 2850 cm^{-1}), lipid (around 1700–1740 cm^{-1}), protein (1650 and 1500 cm^{-1}), and carbohydrates and chitin (from 1200 to 900 cm^{-1}), with a similar absorbance.

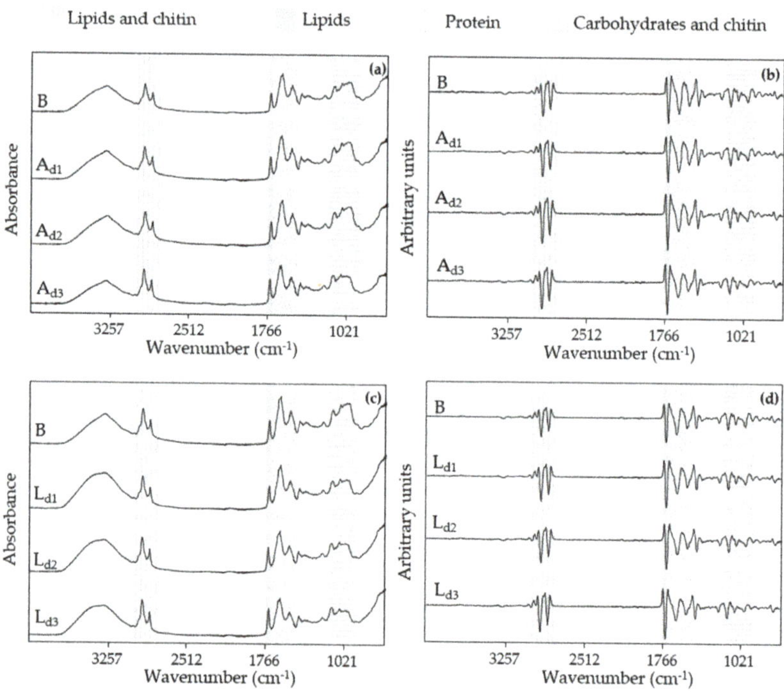

Figure 3. Raw MIR spectra (4000 to 800 cm^{-1}) from doughs with (**a**) different concentrations of *A. diaperinus* (A$_{d1}$, A$_{d2}$, and A$_{d3}$) and (**b**) the transformed spectra (vector length normalized and second derivative, 13 points). Graphs (**c**) show the raw mid-infrared spectra of different concentrations of *L. migratoria* (L$_{d1}$, L$_{d2}$, and L$_{d3}$) and (**d**) the transformed spectra. Abbreviations used: B, blank dough with 0% insect powder; A$_{d1}$, dough with 4.6% of *A. diaperinus* powder; A$_{d2}$, dough with 9.3% of *A. diaperinus* powder; A$_{d3}$, dough with 13.9% of *A. diaperinus* powder; L$_{d1}$, dough with 4.6% of *L. migratoria* powder; L$_{d2}$, dough with 9.3% of *L. migratoria* powder; L$_{d3}$, dough with 13.9% of *L. migratoria* powder.

All this information was important to decide the spectral region selected for the building up of the SIMCA models.

First of all, the region from 4000 to 3000 cm^{-1} was excluded from the model due to the high moisture content of the dough samples, trying to avoid discriminations based on water content. Furthermore, the IR bands between 2700 and 1800 cm^{-1} were also omitted to reduce the noise impact in the spectral analysis caused by the crystal.

Moreover, for all the formulations, the amount of insect powder added was increasing at the same time that the proportion of chickpea flour was decreasing due to the substitution of ingredients shown in Table 1. As can be seen in the raw spectra of the chickpea flour in Figure 2, the signal in the carbohydrate region is quite intense compared to the other ingredients. Therefore, since the aim of our research was to determine if it was possible to discriminate the presence of insect powder in the designed product and in view of the fact that the region from 1200 to 800 cm^{-1} was strongly influenced by the variable quantities of chickpea flour, it was considered more appropriate to exclude this region from the model. Finally, the spectral region chose for the data analysis was the one between 3000 to 2700 cm^{-1} and 1800 to 1200 cm^{-1}.

3.2. Discrimination and Classification of Doughs by ATR-FT-MIR Combined with SIMCA

Initially, two different four-class SIMCA models were built up to obtain classification models to discriminate the blank and the doughs enriched with different amounts of *A. diaperinus* (A_{d1}, A_{d2}, and A_{d3}) or *L. migratoria* (L_{d1}, L_{d2}, and L_{d3}) powders and obtain information about their biochemical differences.

To obtain these models, the original variables are replaced by linear combinations of the same variables called factors, helping to reduce the dimensionality of the data without losing information [29]. The number of factors was chosen to achieve a minimum of 90% of the variance in each class of both models (Table 3).

Table 3. The number of factors, their cumulative variance, and the number of outliers used to create 4-class SIMCA models of *A. diaperinus* and *L. migratoria* dough.

Model	Class	Factor 1 (%)	Factor 2 (%)	Factor 3 (%)	Outliers
4-class	B	82.2	92.7	95.0	0
SIMCA	A_{d1}	81.2	92.5	96.3	5
A. diaperinus	A_{d2}	92.0	95.8	98.0	3
dough model	A_{d3}	91.7	97.0	98.5	5
4-class	B	82.2	95.7	95.0	0
SIMCA	L_{d1}	86.0	91.7	95.7	4
L. migratoria	L_{d2}	96.9	98.2	99.1	6
dough model	L_{d3}	75.0	85.7	91.9	5

Abbreviations used: B, blank dough with 0% insect powder; A_{d1}, dough with 4.6% of *A. diaperinus* powder; A_{d2}, dough with 9.3% of *A. diaperinus* powder; A_{d3}, dough with 13.9% of *A. diaperinus* powder; L_{d1}, dough with 4.6% of *L. migratoria*; L_{d2}, dough with 9.3% of *L. migratoria*; L_{d3}, dough with 13.9% of *L. migratoria*.

SIMCA class projection is a three-dimensional scatter plot that gives information about the similarity of the spectral data (Figure 4). Each point in the class projection represents a spectrum that belongs to a cluster (represented with an ellipse) with 95% of confidence. Those groups that seem closer will be more similar to each other unlike those that will appear further away, which will be much different [30].

The transformed spectra (3000–2700 cm^{-1} and 1800 to 1200 cm^{-1} region) of doughs enriched with *A. diaperinus* powder (Figure 4a) or *L. migratoria* (Figure 4b) showed non-overlapping and well-separated clusters allowing accurate dough classification for each type. SIMCA's misclassification algorithm for doughs enriched with *A. diaperinus* or *L. migratoria* powders showed zero misclassifications, indicating that the training set for each model was homogeneous, and all samples were correctly classified into their assigned classes.

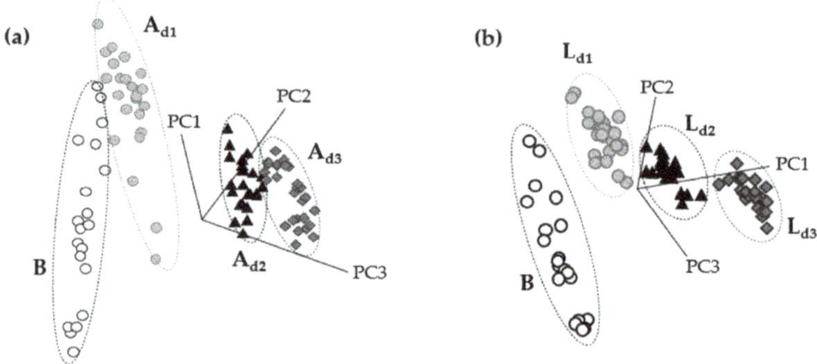

Figure 4. SIMCA class projection plots of (**a**) 4-class SIMCA *A. diaperinus* dough model and (**b**) 4-class SIMCA *L. migratoria* dough model. Abbreviations used: B, blank dough with 0% insect powder; A_{d1}, dough with 4.6% of *A. diaperinus* powder; A_{d2}, dough with 9.3% of *A. diaperinus* powder; A_{d3}, dough with 13.9% of *A. diaperinus* powder.

Discrimination power plots (Figure 5) show which are the major IR bands that contributed to the development of the classification models. Despite the fact both models were made with distinct insect species, similar IR bands were responsible for the differentiation of the clusters. As can be seen in Figure 5, the IR bands mainly responsible for the discrimination of the models were related to the stretching of C-H bonds from methyl groups associated with lipid content (2959 cm^{-1} and 2877 cm^{-1}), the C=O stretching of esters of lipids (1777 cm^{-1}), amide I groups most likely from proteins and chitin (1647 cm^{-1} till 1546 cm^{-1}), and bending vibrations of the CH$_2$ and CH$_3$ aliphatic groups (1427 cm^{-1}) that can be related to polysaccharides and sugars [31,32].

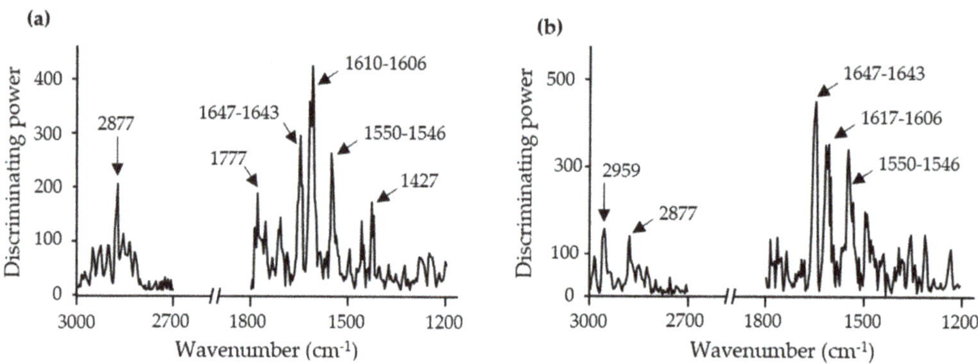

Figure 5. Soft independent modeling of class analogy discriminating power plots of FTIR spectroscopy spectra (3000–2700 cm^{-1} and 1800–1200 cm^{-1}) from (**a**) 4-class SIMCA *A. diaperinus* dough model and (**b**) 4-class SIMCA *L. migratoria* dough model.

Interclass distance (ICD) is a ratio of Euclidean distances between two classes that indicate the similarities and/or dissimilarities between them [33]. ICD values above 3 are considered significant to discriminate two clusters of samples as distinct classes [34]. As shown in Table 3, ICD values between the clusters of dough enriched with *A. diaperinus* powder increased with the amount of insect powder added, ranging from 3.1 to 12.2. In the case of dough enriched with *L. migratoria*, the same tendency is observed, showing increasing ICD values with the amount of insect powder added from 5.8 to 13.6 (Table 4).

The fact that ICD values increased with the amount of insect powder added shows that this ingredient is responsible for the dough classification [35].

Table 4. Soft independent modeling of class analogy interclasses distances from the 4-class SIMCA *A. diaperinus* doughs model and the 4-class SIMCA *L. migratoria* doughs model.

A. diaperinus	B	A_{d1}	A_{d2}	A_{d3}
B	0.0			
A_{d1}	3.1	0.0		
A_{d2}	8.7	4.5	0.0	
A_{d3}	12.2	7.2	3.1	0.0
L. migratoria	B	L_{d1}	L_{d2}	L_{d3}
B	0.0			
L_{d1}	5.8	0.0		
L_{d2}	8.4	3.0	0.0	
L_{d3}	13.6	7.4	3.9	0.0

Abbreviations used: B, blank dough with 0% insect powder; A_{d1}, dough with 4.6% of *A. diaperinus* powder; A_{d2}, dough with 9.3% of *A. diaperinus* powder; A_{d3}, dough with 13.9% of *A. diaperinus* powder; L_{d1}, dough with 4.6% of *L. migratoria*; L_{d2}, dough with 9.3% of *L. migratoria*; L_{d3}, dough with 13.9% of *L. migratoria*.

Model validation using an independent set of spectra (five spectra not included in the SIMCA model from each class) showed a 100% correct classification for each type of dough in both models tested. The specificity of each model was also evaluated by making predictions of the *L. migratoria* spectra of each class into the four-class SIMCA *A. diaperinus* dough model and by using this one to classify the spectra (Table 5). The same procedure was used to evaluate the *L. migratoria* model. These results confirmed the ability of the models developed to discriminate dough samples depending on the species of insect used (*A. diaperinus* and *L. migratoria*) and the quantity of insect powder added [36].

Table 5. Dough model predictions of *A. diaperinus* SIMCA model using *L. migratoria* spectra and *L. migratoria* SIMCA model using *A. diaperinus* spectra.

Model	Dough	No. Spectra	Best Class	Next Best	Not Classified
4-class	L_{d1}	26	0%	0%	100%
SIMCA	L_{d2}	24	0%	0%	100%
A. diaperinus	L_{d3}	25	0%	0%	100%
4-class	A_{d1}	25	0%	0%	100%
SIMCA	A_{d2}	27	0%	0%	100%
L. migratoria	A_{d3}	25	0%	0%	100%

Abbreviations used: B, blank dough with 0% insect powder; A_{d1}, dough with 4.6% of *A. diaperinus* powder; A_{d2}, dough with 9.3% of *A. diaperinus* powder; A_{d3}, dough with 13.9% of *A. diaperinus* powder; L_{d1}, dough with 4.6% of *L. migratoria*; L_{d2}, dough with 9.3% of *L. migratoria*; L_{d3}, dough with 13.9% of *L. migratoria*.

It was also important to study if it was possible to differentiate between the type of insects used to make the doughs. For this purpose, a six-class SIMCA model was created using the IR data of doughs with different concentrations of *A. diaperinus* and *L. migratoria* powders. The number of factors chosen for this model as well as their cumulative variance is shown in Table 6.

The class projection plot (Figure 6) showed distinctive clustering patterns and six well-defined classes, which are closer to those with a low amount of insect powder and far away from the clusters with the highest quantity of insect powder.

Table 6. The number of factors, their cumulative variance, and the number of outliers used to create a 6-class SIMCA model of *A. diaperinus* vs. *L. migratoria* dough.

Model	Class	Factor 1 (%)	Factor 2 (%)	Factor 3 (%)	Outliers
6-class SIMCA *A. diaperinus* vs. *L. migratoria* dough model	L_{d1}	86.0	91.7	95.7	4
	A_{d1}	83.9	94.7	96.6	6
	L_{d2}	96.7	98.2	99.1	7
	A_{d2}	92.0	95.8	98.0	6
	L_{d3}	75.9	85.7	91.9	5
	A_{d3}	91.7	96.9	98.5	5

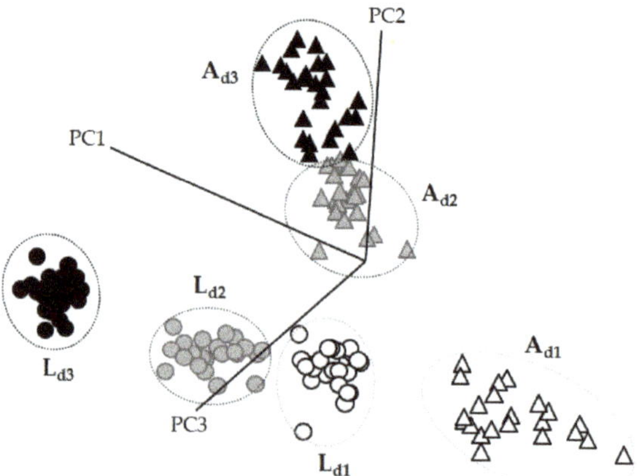

Figure 6. Soft independent modeling of class analogy class projection of FTIR spectroscopy spectra (3000–2700 cm^{-1} and 1800–1200 cm^{-1}) from a 6-class SIMCA *A. diaperinus* vs. *L. migratoria* dough model. Abbreviations used: A_{d1}, dough with 4.6% of *A. diaperinus* powder; A_{d2}, dough with 9.3% of *A. diaperinus* powder; A_{d3}, dough with 13.9% of *A. diaperinus* powder; L_{d1}, dough with 4.6% of *L. migratoria*; L_{d2}, dough with 9.3% of *L. migratoria*; L_{d3}, dough with 13.9% of *L. migratoria*.

The discriminating power plot (Figure 7) showed that, as in the previous models, the main discrimination was related to lipids (C=O stretching of esters from lipid content at 2877 cm^{-1} and 1722 cm^{-1} to 1710 cm^{-1}), protein or/and chitin content (amide I from proteins and/or chitin at 1606 cm^{-1} and 1535 cm^{-1}), and polysaccharides (1461 cm^{-1}). This reinforces the idea that these components might be important factors in the discrimination of different insect species.

Moreover, the maximum discriminating power obtained in the *A. diaperinus* and *L. migratoria* dough models was almost triple that obtained in the *A. diaperinus* vs. *L. migratoria* dough model (Figure 7). This information reveals that both types of dough display remarkably similar composition despite being made with different insect powders.

The ICD values observed in the model that compare the different insect doughs exhibit the same trend as the individual models (see Table 4). The ICD value increases with the insect powder content in the dough regardless of the type of insect powder used (Table 7).

The validation of the model was carried out using a new set of spectra, five for each class. A 100% correct classification was obtained from all the spectra in all the tested classes. Therefore, the capability of this model to discriminate between *A. diaperinus* and *L. migratoria* powders in a food matrix was corroborated.

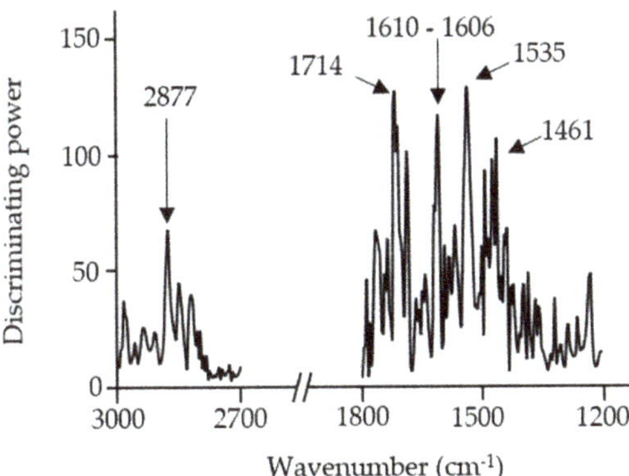

Figure 7. Soft independent modeling of class analogy discriminating power plot of FTIR spectroscopy spectra (3000–2700 cm^{-1} and 1800–700 cm^{-1}) from a 6-class SIMCA *A. diaperinus* vs. *L. migratoria* dough model.

Table 7. Soft independent modeling of class analogy (SIMCA) interclasses distances from a 6-class SIMCA *A. diaperinus* vs. *L. migratoria* doughs model.

Dough	L_{d1}	A_{d1}	L_{d2}	A_{d2}	L_{d3}	A_{d3}
L_{d1}	0.0					
A_{d1}	3.0	0.0				
L_{d2}	3.0	4.7	0.0			
A_{d2}	3.3	4.3	3.8	0.0		
L_{d3}	7.4	9.3	3.9	6.7	0.0	
A_{d3}	6.4	7.3	5.1	3.1	5.0	0.0

Abbreviations used: A_{d1}, dough with 4.6% of *A. diaperinus* powder; A_{d2}, dough with 9.3% of *A. diaperinus* powder; A_{d3}, dough with 13.9% of *A. diaperinus* powder; L_{d1}, dough with 4.6% of *L. migratoria*; L_{d2}, dough with 9.3% of *L. migratoria*; L_{d3}, dough with 13.9% of *L. migratoria*.

3.3. Spectral Information of 3D-Printed Snacks

Another objective proposed in this research was to study if the insect powder could be discriminated in the final product after being extruded by a 3D printer and going through a baking process. Figure 8 shows the raw spectra of 3D-printed baked snacks made with different amounts of *A. diaperinus* powder.

Thermal treatments, such as baking, are commonly used processes in the food industry and in food preparation [37]. Some of these methods can help in elongating the shelf-life of the processed products and improve the digestibility and bioavailability of proteins. Nevertheless, thermal treatments can negatively affect a wide diversity of molecules (e.g., lipid oxidation, protein denaturation, and vitamin solubilization) [38,39].

When comparing the raw spectra of these snacks with the raw spectra of the corresponding doughs in Figure 2, we can observe that the broad IR band from 3000 to 3500 cm^{-1} was reduced. This IR band range is linked to the stretching of O-H bonds, and its decreasing signal can be related to the water loss reported during the baking process [40].

The oxidation of unsaturated lipids is an autocatalytic reaction enhanced by thermal processes [41]. However, the signal on the IR bands around 2900, 2850, and 1740 cm^{-1}, most likely related to the presence of lipid, seems to remain similar to the one shown in the dough spectra.

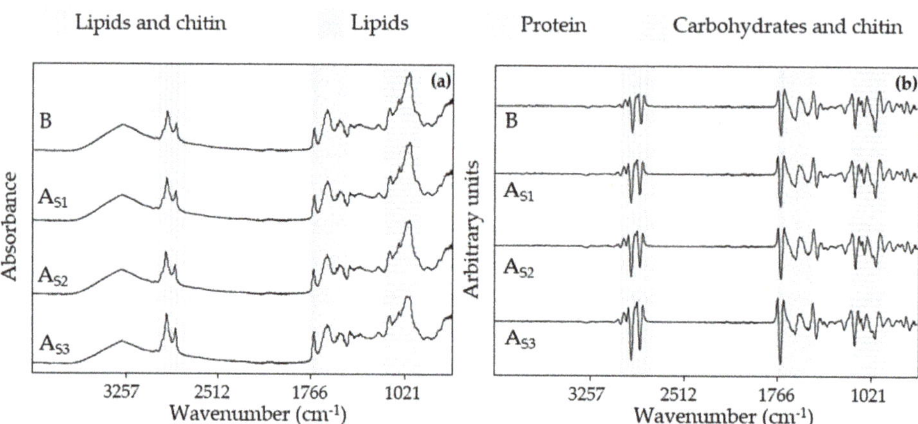

Figure 8. Raw mid-infrared spectra (4000 to 800 cm^{-1}) from the snacks made with *A. diaperinus* powder (**a**) and the transformed spectra (MSA and second derivative, 13 points) (**b**). Abbreviations used: Bs, blank snack with 0% insect powder; As$_1$, snack with 7.2% of *A. diaperinus* powder; As$_2$, snack with 14.4% of *A. diaperinus* powder; As$_3$, snack with 21.9% of *A. diaperinus* powder.

Moreover, another spectrum region that showed a distinct signal was the region at 1650 and 1500 cm^{-1}, presumably from protein or chitin content. A plausible explanation for this fact could be the denaturation of proteins due to the heating process at high temperatures [42]. Studies on protein denaturation in milk-derived products reveal that certain regions of the spectrum related to protein content (1700–1695 cm^{-1} (aggregated β-sheets), 1645 cm^{-1} (random structure), and 1609 cm^{-1} (side chains)) have been affected by heat treatments, altering the spatial conformation of the proteins and, as a consequence, the intensity of the related IR bands [43]. Another approach for this variation on the IR bands could be the presence of Maillard reaction products. Maillard reaction is a non-enzymatic reaction, binding amino components and reducing sugars with covalent bonds, obtaining aromatic compounds as a result [44].

Finally, a characteristic IR band at 1100–900 cm^{-1}, probably from carbohydrates, can be observed. As has been commented before, thermal treatments can lead to changes in different nutrients. In the case of carbohydrates, these can be involved in several complex reactions. One of these reactions is the hydrolyzation of long carbohydrates, such as starch, obtaining reduced sugars and molecules that can enhance other reactions, such as the above-mentioned Maillard reaction. Regarding the starch content, this macromolecule can also undergo a process called gelatinization, changing its structure from ordered to disordered, affecting its solubilization. Furthermore, free sugars can caramelize due to the dehydration caused by the thermal treatment, contributing to the browning process [42,45].

All these changes produced by the aforementioned chemical reactions were reflected in the raw spectra of the snacks, which were used afterward to explain the origin of the IR regions responsible for the discrimination of the model created.

3.4. Discrimination and Classification of Snacks by ATR-FT-MIR Combined with SIMCA

A four-class SIMCA model was created to assess the differences between snacks made with an increasing amount of *A. diaperinus* powder. A total of two factors were selected to create this model, achieving a cumulative variance higher than 90% in all the classes of the model (Table 8).

In the case of the class projection of the *A. diaperinus* snacks, the model showed well-defined clusters. A two-axis plot was used to represent this model since only two factors were needed to discriminate the different classes (Figure 9).

Table 8. The number of factors, their cumulative variance, and the number of outliers used to create a 4-class SIMCA model of *A. diaperinus* snacks.

Model	Class	Factor 1 (%)	Factor 2 (%)	Outliers
4-class SIMCA *A. diaperinus* snack model	Bs	97.2	98.7	1
	As$_1$	97.0	97.9	6
	As$_2$	89.5	93.5	4
	As$_3$	93.8	96.9	0

Abbreviations used: Bs, blank snack with 0% insect powder; As$_1$, snack with 7.2% of *A. diaperinus* powder; As$_2$, snack with 14.4% of *A. diaperinus* powder; As$_3$, snack with 21.9% of *A. diaperinus* powder.

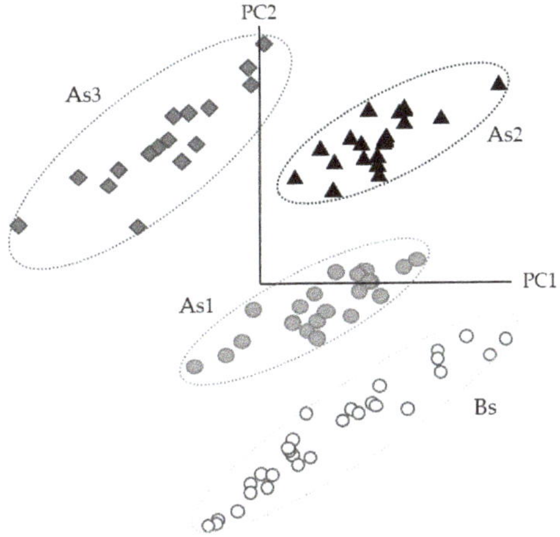

Figure 9. Soft independent modeling of class analogy class projection of FTIR spectroscopy spectra (3000–2700 cm^{-1} and 1800–1200 cm^{-1}) from a 4-class SIMCA *A. diaperinus* snack model. Abbreviations used: Bs, blank snack with 0% insect powder; As$_1$, snack with 7.2% of *A. diaperinus* powder; As$_2$, snack with 14.4% of *A. diaperinus* powder; As$_3$, snack with 21.9% of *A. diaperinus* powder.

The discriminating power plot of the *A. diaperinus* snack model (Figure 10) showed major discrimination of the IR band at 1647 cm^{-1}, from amide I groups most likely from protein and chitin presence, and the IR bands at 1707, 1699, and 1610 cm^{-1}, probably linked to different structural configurations of proteins affected by the thermal treatment. This fact can also explain the decrease in the discriminating power from the IR band at 1505 cm^{-1}.

All the classes exhibited a good classification, showing ICD values over 3. In addition, this model shows the same trend as the previous models shown before of *A. diaperinus* or *L. migratoria* dough, increasing the ICD value with the increase in insect powder concentration (Table 9).

Five spectra from each class, that were not previously included, were used to validate the model. Each class obtained a 100% correct classification of the new spectra showing the capability of the model created to discriminate different baked snacks based on the concentration of *A. diaperinus* powder used in the formulation.

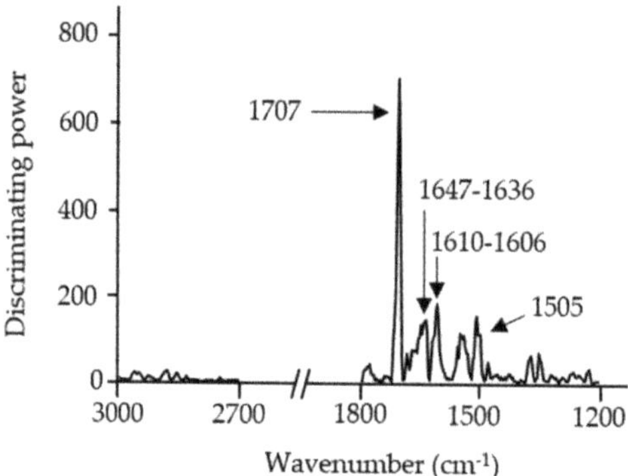

Figure 10. Soft independent modeling of class analogy discriminating power plot of FTIR spectroscopy spectra (3000–2700 cm^{-1} and 1800–1200 cm^{-1}) from a 4-class SIMCA *A. diaperinus* snack model.

Table 9. Soft independent modeling of class analogy interclass distances from a 4-class SIMCA *A. diaperinus* snack model. Abbreviations used: Bs, blank snack with 0% insect powder; As$_1$, snack with 7.2% of *A. diaperinus* powder; As$_2$, snack with 14.4% of *A. diaperinus* powder; As$_3$, snack with 21.9% of *A. diaperinus* powder.

Snack	Bs	As$_1$	As$_2$	As$_3$
B	0.0			
As$_1$	3.3	0.0		
As$_2$	5.9	3.0	0.0	
As$_3$	8.2	5.1	3.3	0.0

3.5. PLSR of Insect Powder Concentration in Doughs and Snacks

Based on the information obtained from the SIMCA models, a partial least square regression (PLSR) analysis was applied to study the potential of predicting the concentration of insect powder added to enrich the doughs and the snacks. PLSR models were built up using the MIR region between 3000–2700 cm^{-1} and 1800–1200 cm^{-1} (Figure 11).

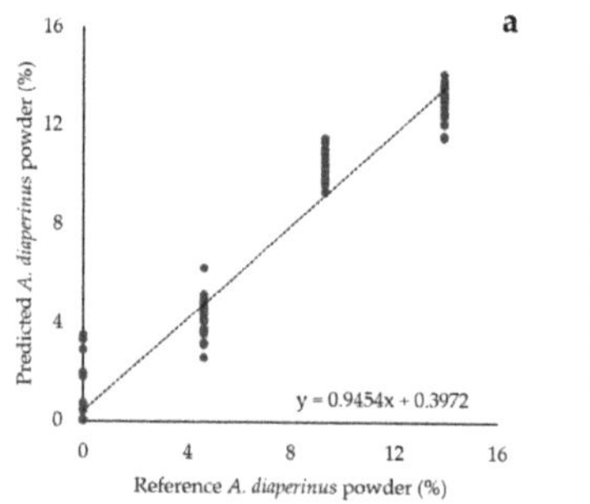

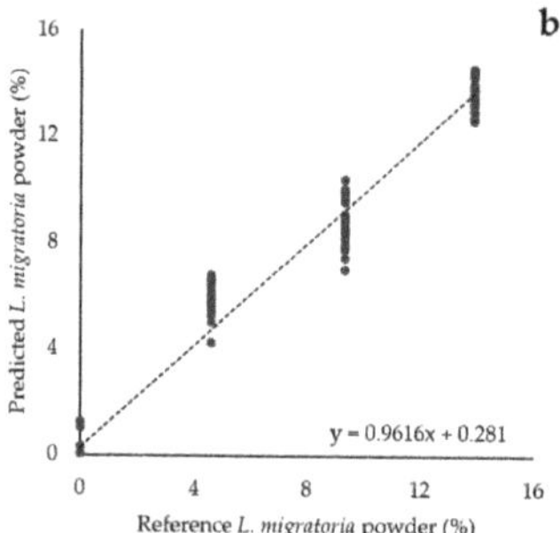

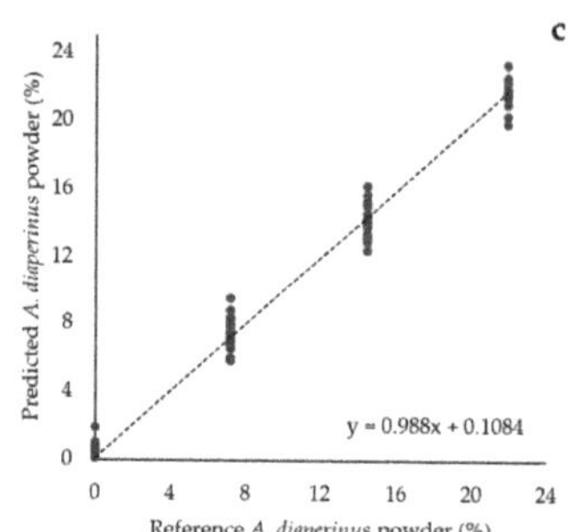

Figure 11. Modeling equations of the calibrated partial least square regressions (PLSR) from the models (**a**) A$_d$ PLSR, *A. diaperinus* dough PLSR model; (**b**) L$_d$ PLSR, *L. migratoria* dough model; and (**c**) A$_s$ PLSR, *A. diaperinus* 3D-printed baked snacks PLSR model.

The optimal number of factors was chosen based on a cumulative variance of the factors higher than 90%, a low standard error of prediction (SEP), and the lowest jaggedness possible. The SEP estimates the expected error of predicting an unknown sample, and the jaggedness is a measurement that quantifies the relevance of the noise compared to the overall signal [46]. A low number of factors can lead to a poor explanation of the data, obtaining high SEP and jaggedness values. On the other hand, a very large number of factors end up overfitting the data, causing the algorithm to perform inaccurately with new data sets [47]. In this case, the SEP will keep reducing its value but the jaggedness value will increase revealing a high impact of the noise in the model. Following these criteria, two factors were chosen for the PLSR models of *A. diaperinus* doughs (A_d PLSR), *L. migratoria* doughs (L_d PLSR), and the *A. diaperinus* snacks model (A_S PLSR). It is also important to mention that all the PLSR models obtained a high R^2, showing an excellent prediction accuracy (Table 10).

Table 10. Parameters associated with each partial least square regression (PLSR) model. Abbreviations used: *n* refers to the number of spectra used to create the model; A_d PLSR, *A. diaperinus* dough PLSR model; L_d PLSR, *L. migratoria* dough model; and A_s PLSR, *A. diaperinus* 3D-printed baked snacks PLSR model.

Model	Factors	*n*	Cumulative Variance (%)	SEP	R^2_{val}	SEC	R^2_{cal}
A_d PLSR	2	93	96.2	1.24	0.970	1.21	0.972
L_d PLSR	2	91	95.6	1.08	0.978	1.06	0.979
A_s PLSR	2	89	97.8	0.90	0.994	0.88	0.994

The regression vector was taken into account in order to find out which were the most remarkable wavenumbers used to develop the PLSR model. The regression vector is the weighted addition, depending on the variance, of each of the wavelengths comprised in the model. The variable that contributes significantly to the prediction of the sample has a higher coefficient compared to those that do not.

In the case of the PLSR dough models, the regression vector had a similar profile (Figure 12). Both models showed that the IR bands that contributed most to the prediction of the percentage of insect powder were 1647 cm^{-1} and 1610 to 1606 cm^{-1}, linked to amide I groups probably related to protein and chitin presence. Moreover, the IR band at 2877 cm^{-1}, associated with the stretching of C-H bonds from methyl groups, most likely from lipid presence, was also detected. All these regions from the spectra were exhibited previously in the discriminating power from the *A. diaperinus* and *L. migratoria* SIMCA dough models explained above.

Prediction models from *A. diaperinus* dough and *L. migratoria* dough exhibited similar IR bands at 2922 and 2855 cm^{-1}, associated with the stretching of C-H bonds from methyl groups (most likely from lipid presence), and 1535 cm^{-1}, presumably from protein and chitin presence (amide I groups).

The same effect was found in the regression vector of the 3D-printed baked snack (Figure 13). The regression vector obtained from the PLSR snack model showed similar IR bands to those exhibited in the discriminating power of the *A. diaperinus* SIMCA snack model in Section 3.4. In the case of the regression vector, the IR bands with the highest contribution in the prediction model were probably related to protein, in particular to different secondary structures, such as aggregated β-sheets (1700–1695 cm^{-1}), random structures (1645 cm^{-1}), and exposed side chains (1609 cm^{-1}), as a result of the thermal treatment during the baking process.

To properly evaluate if the IR bands present in the regression vectors were linked to the insect powder or/and chickpea flour, it was necessary to compare them not only with the pretreated spectra of ingredients (Figure 2) but also with the doughs and snacks spectral information (Figures 3 and 8). In the case of the IR bands related to lipids (1740 cm^{-1}) and lipids and chitin (the region between 3000 and 2700 cm^{-1}), a slight increase in the signal was detected in the doughs as the amount of insect powder increased (Figure 3). Comparing to the transform spectra of raw ingredients (Figure 12c), the amount of lipids present in the chickpea flour was lower than that in both insect powders. Thus, the differentiation by the lipid region in the PLSR model was mainly related to the insect powder added to the dough. However, in the case of the protein region of the pretreated spectra from the doughs (between 16,500 and 1500 cm^{-1}), most of the IR bands showed increasing absorbances while others exhibited a decreasing signal (Figure 3a,c). The pretreated spectra of chickpea flour and *A. diaperinus* and *L. migratoria* powders also showed IR bands at this region (Figure 12c). It is also known that different secondary structures of proteins absorb at different wavenumbers [48]. Thus, the differences in the absorbance of the protein region were not only due to the variation of the chickpea flour/insect powder ratio but also due to the type of proteins added, the bonds that conform them, and the media (dough or snack).

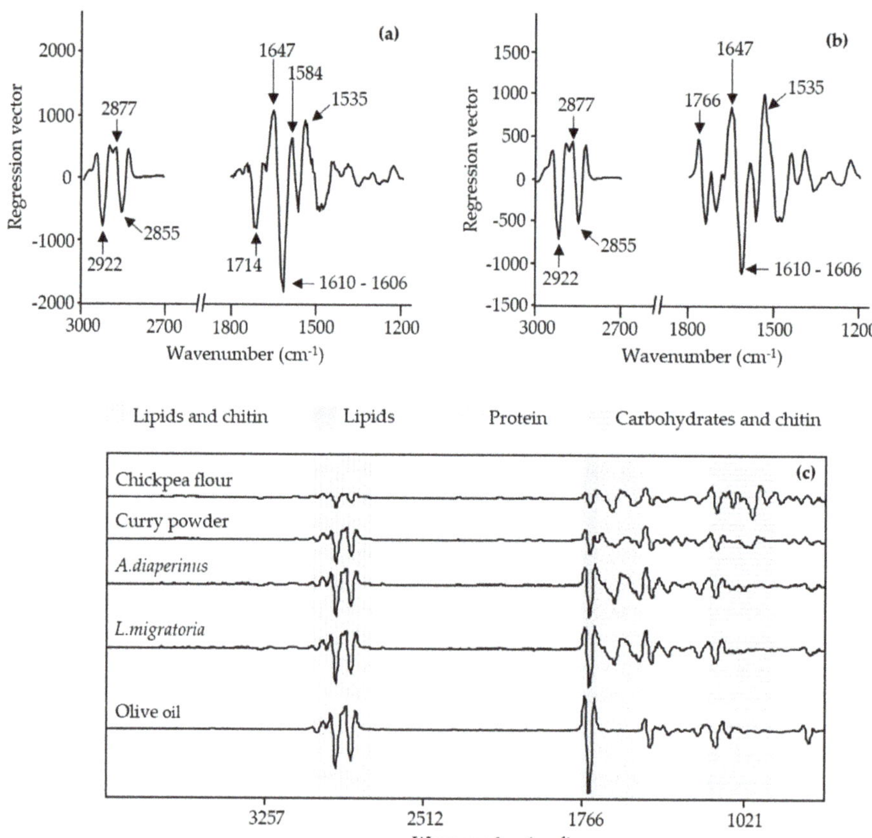

Figure 12. PLSR regression vector plot of FTIR spectroscopy spectra (3000–2700 cm^{-1} and 1800–1200 cm^{-1}) from an (**a**) *A. diaperinus* dough prediction model, (**b**) *L. migratoria* dough prediction model, and (**c**) the transformed spectra (vector length normalized and transformed using second derivative polynomial-fit Sa-vitzky–Golay function (13 points)) of the raw ingredients used for the dough preparation.

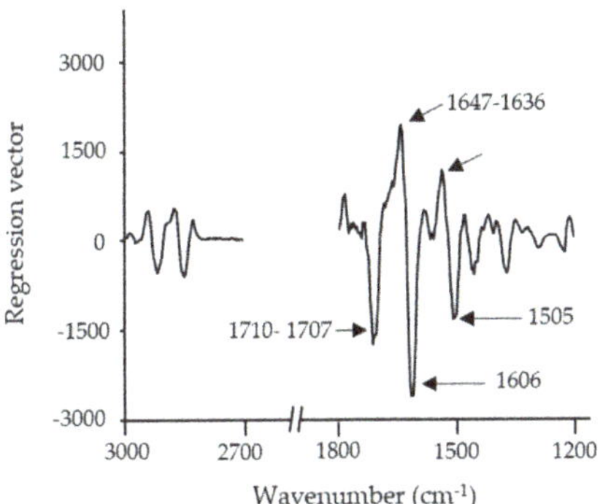

Figure 13. PLSR regression vector plot of FTIR spectroscopy spectra (3000–2700 cm^{-1} and 1800–1200 cm^{-1}) from an *A. diaperinus* snack model.

In the literature, a collection of models created with the PLSR algorithm using MIR and NIR spectral data have been successfully used to predict the concentration or amount of ingredients and products used by the food industry [49,50]. A portable FT-MIR spectrometer combined with multivariate analysis can be used for routine controls for authentication of standardization of quality control of *A. diaperinus* and *L. migratoria* products both during the fabrication process or at the end of the production line.

4. Conclusions

ATR-FT-MIR combined with multivariate analysis can be used as a rapid technique to discriminate edible insect powders used as ingredients for doughs and 3D-printed snacks. Moreover, using PLSR analysis, calibration models can be built up to easily predict the concentration of insect powder present in doughs and snacks.

Further work needs to be carried out to determine the feasibility of this method for detecting insect powders in other food matrices, either raw or cooked.

Supplementary Materials: The following are available online at https://www.mdpi.com/article/10.3390/foods10081806/s1, Figure S1: Particle size distribution of commercial and milled *A. diaperinus* powder; and Figure S2: Particle size distribution of milled *L. migratoria* powder.

Author Contributions: Conceptualization, S.d.L.-C., A.S., and T.S.; methodology, N.G.-G., J.M.-C., J.W., C.G., M.F., and S.d.L.-C.; validation, A.S.; T.S.; C.G., M.F., and S.d.L.-C.; formal analysis, N.G.-G., J.M.-C., and S.d.L.-C.; investigation, N.G.-G.; resources, C.G., M.F., S.d.L.-C., and C.B.; data curation, N.G.-G.; writing—original draft preparation, N.G.-G.; writing—review and editing, C.G., M.F., S.d.L.-C., A.S., and T.S.; visualization, N.G.-G.; supervision, S.d.L.-C., A.S., and T.S.; project administration, C.B., C.G., M.F., and S.d.L.-C.; funding acquisition, C.B., C.G., M.F., and S.d.L.-C. All authors have read and agreed to the published version of the manuscript.

Funding: This research was funded by Ministerio de Economía i Competitividad (CTQ 2014-54520-P), Ministerio de Ciencia e Innovación (PGC2018-097095-B-I00), and Agencia Estatal de Investigación, Fondo Social Europeo (FSE) and Iniciativa de Empleo Juvenil (PEJ2018-004192-A).

Institutional Review Board Statement: Not applicable.

Informed Consent Statement: Not applicable.

Conflicts of Interest: The authors declare no conflict of interest.

References

1. United Nations: Department of Economic and Social Affairs: Population Division. *World Population Prospects 2019: Highlights*; United Nations: New York, NY, USA, 2019.
2. Production, F.; Cervantes-Godoy, D.; Dewbre, J.; PIN; Amegnaglo, C.J.; Soglo, Y.Y.; Akpa, A.F.; Bickel, M.; Sanyang, S.; Ly, S.; et al. *The Future of Food and Agriculture: Trends and Challenges*; Food and Agriculture Organization of the United Nations: Rome, Italy, 2017; Volume 4, ISBN 1815-6797.
3. Grau, T.; Vilcinskas, A.; Joop, G. Sustainable farming of the mealworm Tenebrio molitor for the production of food and feed. *A J. Biosci.* **2017**, *72*, 337–349. [CrossRef]
4. Van Huis, A.; Van Itterbeeck, J.; Klunder, H.; Mertens, E.; Halloran, A.; Muir, G.; Vantomme, P. *Edible Insects. Future Prospects for Food and Feed Security*; Food and Agriculture Organization of the United Nations: Rome, Italy, 2013; Volume 171, ISBN 9789251075951.
5. Van Broekhoven, S.; Oonincx, D.G.A.B.; Van Huis, A.; Loon, J.J.A. Van Growth performance and feed conversion efficiency of three edible mealworm species (Coleoptera: Tenebrionidae) on diets composed of organic by-products. *J. Insect Physiol.* **2015**, *73*, 1–10. [CrossRef] [PubMed]
6. Premalatha, M.; Abbasi, T.; Abbasi, T.; Abbasi, S.A. Energy-efficient food production to reduce global warming and ecodegradation: The use of edible insects. *Renew. Sustain. Energy Rev.* **2011**, *15*, 4357–4360. [CrossRef]
7. Nowak, V.; Persijn, D.; Rittenschober, D.; Charrondiere, U.R. Review of food composition data for edible insects. *Food Chem.* **2016**, *193*, 39–46. [CrossRef]
8. Rumpold, B.A.; Schlüter, O.K. Nutritional composition and safety aspects of edible insects. *Mol. Nutr. Food Res.* **2013**, *57*, 802–823. [CrossRef] [PubMed]
9. Ribeiro, J.C.; Cunha, L.M.; Sousa-Pinto, B.; Fonseca, J. Allergic risks of consuming edible insects: A systematic review. *Mol. Nutr. Food Res.* **2018**, *62*, 1700030. [CrossRef] [PubMed]
10. Garino, C.; Zagon, J.; Braeuning, A. Insects in food and feed—Allergenicity risk assessment and analytical detection. *EFSA J.* **2019**, *17*, e170907. [CrossRef] [PubMed]
11. Turck, D.; Castenmiller, J.; De Henauw, S.; Hirsch-Ernst, K.I.; Kearney, J.; Maciuk, A.; Mangelsdorf, I.; McArdle, H.J.; Naska, A.; Pelaez, C.; et al. Safety of dried yellow mealworm (Tenebrio molitor larva) as a novel food pursuant to Regulation (EU) 2015/2283. *EFSA J.* **2021**, *19*, e06343. [CrossRef]
12. International Platform of Insects for Food and Feed (IPIFF). Edible Insects on the European Market. Available online: https://ipiff.org/wp-content/uploads/2020/06/10-06-2020-IPIFF-edible-insects-market-factsheet.pdf (accessed on 4 August 2021).
13. European Commission. Approval of First Insect as Novel Food. Available online: https://ec.europa.eu/food/safety/novel_food/authorisations/approval-first-insect-novel-food_en (accessed on 27 May 2021).
14. Danezis, G.P.; Tsagkaris, A.S.; Camin, F.; Brusic, V.; Georgiou, C.A. Trends in Analytical Chemistry Food authentication: Techniques, trends & emerging approaches. *Trends Anal. Chem.* **2016**, *85*, 123–132. [CrossRef]
15. Reed, J.D.; Krueger, C.G.; Vestling, M.M. MALDI-TOF mass spectrometry of oligomeric food polyphenols. *Phytochemistry* **2005**, *66*, 2248–2263. [CrossRef]
16. Ulrich, S.; Kühn, U.; Biermaier, B.; Piacenza, N.; Schwaiger, K.; Gottschalk, C.; Gareis, M. Direct identification of edible insects by MALDI-TOF mass spectrometry. *Food Control* **2017**, *76*, 96–101. [CrossRef]
17. Rohman, A.; Sismindari; Erwanto, Y.; Che Man, Y.B. Analysis of pork adulteration in beef meatball using Fourier transform infrared (FTIR) spectroscopy. *Meat Sci.* **2011**, *88*, 91–95. [CrossRef]
18. Rodriguez-Saona, L.E.; Allendorf, M.E. Use of FTIR for Rapid Authentication and Detection of Adulteration of Food. *Annu. Rev. Food Sci. Technol.* **2011**, *2*, 467–483. [CrossRef]
19. Mellado-Carretero, J.; García-Gutiérrez, N.; Ferrando, M.; Güell, C.; García-Gonzalo, D.; De Lamo-Castellví, S. Rapid discrimination and classification of edible insect powders using ATR-FTIR spectroscopy combined with multivariate analysis. *J. Insects Food Feed* **2019**, *6*, 141–148. [CrossRef]
20. Suhandy, D.; Yulia, M. Peaberry coffee discrimination using UV-visible spectroscopy combined with SIMCA and PLS-DA. *Int. J. Food Prop.* **2017**, *20*, S331–S339. [CrossRef]
21. Mertens, B.; Thompson, M.; Fearn, T. Principal component outlier detection and SIMCA: A synthesis. *Analyst* **1994**, *119*, 2777–2784. [CrossRef]
22. Afseth, N.K.; Segtnan, V.H.; Marquardt, B.J.; Wold, J.P. Raman and near-infrared spectroscopy for quantification of fat composition in a complex food model system. *Appl. Spectrosc.* **2005**, *59*, 1324–1332. [CrossRef] [PubMed]
23. Indrayanto, G.; Rohman, A. The Use of FTIR Spectroscopy Combined with Multivariate Analysis in Food Composition Analysis. In *Spectroscopic Techniques & Artificial Intelligence for Food and Beverage Analysis*; Springer: Singapore, 2020; pp. 25–51. ISBN 978-981-15-6495-6.
24. Kumar, Y.; Sharanagat, V.S.; Singh, L.; Mani, S. Effect of germination and roasting on the proximate composition, total phenolics, and functional properties of black chickpea (Cicer arietinum). *Legum. Sci.* **2020**, *2*, e20. [CrossRef]
25. Cárdenas, G.; Cabrera, G.; Taboada, E.; Miranda, S.P. Chitin characterization by SEM, FTIR, XRD, and 13C cross polarization/mass angle spinning NMR. *J. Appl. Polym. Sci.* **2004**, *93*, 1876–1885. [CrossRef]
26. Rohman, A.; Man, Y.B.C. Fourier transform infrared (FTIR) spectroscopy for analysis of extra virgin olive oil adulterated with palm oil. *Food Res. Int.* **2010**, *43*, 886–892. [CrossRef]

27. Majtán, J.; Bíliková, K.; Markovič, O.; Gróf, J.; Kogan, G.; Šimúth, J. Isolation and characterization of chitin from bumblebee (*Bombus terrestris*). *Int. J. Biol. Macromol.* **2007**, *40*, 237–241. [CrossRef]

28. Sun, B.; Rahman, M.M.; Tar'an, B.; Yu, P. Determine effect of pressure heating on carbohydrate related molecular structures in association with carbohydrate metabolic profiles of cool-climate chickpeas using Globar spectroscopy. *Spectrochim. Acta Part A Mol. Biomol. Spectrosc.* **2018**, *201*, 8–18. [CrossRef]

29. Meza-Márquez, O.G.; Gallardo-Velázquez, T.; Osorio-Revilla, G. Application of mid-infrared spectroscopy with multivariate analysis and soft independent modeling of class analogies (SIMCA) for the detection of adulterants in minced beef. *MESC* **2010**, *86*, 511–519. [CrossRef]

30. Tsugawa, H.; Tsujimoto, Y.; Arita, M.; Bamba, T.; Fukusaki, E. GC/MS based metabolomics: Development of a data mining system for metabolite identification by using soft independent modeling of class analogy (SIMCA). *BMC Bioinform.* **2011**, *12*, 1–13. [CrossRef]

31. Talari, A.C.S.; Martinez, M.A.G.; Movasaghi, Z.; Rehman, S.; Rehman, I.U. Advances in Fourier transform infrared (FTIR) spectroscopy of biological tissues. *Appl. Spectrosc. Rev.* **2017**, *52*, 456–506. [CrossRef]

32. Mantilla, S.M.O.; Alagappan, S.; Sultanbawa, Y.; Smyth, H.E.; Cozzolino, D. A Mid Infrared (MIR) Spectroscopy Study of the Composition of Edible Australian Green Ants (*Oecophylla smaragdina*)—A Qualitative Study. *Food Anal. Methods* **2020**, *13*, 1627–1633. [CrossRef]

33. Peren, D.; Menevseoglu, A. A rapid method to detect green pea and peanut adulteration in pistachio by using portable FT-MIR and FT-NIR spectroscopy combined with chemometrics. *Food Control* **2021**, *121*, 107670. [CrossRef]

34. Gómez-de-anda, F.; Dorantes-álvarez, L.; Gallardo-velázquez, T.; Osorio-revilla, G. Determination of Trichinella spiralis in pig muscles using Mid-Fourier Transform Infrared Spectroscopy (MID-FTIR) with Attenuated Total Re fl ectance (ATR) and Soft Independent Modeling of Class Analogy (SIMCA). *MESC* **2012**, *91*, 240–246. [CrossRef]

35. De Lamo-Castellví, S.; Männing, A.; Rodríguez-Saona, L.E. Fourier-transform infrared spectroscopy combined with immunomagnetic separation as a tool to discriminate Salmonella serovars. *Analyst* **2010**, *135*, 2987–2992. [CrossRef] [PubMed]

36. Gelaw, T.K.; Espina, L.; Pagán, R.; García-Gonzalo, D.; De Lamo-Castellví, S. Prediction of Injured and Dead Inactivated Escherichia coli O157:H7 Cells after Heat and Pulsed Electric Field Treatment with Attenuated Total Reflectance Infrared Microspectroscopy Combined with Multivariate Analysis Technique. *Food Bioprocess Technol.* **2014**, *7*, 2084–2092. [CrossRef]

37. Kwok, K.C.; Liang, H.H.; Niranjan, K. Optimizing conditions for thermal processes of soy milk. *J. Agric. Food Chem.* **2002**, *50*, 4834–4838. [CrossRef] [PubMed]

38. Melgar-Lalanne, G.; Hernández-Álvarez, A.J.; Salinas-Castro, A. Edible Insects Processing: Traditional and Innovative Technologies. *Compr. Rev. Food Sci. Food Saf.* **2019**. [CrossRef] [PubMed]

39. Byler, D.M.; Purcell, J.M. FTIR Examination Of Thermal Denaturation And Gel-Formation In Whey Proteins. *SPIE Digit. Libr.* **1989**, *1145*, 415. [CrossRef]

40. Sablani, S.S.; Marcotte, M.; Baik, O.D.; Castaigne, F. Modeling of simultaneous heat and water transport in the baking process. *LWT Food Sci. Technol.* **1998**, *31*, 201–209. [CrossRef]

41. Frankel, E.N. Free radical oxidation. *Lipid Oxid.* **2012**, 15–24. [CrossRef]

42. Fox, P.F.; Uniacke-Lowe, T.; McSweeney, P.L.H.; O'Mahony, J.A. Heat-Induced Changes in Milk. In *Dairy Chemistry and Biochemistry*; Springer International Publishing: Cham, Switzerland, 2015; pp. 345–375.

43. Markoska, T.; Huppertz, T.; Grewal, M.K.; Vasiljevic, T. Structural changes of milk proteins during heating of concentrated skim milk determined using FTIR. *Int. Dairy J.* **2019**, *89*, 21–30. [CrossRef]

44. Kröncke, N.; Grebenteuch, S.; Keil, C.; Demtröder, S.; Kroh, L.; Thünemann, A.F.; Benning, R.; Haase, H. Effect of different drying methods on nutrient quality of the yellow mealworm (*Tenebrio molitor* L.). *Insects* **2019**, *10*, 84. [CrossRef]

45. Carl Hoseney, R. Chemical changes in carbohydrates produced by thermal processing. *J. Chem. Educ.* **1984**, *61*, 308–312. [CrossRef]

46. Mevik, B.H.; Cederkvist, H.R. Mean squared error of prediction (MSEP) estimates for principal component regression (PCR) and partial least squares regression (PLSR). *J. Chemom.* **2004**, *18*, 422–429. [CrossRef]

47. Infometrix Inc. *Pirouette Multivariate Data Analysis Software*; Infometrix Inc.: Washington, DC, USA, 2011.

48. Byler, D.M.; Susi, H. Examination of the secondary structure of proteins by deconvolved FTIR spectra. *Biopolymers* **1986**, *25*, 469–487. [CrossRef]

49. Candoğan, K.; Altuntas, E.G.; İğci, N. Authentication and Quality Assessment of Meat Products by Fourier-Transform Infrared (FTIR) Spectroscopy. *Food Eng. Rev.* **2021**, *13*, 66–91. [CrossRef]

50. Cebi, N.; Taylan, O.; Abusurrah, M.; Sagdic, O. Detection of orange essential oil, isopropyl myristate, and benzyl alcohol in lemon essential oil by ftir spectroscopy combined with chemometrics. *Foods* **2021**, *10*, 27. [CrossRef] [PubMed]

Article

Development of a DNA Metabarcoding Method for the Identification of Bivalve Species in Seafood Products

Kristina Gense [1], Verena Peterseil [2], Alma Licina [3], Martin Wagner [1,4], Margit Cichna-Markl [5], Stefanie Dobrovolny [2] and Rupert Hochegger [2,*]

1   Austrian Competence Centre for Feed and Food Quality, Safety and Innovation, FFoQSI GmbH, Technopark 1, 3430 Tulln an der Donau, Austria; kristina.gense@ffoqsi.at (K.G.); martin.wagner@ffoqsi.at (M.W.)
2   Austrian Agency for Health and Food Safety (AGES), Institute for Food Safety, Department of Molecular Biology and Microbiology, Spargelfeldstr. 191, 1220 Vienna, Austria; verena.peterseil@ages.at (V.P.); stefanie.dobrovolny@ages.at (S.D.)
3   LVA GmbH, Magdeburggasse 10, 3400 Klosterneuburg, Austria; alma.licina@lva.at
4   Department for Farm Animals and Veterinary Public Health, Institute of Milk Hygiene, University of Veterinary Medicine, Veterinärplatz 1, 1210 Vienna, Austria
5   Department of Analytical Chemistry, Faculty of Chemistry, University of Vienna, Währinger Straße 38, 1090 Vienna, Austria; margit.cichna@univie.ac.at
*   Correspondence: rupert.hochegger@ages.at

Abstract: The production of bivalve species has been increasing in the last decades. In spite of strict requirements for species declaration, incorrect labelling of bivalve products has repeatedly been detected. We present a DNA metabarcoding method allowing the identification of bivalve species belonging to the bivalve families Mytilidae (mussels), Pectinidae (scallops), and Ostreidae (oysters) in foodstuffs. The method, developed on Illumina instruments, targets a 150 bp fragment of mitochondrial 16S rDNA. We designed seven primers (three primers for mussel species, two primers for scallop species and a primer pair for oyster species) and combined them in a triplex PCR assay. In each of eleven reference samples, the bivalve species was identified correctly. In ten DNA extract mixtures, not only the main component (97.0–98.0%) but also the minor components (0.5–1.5%) were detected correctly, with only a few exceptions. The DNA metabarcoding method was found to be applicable to complex and processed foodstuffs, allowing the identification of bivalves in, e.g., marinated form, in sauces, in seafood mixes and even in instant noodle seafood. The method is highly suitable for food authentication in routine analysis, in particular in combination with a DNA metabarcoding method for mammalian and poultry species published recently.

Keywords: DNA metabarcoding; next generation sequencing; food authentication; bivalves; Mytilidae; Pectinidae; Ostreidae; species identification; mitochondrial 16S rDNA; seafood

Citation: Gense, K.; Peterseil, V.; Licina, A.; Wagner, M.; Cichna-Markl, M.; Dobrovolny, S.; Hochegger, R. Development of a DNA Metabarcoding Method for the Identification of Bivalve Species in Seafood Products. *Foods* **2021**, *10*, 2618. https://doi.org/10.3390/foods10112618

Academic Editor: Maria Castro-Puyana

Received: 28 September 2021
Accepted: 23 October 2021
Published: 28 October 2021

Publisher's Note: MDPI stays neutral with regard to jurisdictional claims in published maps and institutional affiliations.

1. Introduction

Bivalves, a class of molluscs, are distributed worldwide. Due to their high content of essential nutrients, their production has steadily been increased over the last three decades [1–5]. Mytilidae (mussels), Pectinidae (scallops), and Ostreidae (oysters) are the most important bivalve families for human consumption. Each of these bivalve families is divided into several genera comprising a high number of species [6]. In 2019, 1.03 million tons of mussels, scallops, and oysters were caught in nature and 10.25 million tons were cultivated in aquaculture, earning a profit of millions of US dollars [7].

In the EU, international and national regulations exist to ensure legal trade in seafood and seafood products. The EU directive 1379/2013 regulates market organization of fishery and aquaculture products, including correct declaration of seafood [8]. To comply with legal regulations, labels must include both the local trade name in the official language(s) and the correct scientific Latin name [8,9]. Correct labelling of seafood products is important for traceability issues, protection of endangered species, mitigation of illegal fishing, and for individual reasons of end consumers [10,11]. Regardless of clear and strict requirements for species declaration, incorrect labelling of bivalve products has repeatedly been detected in Europe [12–17]. In German and Swiss studies, more than half of the products declared to contain "Jakobsmuschel" (or "Jacobsmuschel") were labelled incorrectly [15,18,19]. Although the German name "Jakobsmuschel" (or "Jacobsmuschel") may only be used for scallop species belonging to the genus *Pecten*, species of other genera (particularly *Placopecten* and *Mizuhopecten*) were identified in these products.

For authentication of seafood products, laboratories may choose from a variety of methodologies. In the case of bivalves, morphological characteristics such as shell, color, and size may allow correct species classification. However, after shell removal or mechanical processing, classification by morphology may be hampered or even be impossible [16,20]. Recently, matrix-assisted laser desorption ionization time of flight mass spectrometry (MALDI-TOF MS) has been shown to be suitable for accurate species identification of scallops [19]. However, since MALDI-TOF MS instruments are rather expensive and do not allow high-throughput analysis, this methodology is less applicable for routine analyses.

To date, DNA-based methods are considered most suitable for the identification of seafood species, even in highly processed food products [21–23]. Due to its high copy number and robustness, mitochondrial DNA (mtDNA) is frequently preferred over genomic DNA [24,25]. The mtDNA regions most commonly used for species identification are cytochrome c oxidase subunit I (COI), cytochrome b (cyt b), and 16S ribosomal DNA (16S rDNA) [15,26–33]. Compared to other seafood, e.g., fish, crustaceans, and cephalopods, (real-time) polymerase chain reaction (PCR) assays for bivalve species are limited in number [18,32,34–41]. The disadvantage of (real-time) PCR is that for each target species, a specific primer (probe) system is required [18,31,33,36,39–43].

A powerful alternative is DNA barcoding, aiming at detecting a broader range of species by using universal primer systems [22,26,34,44]. DNA barcodes commonly contain conserved regions at both ends, serving as binding sites for universal primers, and a variable part in between the primer binding sites, for differentiation between the species of interest [34,45]. DNA barcodes of approximately 600 base pairs (bp) in length have been found to be suitable for the analysis of highly processed food products [22,26,27,34,44,46–48]. In conventional DNA barcoding, PCR products obtained by amplifying the selected DNA barcode region are then subjected to Sanger sequencing [22,34,44,49,50]. However, sample throughput of Sanger sequencing is limited since samples are sequenced one by one. A much more efficient approach is to combine DNA barcoding with next-generation sequencing (NGS) technologies [22,26,34]. So-called DNA metabarcoding allows the identification of multiple species in multiple food samples in one and the same sequencing run [45,46,51–54]. The suitability of DNA metabarcoding for the analysis of ultra-processed food products has already been demonstrated, e.g., for the detection of mammals in sausages or insects in bars [47,48].

In this study, we present a DNA metabarcoding method allowing the differentiation between species from three bivalve families, Pectinidae, Ostreidae, and Mytilidae, in raw and processed food products to detect food adulteration. The method was developed on the Illumina MiSeq® (San Diego, CA, USA) and iSeq® (San Diego, CA, USA) platforms due to their low error rates compared to other NGS platforms [55].

2. Materials and Methods

2.1. Sample Collection and Storage

A total of 86 commercial food products were collected from regional supermarkets, fish markets, and delicacy shops in Austria from summer 2018 until winter 2020 (Supplementary Table S1). Samples were either fresh, deep-frozen, or in processed condition. Each sample was given a specific ID number, with the letter "O" referring to oysters, "S" to scallops, "M" to mussels, and "Mi" to mixed-species seafood. Samples were stored at −20 °C until DNA extraction.

Eleven out of the 86 samples ("reference samples"), comprising three mussel, six scallop, and two oyster species (see Table 1), were used for method development. Identity of bivalve species in these reference samples (samples M12, M13 and M27 for mussels; samples S42, S46, S47, S49, S50, and S55 for scallops; samples O2 and O3 for oysters; Supplementary Table S1) was verified by subjecting DNA extracts to Sanger sequencing (Microsynth, Balgach, Switzerland) and matching the sequences against the public databases provided by the National Center for Biotechnology Information (NCBI, Bethesda, MD, USA). For Sanger sequencing, the forward and reverse primers listed in Table 2 were used.

Table 1. Bivalve species used for development of the DNA metabarcoding method.

Scientific Name	Commercial Name (German)	Commercial Name (English)
Mytilidae	Miesmuscheln	Mussels
Mytilus edulis	Gemeine Miesmuschel	Blue mussel
Mytilus galloprovincialis	Mittelmeer-Miesmuschel	Mediterranean mussel
Perna canaliculus	Neuseeland-Miesmuschel	New Zealand green-lipped mussel
Pectinidae	Kammmuscheln	Scallops
Placopecten magellanicus	Atlantischer Tiefseescallop	Atlantic deep-sea scallop
Mizuhopecten yessoensis	Japanische Kammmuschel	Yesso scallop
Pecten jacobaeus	Jakobsmuschel	Great scallop
Zygochlamys patagonica	Patagonische Kammmuschel	Patagonian scallop
Argopecten purpuratus	Purpur-Kammmuschel	Purple scallop
Aequipecten opercularis	Kleine Pilgermuschel	Queen scallop
Ostreidae	Austern	Oysters
Magallana gigas	Pazifische Felsenauster	Pacific oyster
Ostrea edulis	Europäische Auster	European flat oyster

Table 2. Primers designed in this study.

Name	Sequence $5'{\rightarrow}3'$
mussel	
For_Mu	CCTTTTGCATAAGGGTTTTTCAAG
Rev1_Mu	CGAATAGTATCTAGCCGCCATTC
Rev2_Mu	GCAAATAGCATATCACTTTCACCTC
scallop	
For_Mu	TGCTAAGGTAGCTAAATTATGGCC
Rev_Mu	CTTCACGGGGTCTTCTCGTC
oyster	
For_Mu	GGTAGCGAAATTCCTTGCCTT
Rev_Mu	AAAGTTGCACGGGGTCTT
overhang	
Forward	TCGTCGGCAGCGTCAGATGTGTATAAGAGACAG
Reverse	GTCTCGTGGGCTCGGAGATGTGTATAAGAGACAG

2.2. DNA Extraction and Quantification

Raw material was cut into smaller pieces or homogenized. To 2.0 g of each sample, 10 mL of a hexadecyltrimethylammonium bromide (CTAB) buffer was added. After addition of 80 µL proteinase K, the mixture was incubated on an Intelli-Mixer™ RM2 (LTF Labortechnik, Wasserburg, Germany) overnight at 50 °C.

For DNA isolation, a commercial kit (Maxwell® 16 FFS Nucleic Acid Extraction System Custom-Kit, Promega, Madison, WI, USA) was used according to the manufacturer's instructions. DNA concentration was determined fluorometrically (Qubit® 2.0 fluorometer, Thermo Fisher Scientific, Waltham, MA, USA). For higher concentrations, the Qubit® dsDNA broad range assay kit (2 to 1000 ng) was used, and for lower concentrations, the Qubit® dsDNA high-sensitivity assay kit (0.2 to 100 ng) was used. DNA purity was assessed from the ratio of the absorbance at 260 and 280 nm (QIAxpert spectrophotometer, software version 2.2.0.21, Qiagen, Hilden, Germany). DNA extracts were stored at −20 °C until further use.

2.3. DNA Extract Mixtures

Ternary DNA extract mixtures were prepared by mixing DNA extracts (DNA concentration 5 ng/µL) from *Pecten* spp., *Magallana gigas* and *Mytilus galloprovincialis*, representing the three bivalve families Pectinidae, Ostreidae, and Mytilidae, respectively. Individual DNA extracts were mixed in a ratio of 98.0:1.5:0.5 (*v/v/v*).

In addition, DNA extract mixtures consisting of DNA from species belonging to one bivalve family were prepared. In these mixtures, DNA from one species was present as the main component, DNA from the other species as minor components (1.0% each). Since only two oyster species were available, the DNA extract mixture representing the bivalve family Ostreidae contained the closely related scallop (*Placopecten magellanicus*) as a major component (98.0%) and DNA from the two oyster species as minor components (1.0% each).

In addition to mixtures consisting of DNA from bivalve species only, a DNA extract mixture containing another mollusc species was prepared. DNA extract from a squid species (*Sepiella inermis*) was chosen as the main component (97.0%) and DNA from the bivalve species *Placopecten magellanicus*, *Ostrea edulis* and *Perna canaliculus* was present as minor components (1.0% each).

2.4. Reference Sequences

A 150 bp fragment of the mitochondrial 16S rDNA gene was used as a DNA barcode. Reference sequences for commonly consumed bivalve species and some exotic seafood species, that are permitted for consumption in Austria ("Codex Alimentarius Austriacus" chapter B35, [56]), were downloaded from the NCBI databases (Supplementary Table S2) by using CLC Genomics Workbench software (version 10.1.1, Qiagen, Hilden, Germany). If available, complete reference sequences from the RefSeq database were preferentially downloaded due to their reliability. In case complete reference sequences were not available, all DNA sequences of the mitochondrial 16S rDNA available for one and the same species, submitted by individual scientists, were aligned and checked for similarity and unidentified nucleotides. Subsequently, the DNA sequence with the highest quality (e.g., without unknown nucleotides, full-length of the DNA barcode) was chosen as a reference sequence.

2.5. Primer Systems

Primers were designed manually on a multiple DNA sequence alignment of the mitochondrial 16S rDNA of approximately 90 bivalve species using the CLC Genomics Workbench software (version 10.1.1, Qiagen, Hilden, Germany). The designed primers were checked for their physical and structural properties (e.g., formation of dimers, secondary structure, annealing temperature) using Oligo Calc, the OligoAnalyzer Tool provided by Integrated DNA Technologies (IDT, Coralville, IA, USA) and the online product descriptions from TIB Molbiol (Berlin, Germany). The primers, listed in Table 2, were synthesized by TIB Molbiol. Table 2 also shows the Illumina overhang adapter sequences which were linked to the target-specific primers.

All in-house-designed primers were tested in real-time PCR with DNA extracted from the eleven reference samples. During optimization, the following PCR conditions/parameters were kept constant and applied as published previously: DNA input amount of 12.5 ng, 'ready-to-use' HotStarTaq Master Mix Kit, annealing temperature (62 °C), 25 cycles [47].

Only one variable, the addition of magnesium chloride solution, was modified (addition of 1.5 or 3 mM $MgCl_2$). Real-time PCR reactions were carried out using a fluorescent intercalating dye (EvaGreen® (20x in water)) in strip tubes or in 96-well plates, depending on the thermocycler used, the Rotor-Gene Q (Qiagen, Hilden, Germany) or the LightCycler® 480 System (Roche, Penzberg, Germany), respectively. The total volume of the PCR reactions was 25 µL, consisting of 22.5 µL reaction mix and 2.5 µL of template DNA (diluted DNA samples (5 ng/µL)) or water as negative control. In the reaction mix, the HotStarTaq Master Mix Kit (Qiagen, Hilden, Germany) was used at a final concentration of 1x and the final concentration of primers was 0.2 µM, except the forward primer for mussels (0.4 µM). PCR cycling conditions were 15 min initial denaturation at 95 °C, 25 cycles at 95 °C, 62 °C and 72 °C for 30 s each, and a final elongation for 10 min at 72 °C. The primer pairs for mussels, scallops, and oysters with and without Illumina overhang adapter sequences were first used in singleplex PCR assays. Then, the seven primers (three forward and four reverse primers) listed in Table 2 were combined in a triplex assay. The identity of the PCR products was confirmed by melting curve analysis and/or agarose gel electrophoresis.

2.6. Library Preparation and NGS

In general, samples were sequenced by using either the MiSeq® or the iSeq® platform (Illumina, San Diego, CA, USA). DNA extracts were diluted to a DNA concentration of 5 ng/µL. Extracts with a DNA concentration < 5 ng/µL were used undiluted.

DNA library preparation was performed according to Dobrovolny et al. [47] with minor modifications (excess of $MgCl_2$, final concentration 3 mM; average library size: 278 bp; diluted libraries of the iSeq® system were denatured automatically on the instrument).

For the MiSeq® and iSeq® platform, the DNA library was adjusted to 4 and 1 nM, respectively, with 10 mM Tris-HCl, pH 8.6. After pooling individual DNA libraries (5 µL MiSeq®, 7 µL iSeq®), the DNA concentration was determined using Qubit® 2.0 fluorimeter.

All sequencing runs were performed using either the MiSeq® Reagent Kit v2 (300-cycles) or the iSeq® 100 i1 Reagent v2 (300-cycles) with a final loading concentration of 8 pM. The pooled DNA libraries contained a 5% PhiX spike-in.

Reference samples were sequenced in six replicates (three sequencing runs, two replicates per run), while DNA extract mixtures were sequenced in nine replicates (three sequencing runs, three replicates per run). Commercial food products were sequenced in triplicates (three sequencing runs, one replicate per run) and food products were sequenced at least once by using either the MiSeq® or the iSeq® platform.

2.7. NGS Data Analysis Using Galaxy

After paired-end sequencing, the resulting FastQ files, generated by the instrument control software, were used as input for data analysis. The sequencing output in FastQ format was then processed with an analysis pipeline as described previously by using Galaxy (version 19.01) [47]. The published amplicon analysis workflow was modified as follows: the target-specific primers were trimmed from both ends using the tool Cutadapt and reads were not clustered into Operational Taxonomic Units (OTUs) [57]. Completely identical sequences were collapsed into a single representative sequence with the tool Dereplicate to minimize the number of reads, and then compared against a customized database for bivalves (Supplementary Table S2) using BLASTn [58].

3. Results and Discussion

3.1. Barcode Region and Primer Systems

We aimed to develop a DNA metabarcoding method allowing the differentiation between species belonging to the bivalve families Pectinidae, Ostreidae, and Mytilidae. To be applicable in routine analysis, the method should allow identifying the economically most important bivalve species in raw and highly processed food products.

We started with searching for appropriate DNA barcode regions of about 150 bp in length, containing conserved parts at the ends and a variable part in between. Potential

DNA barcode regions were found in the mitochondrial DNA, especially the mitochondrial 16S rDNA. Several metabarcoding studies have shown that the sequences of the 16S rDNA gene are suitable as barcodes for species identification. Since we have already used a barcode region of the mitochondrial 16S rDNA to identify mammals and poultry [47], this marker gene was chosen as the DNA barcode for our assay.

Since the DNA metabarcoding method for bivalves should be compatible with the DNA metabarcoding method for mammalian and poultry species published recently [47], the primers should anneal at the same temperature (62 °C). In addition, the PCR cycle number should be limited to 25 and DNA libraries should be sequenced with Illumina reagent kits in the 300-cycle format. Due to high sequence variability between closely related bivalve species, none of the primer sets designed enabled obtaining a PCR product for each of the bivalve species of interest. Thus, we continued by designing three primer sets, one for each of the three bivalve families, Pectinidae, Ostreidae, and Mytilidae. Primer pairs consisting of one forward and one reverse primer allowed amplifying the DNA barcode region in scallop and oyster species (Table 2). However, in the case of mussels, a primer set consisting of one forward primer and two reverse primers (Table 2) was necessary to obtain a PCR product for the mussel species listed in Table 1. Figure 1 shows an alignment of selected DNA barcode sequences for the commercially most relevant bivalve species. The alignment of the 90 bivalve species is shown in Supplementary Figure S1. Blue, green, and red bars indicate the binding sites of the primers for Pectinidae, Ostreidae and Mytilidae, respectively. With the three primer sets, PCR products differing in at least one base should be obtained for all bivalve species of interest.

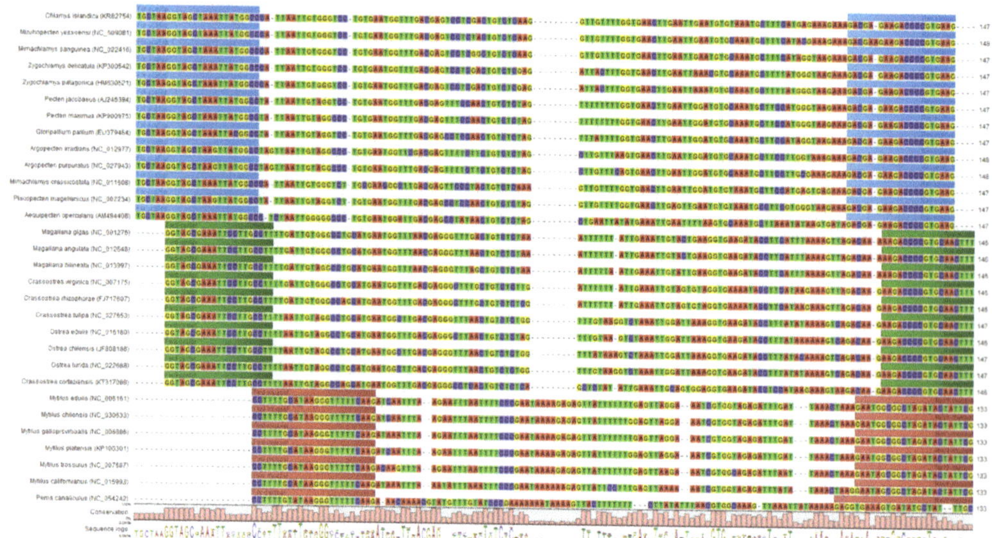

Figure 1. Multi-species sequence alignment of the mitochondrial 16S rDNA barcoding region for bivalve species. Colored bars indicate the binding sites of the primer sets for scallops (blue), oysters (green), and mussels (red, CLC Genomics Workbench software version 10.1.1, Qiagen, Hilden, Germany).

Further sequence alignments indicated that the DNA barcode region selected does not allow distinguishing between all species of the following genera: *Chlamys* spp., *Euvola* spp., *Pecten* spp., *Crassostrea* spp., *Magallana* spp., *Ostrea* spp. and *Saccostrea* spp. These species cannot be distinguished: *Chlamys rubida* and *Chlamys behringiana*; *Pecten albicans*, *Pecten fumatus*, *Pecten jacobaeus*, *Pecten keppelianus*, *Pecten novaezelandiae*, *Pecten sulcicostatus*, *Crassostrea hongkongensis*, and *Crassostrea rivularis*; *Ostrea angelica* and *Ostrea lurida*; as well as *Ostrea permollis* and *Ostrea puelchana*; and *Saccostrea echinata*, *Saccostrea glomerata*, and

Saccostrea mytiloides. In addition, two mussel species, *Mytilus platensis* and *Mytilus chilensis*, can also not be distinguished (for *Mytilus platensis* only one DNA sequence entry was in the public databases provided by NCBI). However, differentiation at the genus level (*Chlamys* spp., *Pecten* spp., *Crassostrea* spp., *Ostrea* spp., *Mytilus* spp.) is sufficient according to the "Codex Alimentarius Austriacus" chapter B35 [56].

When we tested the primers in singleplex PCR assays, for each of the reference samples a PCR product of about 150 bp in length was obtained by increasing the concentration of the forward primer for mussels to 0.4 μM and keeping the concentration of the other six primers at 0.2 μM. In addition, we tested whether the seven primers could be combined to a triplex system. PCR products for the bivalve species of interest were obtained in one and the same vial by increasing the $MgCl_2$ concentration to a final concentration of 3 mM. Thus, we achieved our objective to perform the triplex PCR assay in combination with the previously published DNA metabarcoding assay for mammalian and poultry species [47].

3.2. Library Preparation, Pooling of Libraries, and Sequencing

Library preparation, pooling of 5 or 7 μL per normalized DNA library, and the sequencing process were performed as described previously [47]. However, in case of the pooling process, all DNA libraries were mixed in equal volumes as recommended by the manufacturer's instruction. In our previous study, different volumes from individual DNA libraries were taken to achieve sufficient sequencing depth for minor components. For sample pooling to the maximum of 96 libraries, more than 100,000 NGS reads per sample were expected to be obtained using the 300-cycle MiSeq® Reagent Kit v2.

Sequencing runs were performed in triplicate and the average run metrics were as follows: cluster density (969 K/mm^2) on the flow cell, cluster passing filter (70.22%) as well as the Q-scores (Q30) for read 1 and read 2 were 92.6% and 89.28%, respectively. A total of 5.02% of the total reads were identified as PhiX control sequences with an error rate of 1.49%.

3.3. Analysis of DNA Extracts from Reference Samples

PCR products were obtained for each of the reference samples and sequencing results for those samples are summarized in Table 3. The table shows mean values of the total number of raw reads, the total number of reads that passed the analysis pipeline in Galaxy as well as the total number and percentage of reads that were assigned correctly to the eleven species (based on six replicates).

No significant differences were observed in the total number of reads (before data analysis process) between these species, except *Mytilus galloprovincialis* (162843), *Perna canaliculus* (169631), and *Mytilus edulis* (134500). With the exception of *Perna canaliculus*, >70% of the reads passed the amplicon analysis workflow. All three mussel species, six scallop species and two oyster species could be identified with this workflow at a high rate (>97.5%), except *Mytilus edulis*.

3.4. Analysis of DNA Extract Mixtures

Six ternary DNA extract mixtures were analyzed containing the DNA of the three bivalve families Pectinidae, Ostreidae, and Mytilidae in ratios of 98.0:1.5:0.5 (*v/v/v*). The composition of the DNA extract mixtures and the results obtained by DNA metabarcoding are summarized in Table 4. The total number of raw reads *ranged* from 80856 to 159,737 and the reads that passed the workflow were in the range from 65961 to 147196. For the main components (98.0%), the number of reads assigned correctly ranged from 62434 to 140147. In addition, both minor components (1.5% and 0.5%) could be identified. The number of reads assigned correctly was in the range from 1710 to 4356 and 555 to 1478, respectively.

Table 3. Results for DNA extracts from reference samples. Numbers are mean values ($n = 6$, three sequencing runs, two replicates per run).

| Sample ID | Declaration on the Product | | Species Identified | Total Number of Raw Reads | Total Number of Reads Passing the Workflow | Number of Reads Assigned Correctly | Percentage of Reads Assigned Correctly (%) |
	Scientific/Latin Name	Product Description [Eng]					
O2	*Ostrea edulis*	Oyster	*Ostrea edulis*	78559	63491	61875	97.46
O3	*Crassostrea gigas* *	Oyster	*Magallana gigas* *	76143	65389	64125	98.07
M12	*Mytilus galloprovincialis*	Blue Mussel	*Mytilus galloprovincialis*	162843	150678	149315	99.09
M13	*Perna canaliculus*	New Zealand green-lipped mussel	*Perna canaliculus*	169631	104861	103350	98.56
M27	*Mytilus edulis*	Mussels in marinade	*Mytilus edulis*	134500	120686	105024	87.02
S42	*Mizuhopecten yessoensis*	Yesso scallop	*Mizuhopecten yessoensis*	75927	58069	57058	98.26
S46	*Pecten jacobaeus*	Great scallop	*Pecten* spp.	79472	61484	60514	98.42
S47	*Zygochlamys patagonica*	Scallop "á la Bretonne"	*Zygochlamys patagonica*	77747	59245	58429	98.62
S49	*Placopecten magellanicus*	Great scallop	*Placopecten magellanicus*	79131	61531	60886	98.95
S50	*Argopecten purpuratus*	Pacific scallop	*Argopecten purpuratus*	77383	55455	54588	98.44
S55	*Aequipecten opercularis*	Scallop in sauce	*Aequipecten opercularis*	79141	56064	55800	99.53

* former nomenclature, synonym for *Magallana gigas*.

Table 4. Results for ternary DNA extract mixtures representing the three bivalve families of interest. DNA extracts (5 ng/μL) were mixed in a ratio of 98.0:1.5:0.5 ($v/v/v$). Numbers are mean values ($n = 9$, three sequencing runs, three replicates per run).

| Proportion | | | Total Number of Raw Reads | Total Number of Reads Passing the Workflow | Reads Assigned Correctly | | | | | |
Species 1 (98%)	Species 2 (1.5%)	Species 3 (0.5%)			Species 1	(%)	Species 2	(%)	Species 3	(%)
Magallana gigas	*Mytilus galloprovincialis*	*Pecten* spp.	80856	69506	66430	95.57	1985	2.86	658	0.95
Magallana gigas	*Pecten* spp.	*Mytilus galloprovincialis*	89552	76669	73114	95.36	2182	2.85	894	1.17
Pecten spp.	*Magallana gigas*	*Mytilus galloprovincialis*	88971	69682	66291	95.13	1710	2.45	922	1.32
Pecten spp.	*Mytilus galloprovincialis*	*Magallana gigas*	84085	65961	62434	94.65	2281	3.46	555	0.84
Mytilus galloprovincialis	*Pecten* spp.	*Magallana gigas*	159737	147196	140147	95.21	4356	2.96	1478	1.00
Mytilus galloprovincialis	*Magallana gigas*	*Pecten* spp.	147443	136629	130986	95.87	3304	2.42	1156	0.85

In addition, we analyzed three DNA extract mixtures consisting of DNA from species belonging to one bivalve family (Table 5). The mixtures contained DNA from a scallop or mussel species, respectively. DNA from other bivalve species was present in a proportion of 1.0% each. Both species being present as main components, *Placopecten magellanicus* and *Perna canaliculus*, could be identified, with the number of reads assigned correctly ranging from 58156 to 77483. However, quite different numbers of reads were correctly assigned to the minor components, ranging from 626 (*Mizuhopecten yessoensis*) to 50,391 (*Mytilus galloprovincialis*). *Aequipecten opercularis* was the only minor component that could not be detected.

Table 5. Results for DNA extract mixtures representing one bivalve family. DNA from minor components was present in a proportion of 1% each. In addition, results for a DNA extract mixture containing DNA from a squid species (*Sepiella inermis*) as main component (97.0%) and DNA from three bivalve species (1% each) is shown. Numbers are mean values (*n* = 9, three sequencing runs, three replicates per run).

Main Component	Minor Component (1.0% Each)	Total Number of Raw Reads	Total Number of Reads Passed the Workflow	Reads Assigned Correctly	Percentage of Reads Assigned Correctly (%)
Placopecten magellanicus				58156	88.86
	Mizuhopecten yessoensis			626	0.96
	Pecten spp.	83526 *	65446	817	1.25
	Zygochlamys patagonica			4534	6.93
	Argopecten purpuratus			663	1.01
	Aequipecten opercularis			35	0.05
Placopecten magellanicus				63628	95.41
	Magallana gigas	84282 *	66691	1298	1.95
	Ostrea edulis			1088	1.63
Perna canaliculus				77483	60.12
	Mytilus galloprovincialis	179227 *	128882	50391	39.10
	Mytilus edulis			824	0.64
Sepiella inermis					
	Placopecten magellanicus			31424	51.17
	Ostrea edulis	78467	61415	28162	45.86
	Perna canaliculus			806	1.31

* Number of values (*n* = 6, three sequencing runs, two replicates per run).

We analyzed a further DNA extract mixture containing DNA from the squid species *Sepiella inermis* as main component (97.0%) and DNA from the bivalve species *Placopecten magellanicus*, *Ostrea edulis*, and *Perna canaliculus* as minor components (1.0% each). As expected, in this mixture, the main component could not be detected because the primers are not suitable for amplification of the target region for *Sepiella inermis*. 31424, 28162, and 806 reads, respectively, were assigned correctly to the three bivalve species.

In our previous metabarcoding study [47], individual DNA libraries were pooled in different ratios to achieve sufficient sequencing depth for minor components. The present study demonstrates, that minor components down to a proportion of 0.5% could be identified and differentiated although DNA libraries were pooled by mixing them in equal volumes. DNA extracts from reference samples and DNA extract mixtures most frequently resulted in less than 100,000 reads. However, for all samples on average > 75000 raw reads were obtained, which turned out to be sufficient for reliable species identification.

3.5. Analysis of Commercial Seafood Samples

In order to investigate the applicability of the DNA metabarcoding method to foodstuffs, DNA extracts from 75 commercial food products were analyzed. According to declaration, eight samples (O1 and O4–O10) contained oyster species, 27 samples (M11, M14–M26, and M28–M40) mussel species, 15 samples (S41, S43–45, S48, S51–S55, and S56–S61) scallop species and 25 samples (Mi62–Mi86) were mixed-species seafood products (Table 6). The ingredient list of 30 out of 75 food products did not give any information on the bivalve species. A total of 39 samples were declared to contain "*Crassostrea gigas*", "*Mytilus galloprovincialis*", "*Mytilus chilensis*", "*Mytilus edulis*", "*Zygochlamys patagonica*", "*Chlamys opercularis*", "*Placopecten magellanicus*", "*Pecten maximus*", or "*Patinopecten yessoensis*". The remaining samples (*n* = 6) were labelled with "*Mytilus* spp." or "*Pecten* spp.".

Table 6. Results obtained for commercial seafood samples. Samples listed above the double line were sequenced with the MiSeq® (three sequencing runs, one replicate per run, numbers are mean values); samples listed below the double line were sequenced either with the MiSeq® or the iSeq®.

Sample ID	Declaration on the Product		Species Identified	Total Number of Raw Reads	Total Number of Reads Passed the Workflow	Reads Assigned Correctly	Percentage of Reads Assigned Correctly (%)
	Scientific/Latin Name	Product Description [Eng]					
O5	*Crassostrea gigas* [4]	Oyster in sunflower oil	*Magallana gigas* [4]	76930 [1]	65728	64369	97.93
O6	*Crassostrea gigas* [4]	Oyster in sunflower oil	*Magallana gigas* [4]	44848 [1]	38547	37610	97.57
O7	*Crassostrea gigas* [4]	Oyster in water	*Magallana gigas* [4]	76247	64917	63700	98.13
O8	not declared	Oyster sauce	*Saccostrea malabonensis* *Magallana bilineata*	14470	11658	5442 4652	46.68 39.91
M23	not declared	Mussel with sherry vinegar	*Mytilus galloprovincialis*	33517	30794	30358	98.58
M25	not declared	Mussel in marinade sauce	*Mytilus galloprovincialis*	163188	151688	150700	99.35
M26	not declared	Grilled blue mussel	*Mytilus galloprovincialis*	163106	151608	150433	99.23
M29	*Mytilus galloprovincialis*	Blue mussel in tomato sauce	*Mytilus galloprovincialis* *Mytilus edulis*	153435	140475	132354 7937	94.22 5.65
M30	*Mytilus galloprovincialis*	Blue mussel A la mariniere	*Mytilus galloprovincialis* *Mytilus edulis*	185479	171890	170624 1156	99.26 0.67
M31	not declared	Blue mussel in organic marinade	*Mytilus galloprovincialis* *Mytilus edulis*	170303	158379	157015 1267	99.14 0.80
M32	not declared	Marinated blue mussel	*Mytilus galloprovincialis* *Mytilus edulis*	159181	144788	143399 1308	99.04 0.90
M33	*Mytilus chilensis*	Mussel in Escabeche	*Mytilus galloprovincialis* *Mytilus edulis*	167903	151219	118879 31737	78.61 20.99
M34	*Mytilus chilensis*	Mussel	*Mytilus galloprovincialis* *Mytilus edulis*	152112	138768	87964 49601	63.39 35.74
M36	*Mytilus galloprovincialis*	Blue mussel marinated	*Mytilus galloprovincialis* *Mytilus edulis*	176963	163721	162224 1323	99.09 0.81
M37	*Mytilus edulis*	Mussel in honey mustard sauce	*Mytilus galloprovincialis* *Mytilus edulis*	149364	136868	135249 1400	98.82 1.02
M38	not declared	Blue mussel in marinade	*Mytilus galloprovincialis* *Mytilus edulis*	138801	127244	125980 1056	99.01 0.83
S58	not declared	Rillettes de Saint-Jacques	*Aequipecten opercularis* *Mytilus galloprovincialis*	62787	44307	42716 1330	96.41 3.00
S59	not declared	Small scallop in galician sauce	*Aequipecten opercularis*	82550	59722	58296	97.61
Mi62	*Mytilus chilensis*	Seafood mix	*Mytilus galloprovincialis* *Mytilus edulis*	618324	569815	433439 134543	76.07 23.61
Mi63	not declared	Sauce with seafood	*Mytilus edulis* *Mytilus galloprovincialis*	152170	139306	73550 64729	52.80 46.47
Mi64	*Mytilus chilensis* *Mytilus edulis*	Seafood mix	*Mytilus galloprovincialis* *Mytilus edulis*	131285	119350	81590 37211	68.36 31.18
Mi65	not declared	Bouillabaisse Marseille	*Mytilus galloprovincialis* *Mytilus edulis*	157311	143479	138535 4777	96.55 3.33
Mi66	*Mytilus chilensis*	Seafood mix	*Mytilus galloprovincialis* *Mytilus edulis*	152535	140047	92024 47415	65.71 33.86
Mi67	*Mytilus* spp.	Seafood mix	*Mytilus galloprovincialis* *Mytilus edulis*	76544	69081	48275 20459	69.88 29.62
Mi68	*Mytilus galloprovincialis*	Sea fruit salad in sunflower oil	*Mytilus galloprovincialis* *Mytilus edulis*	157861	145671	144468 1046	99.17 0.72
Mi69	*Mytilus chilensis*	Seafood mix	*Mytilus galloprovincialis* *Mytilus edulis*	140227	128007	85679 41686	66.93 32.57
Mi70	not declared	*Sea fruit salad fantasy*	*Mytilus galloprovincialis* *Mytilus edulis*	120677	106674	101121 5413	94.80 5.07
Mi71	not declared	Seafood mix	*Mytilus galloprovincialis* *Mytilus edulis*	160546	147278	79680 66675	54.10 45.27

Table 6. *Cont.*

Sample ID	Declaration on the Product		Species Identified	Total Number of Raw Reads	Total Number of Reads Passed the Workflow	Reads Assigned Correctly	Percentage of Reads Assigned Correctly (%)
	Scientific/Latin Name	Product Description [Eng]					
Mi72	*Mytilus chilensis*	Seafood mix	*Mytilus galloprovincialis* *Mytilus edulis*	160059	146539	91557 54271	62.48 37.03
Mi73	not declared	Seafood mix	*Mytilus edulis* *Mytilus galloprovincialis*	150500	137634	78942 57608	57.36 41.86
Mi74	not declared	Seafood mix	*Mytilus galloprovincialis* *Mytilus edulis*	168841	155701	79035 75612	50.76 48.56
Mi75	not declared	Pizza Frutti di mare	*Mytilus galloprovincialis* *Mytilus edulis*	181822 [1]	172620	95184 71440	55.14 41.39
Mi76	not declared	Paella	*Mytilus galloprovincialis* *Mytilus edulis*	150431	139511	138335 1070	99.16 0.77
Mi77	*Mytilus edulis, Mytilus chilensis*	Paella	*Mytilus galloprovincialis* *Mytilus edulis*	141816	132092	130768 1242	99.00 0.94
Mi78	*Mytilus chilensis*	Seafood all'Olio	*Mytilus galloprovincialis* *Mytilus edulis*	134717	122906	73482 48774	59.79 39.68
Mi79	*Mytilus chilensis*	Seafood mix	*Mytilus galloprovincialis* *Mytilus edulis*	148773	137122	73035 63249	53.26 46.13
Mi80	*Mytilus chilensis*	Seafood mix	*Mytilus galloprovincialis* *Mytilus edulis*	136695	126608	88130 37970	69.61 29.99
Mi81	not declared	Sea fruit salad	*Mytilus galloprovincialis* *Mytilus edulis*	153499	142736	141578 1022	99.19 0.72
Mi82	*Zygochlamys patagonica* *Chlamys opercularis*	Scallop terrine	*Zygochlamys patagonica*	76554	59181	57329	96.87
Mi83	not declared	Terrine of salmon and great scallop	*Pecten* spp.	96596 [1]	76834	75476	98.23
Mi84	*Mytilus chilensis*	Seafood mix	*Mytilus galloprovincialis* *Mytilus edulis*	163885	150852	124468 25916	82.51 17.18
Mi85	not declared	Instant noodle seafood, mild	*Mytilus galloprovincialis*	15409	14118	13750	97.39
Mi86	not declared	Instant noodle seafood, spicy	*Mytilus galloprovincialis*	9787	8892	8473	95.29
O1	*Crassostrea gigas* [4]	Oyster	*Magallana gigas* [4]	139319 [2]	134073	133493	99.57
O4	not declared	Oyster	*Magallana gigas*	46089 [2]	40991	40279	98.26
O9	not declared	Oyster sauce	not evaluable [3]				
O10	not declared	Oyster sauce	not evaluable [3]				
M11	*Mytilus edulis*	Mussel	*Mytilus galloprovincialis*	23766 [2]	22546	22147	98.23
M14	*Mytilus* spp.	Blue mussel	*Mytilus galloprovincialis* *Mytilus edulis*	126880 [2]	119717	79522 39555	66.42 33.04
M15	*Mytilus* spp	Blue mussel	*Mytilus galloprovincialis*	227678	220699	220226	99.79
M16	*Mytilus edulis*	Bouchot mussel	*Mytilus galloprovincialis*	51292 [2]	49604	48832	98.44
M17	not declared	Grilled blue mussel	*Mytilus galloprovincialis* *Mytilus edulis*	9888 [2]	6750	3956 1998	58.61 29.60
M18	*Mytilus chilensis*	Blue mussel	*Mytilus galloprovincialis*	53710 [2]	51670	50733	98.19
M19	not declared	Blue mussel	*Mytilus galloprovincialis*	57238 [2]	54822	53829	98.19
M20	*Mytilus* spp.	Blue mussel	*Mytilus galloprovincialis*	72113 [2]	69576	68969	99.13
M21	*Mytilus edulis*	Mussel	*Mytilus galloprovincialis*	51328 [2]	49908	49459	99.10
M22	*Mytilus galloprovincialis*	Blue mussel	*Mytilus galloprovincialis* *Mytilus edulis*	115950[2]	110777	109262 1466	98.63 1.32
M24	*Mytilus chilensis*	Blue mussel in tomato sauce	*Mytilus galloprovincialis* *Mytilus edulis*	113942 [2]	107150	94449 12505	88.15 11.67
M28	not declared	Dry cat food with green lipped mussel	*Pecten* spp. *Mytilus galloprovincialis* *Perna canaliculus*	128693 [3]	126380	79764 40450 4712	63.11 32.01 3.73
M35	*Mytilus chilensis*	Mussel in tomato sauce	*Mytilus galloprovincialis*	197899 [3]	190771	189540	99.35
M39	*Mytilus chilensis*	Blue mussel	*Mytilus galloprovincialis* *Mytilus edulis*	182612 [3]	175982	96502 75204	54.84 42.73
M40	*Mytilus edulis*	Blue mussel	*Mytilus galloprovincialis*	182958 [3]	179399	178024	99.23

Table 6. *Cont.*

| Sample ID | Declaration on the Product | | Species Identified | Total Number of Raw Reads | Total Number of Reads Passed the Workflow | Reads Assigned Correctly | Percentage of Reads Assigned Correctly (%) |
	Scientific/Latin Name	Product Description [Eng]					
S41	*Placopecten magellanicus*	Deep-sea scallop	*Placopecten magellanicus*	143794 [2]	132140	131583	99.58
S43	*Pecten maximus*	Great scallop	*Mizuhopecten yessoensis*	122156 [2]	113706	113128	99.49
S44	*Pecten* spp.	Great scallop	*Mizuhopecten yessoensis*	2873135 [2]	2718126	2717426	99.97
S45	*Placopecten magellanicus*	Deep-sea scallop	*Placopecten magellanicus*	111673 [2]	107119	106632	99.55
S48	*Patinopecten yessoensis*	Great scallop/ Yesso scallop	*Mizuhopecten yessoensis*	47397 [2]	41076	407873	99.51
S51	not declared	Great scallop	*Placopecten magellanicus*	51565 [2]	45007	44915	99.80
S52	*Patinopecten yessoensis*	Great scallop	*Mizuhopecten yessoensis*	46673 [2]	39769	39627	99.64
S53	*Pecten* spp.	Great scallop	*Mizuhopecten yessoensis*	42857 [2]	36443	35265	96.77
S54	*Placopecten magellanicus*	Great scallop	*Placopecten magellanicus*	55475 [2]	48703	47915	98.38
S56	not declared	Great scallop	*Placopecten magellanicus*	1268169 [3]	1061137	1060653	99.95
S57	*Placopecten magellanicus*	Great scallop	*Pecten* spp.	174497 [3]	171299	170404	99.48
S60	not declared	Deep-sea scallop	*Placopecten magellanicus*	364474 [3]	350953	350869	99.98
S61	*Patinopecten yessoensis*	Great scallop	*Mizuhopecten yessoensis*	159145 [3]	152930	152849	99.95

[1] Mean of two replicates; [2] samples were analyzed with the MiSeq® instrument; [3] samples were analyzed with the iSeq® instrument; [4] former nomenclature, synonym for *Magallana gigas*.

Our results indicate that DNA metabarcoding by targeting the 16S rDNA barcode region of about 150 bp in length is applicable to complex and highly processed foodstuffs. The barcode region could be amplified and sequenced even in products such as Bouillabaisse, Paella, and instant noodle seafood. Oyster sauce was the only sample matrix for which PCR amplification and consequently sequencing failed. Failure of obtaining PCR products for oyster sauce has already been reported by Chin Chin et al. [50], most probably caused by excessive DNA fragmentation due to industrial processing.

Three oyster species (*Saccostrea malabonensis*, *Magallana bilineata*, *Magallana gigas*), three mussel species (*Mytilus galloprovincialis*, *Mytilus edulis*, *Perna canaliculus*), and three scallop species (*Aequipecten opercularis*, *Placopecten magellanicus*, *Pecten* spp.) were detected in food products (O4, O8, M17, M19, M23, M25, M26, M28, M31, M32, M35, M38–M40, S51, S56, S58–S60, Mi63, Mi65, Mi70, Mi71, Mi73–Mi76, Mi81, Mi83, Mi85, and Mi86) although they were not declared on the label.

In each of the six oyster products that could be subjected to sequencing (O1, O4–O8), *Magallana gigas* was identified. *Magallana gigas* is by far the predominant oyster species farmed in the EU [59].

In 21 products (M11, M16, M18, M21, M24, M33–M35, M37, M39, M40, Mi62, Mi64, Mi66, Mi69, Mi72, Mi77–Mi80, and Mi84), the mussel species *Mytilus galloprovincialis* was detected. In addition to *Mytilus galloprovincialis*, *Mytilus edulis* was identified (percentage of reads assigned correctly >1%) in 13 products (M24, M33, M34, M39, Mi62, Mi64, Mi66, Mi69, Mi72, Mi78–Mi80, and Mi84). In four products, *Mytilus edulis* could not be detected although it was declared on the label. *Mytilus galloprovincialis* and *Mytilus edulis* are the two mussel species most frequently cultivated in European mussel farms [59]. In none of the products declared to contain *Mytilus chilensis*, *Mytilus chilensis* was detected. Instead of *Mytilus chilensis*, imported to EU countries from Chile [60], *Mytilus galloprovincialis* and/or *Mytilus edulis* were identified. According to the multi-species sequence alignment shown in Figure 1, the barcode region should allow distinguishing the three Mytilus species.

Placopecten magellanicus and *Patinopecten yessoensis* were listed as ingredients in samples S41, S45, S54, and S57 and samples S48, S52, and S61, respectively. Our results confirmed the presence of these two species, except for sample S57. In sample S43, declared to contain *Pecten maximus*, the species *Mizuhopecten yessoensis* was detected. In sample S44 and S53, declared as *Pecten* spp., the species *Mizuhopecten yessoensis* was also identified. In line with previous studies, most products declared to contain "Jakobsmuschel" did not contain a species of the genus *Pecten* [15,18,19]. Instead, we identified *Placopecten magellanicus* or *Mizuhopecten yessoensis*.

4. Conclusions

The DNA metabarcoding method developed in this study allows the detection of species of Mytilidae (mussels), Pectinidae (scallops), and Ostreidae (oysters), the most important bivalve families for human consumption. By combining three forward and four reverse primers in a triplex PCR assay, the barcode region, a fragment of mitochondrial 16S rDNA, could be amplified in the species of interest.

The applicability of the novel DNA metabarcoding method was investigated by analyzing individual DNA extracts from eleven reference samples, ten DNA extract mixtures and DNA extracts from 75 commercial food products. In each of the eleven reference samples, the bivalve species was identified correctly. In DNA extract mixtures, not only the main component but also the minor components were detected correctly, with just a few exceptions. The analysis of commercial seafood products showed that the DNA metabarcoding method is applicable to complex and processed foodstuffs, allowing the identification of bivalves in, e.g., marinated form, in sauces, in seafood mixes and even in instant noodle seafood.

The DNA metabarcoding method runs on both the MiSeq® and iSeq® instrument of Illumina. Due to the compatibility of PCR and sequencing parameters, the DNA metabarcoding method can be combined with a DNA metabarcoding method for mammalian and poultry species published recently.

5. Patent

This manuscript has been submitted for grant of a European patent (application number: EP21204456.4).

Supplementary Materials: The following are available online at https://www.mdpi.com/article/10.3390/foods10112618/s1, Supplementary Table S1: Declaration, origin and processing condition of the 86 food products, Supplementary Table S2: Sequences included into the reverence database, Supplementary Figure S1: Multi-species sequence alignment of the mitochondrial 16S rDNA barcoding region for the bivalve species of interest.

Author Contributions: Conceptualization, R.H. and S.D.; methodology, K.G., M.C.-M., M.W., R.H., S.D. and V.P.; software, S.D.; formal analysis, K.G. and S.D.; investigation, K.G. and S.D.; resources, K.G.; data curation, K.G. and S.D.; writing—original draft preparation, K.G.; writing—review and editing, A.L., M.C.-M., M.W., R.H., S.D. and V.P.; visualization, K.G.; supervision, M.C.-M., M.W., R.H., S.D. and V.P.; project administration, R.H. and S.D.; funding acquisition, A.L. and V.P. All authors have read and agreed to the published version of the manuscript.

Funding: This research was funded within a research project of the Austrian Competence Centre for Feed and Food Quality, Safety and Innovation (FFoQSI GmbH). The COMET-K1 competence centre FFoQSI is funded by the Austrian ministries BMK, BMDW and the Austrian provinces Lower Austria, Upper Austria and Vienna within the scope of COMET-Competence Centers for Excellent Technologies. The programme COMET is handled by the Austrian Research Promotion Agency FFG.

Data Availability Statement: The datasets generated for this study are available on request to the corresponding author.

Acknowledgments: This research was supported by the Austrian Agency for Health and Food Safety (AGES), Institute for Food Safety Vienna, Department for Molecular Biology and Microbiology and by LVA GmbH in cooperation with the University of Vienna and the University of Veterinary Medicine Vienna.

Conflicts of Interest: The authors declare no conflict of interest.

References

1. Telahigue, K.; Chetoui, I.; Rabeh, I.; Romdhane, M.S.; El Cafsi, M. Comparative fatty acid profiles in edible parts of wild scallops from the Tunisian coast. *Food Chem.* **2010**, *122*, 744–746. [CrossRef]
2. Tan, K.; Ma, H.; Li, S.; Zheng, H. Bivalves as future source of sustainable natural omega-3 polyunsaturated fatty acids. *Food Chem.* **2020**, *311*, 125907. [CrossRef]
3. Manthey-Karl, M.; Lehmann, I.; Ostermeyer, U.; Rehbein, H.; Schröder, U. Meat Composition and Quality Assessment of King Scallops (Pecten maximus) and Frozen Atlantic Sea Scallops (*Placopecten magellanicus*) on a Retail Level. *Foods* **2015**, *4*, 524–546. [CrossRef]
4. Grienke, U.; Silke, J.; Tasdemir, D. Bioactive compounds from marine mussels and their effects on human health. *Food Chem.* **2014**, *142*, 48–60. [CrossRef]
5. Willer, D.F.; Aldridge, D.C. Sustainable bivalve farming can deliver food security in the tropics. *Nat. Food* **2020**, *1*, 384–388. [CrossRef]
6. World Register of Marine Species. Available online: http://www.marinespecies.org/index.php (accessed on 12 December 2020).
7. Food and Agricultural Organization of the United Nations (FAO). Fishery and Aquaculture Statistics. *Global Aquaculture Production. Software for Fishery Statistical Time Series. Version FishStatJ v4.01.4.* Available online: www.fao.org/fishery/statistics/software/fishstatj/en (accessed on 10 June 2021).
8. European Parliament and of the Council. Regulation (EU) No. 1379/2013 of the European Parliament and of the Council of 11 December 2013 on the common organisation of the markets in fishery and aquaculture products, amending Council Regulations (EC) No 1184/2006 and (EC) No 1224/2009 and repealing Council Regulation (EC) No 104/2000. *Off. J. Eur. Union* **2013**, *L 354*, 1–21.
9. European Parliament and of the Council. Regulation (EU) No 1169/2011 of the European Parliament and of the Council of 25 October 2011 on the provision of food information to consumers, amending Regulations (EC) No 1924/2006 and (EC) No 1925/2006 of the European Parliament and of the Council, and repealing Commission Directive 87/250/EEC, Council Directive 90/496/EEC, Commission Directive 1999/10/EC, Directive 2000/13/EC of the European Parliament and of the Council, Commission Directives 2002/67/EC and 2008/5/EC and Commission Regulation (EC) No 608/2004. *Off. J. Eur. Union* **2011**, *L 304*, 18–63.
10. Oceana; Protectin the World´s Oceans. Available online: https://oceana.org/publications/reports/one-name-one-fish-why-seafood-names-matter (accessed on 30 April 2021).
11. Rodríguez, E.M.; Ortea, I. Food authentication of seafood species. In *Proteomics in Food Science: From Farm to Fork*, 3rd ed.; Colgrave, M.L., Ed.; Academic Press: London, UK, 2017; pp. 331–342. ISBN 9780128040072.
12. Parrondo, M.; López, S.; Aparicio-Valencia, A.; Fueyo, A.; Quintanilla-García, P.; Arias, A.; Borrell, Y.J. Almost never you get what you pay for: Widespread mislabeling of commercial "zamburiñas" in northern Spain. *Food Control* **2021**, *120*, 107541. [CrossRef]
13. Giusti, A.; Tosi, F.; Tinacci, L.; Guardone, L.; Corti, I.; Arcangeli, G.; Armani, A. Mussels (*Mytilus* spp.) products authentication: A case study on the Italian market confirms issues in species identification and arises concern on commercial names attribution. *Food Control* **2020**, *118*, 107379. [CrossRef]
14. Harris, D.J.; Rosado, D.; Xavier, R. DNA Barcoding Reveals Extensive Mislabeling in Seafood Sold in Portuguese Supermarkets. *J. Aquat. Food Prod. Technol.* **2016**, *25*, 1375–1380. [CrossRef]
15. Näumann, G.; Stumme, B.; Rehbein, H. Differenzierung von Kammmuscheln durch DNA-Analyse. *Inf. Aus Der Fischereiforschung-Inf. Fish. Res.* **2012**, *59*, 1–7. [CrossRef]
16. Espiñeira, M.; González-Lavín, N.; Vieites, J.M.; Santaclara, F.J. Development of a method for the genetic identification of commercial bivalve species based on mitochondrial 18S rRNA sequences. *J. Agric. Food Chem.* **2009**, *57*, 495–502. [CrossRef]
17. Guardone, L.; Tinacci, L.; Costanzo, F.; Azzarelli, D.; D'Amico, P.; Tasselli, G.; Magni, A.; Guidi, A.; Nucera, D.; Armani, A. DNA barcoding as a tool for detecting mislabeling of fishery products imported from third countries: An official survey conducted at the Border Inspection Post of Livorno-Pisa (Italy). *Food Control* **2017**, *80*, 204–216. [CrossRef]
18. Klapper, R.; Schröder, U. Verification of authenticity: A rapid identification method for commercial scallop species through multiplex real-time PCR. *Food Control* **2021**, *121*, 107574. [CrossRef]
19. Stephan, R.; Johler, S.; Oesterle, N.; Näumann, G.; Vogel, G.; Pflüger, V. Rapid and reliable species identification of scallops by MALDI-TOF mass spectrometry. *Food Control* **2014**, *46*, 6–9. [CrossRef]
20. Fernández, A.; García, T.; Asensio, L.; Rodríguez, M.Á.; González, I.; Céspedes, A.; Hernández, P.E.; Martín, R. Identification of the clam species *Ruditapes decussatus* (Grooved carpet shell), *Venerupis pullastra* (Pullet carpet shell), and *Ruditapes philippinarum* (Japanese carpet shell) by PCR-RFLP. *J. Agric. Food Chem.* **2000**, *48*, 3336–3341. [CrossRef]
21. Galimberti, A.; de Mattia, F.; Losa, A.; Bruni, I.; Federici, S.; Casiraghi, M.; Martellos, S.; Labra, M. DNA barcoding as a new tool for food traceability. *Food Res. Int.* **2013**, *50*, 55–63. [CrossRef]
22. Littlefair, J.E.; Clare, E.L. Barcoding the food chain: From Sanger to high-throughput sequencing. *Genome* **2016**, *59*, 946–958. [CrossRef]
23. Böhme, K.; Calo-Mata, P.; Barros-Velázquez, J.; Ortea, I. Review of Recent DNA-Based Methods for Main Food-Authentication Topics. *J. Agric. Food Chem.* **2019**, *67*, 3854–3864. [CrossRef]
24. Hebert, P.D.N.; Cywinska, A.; Ball, S.L.; deWaard, J.R. Biological identifications through DNA barcodes. *Proc. Biol. Sci.* **2003**, *270*, 313–321. [CrossRef]

25. Teletchea, F.; Maudet, C.; Hänni, C. Food and forensic molecular identification: Update and challenges. *Trends Biotechnol.* **2005**, *23*, 359–366. [CrossRef] [PubMed]
26. Staats, M.; Arulandhu, A.J.; Gravendeel, B.; Holst-Jensen, A.; Scholtens, I.; Peelen, T.; Prins, T.W.; Kok, E. Advances in DNA metabarcoding for food and wildlife forensic species identification. *Anal. Bioanal. Chem.* **2016**, *408*, 4615–4630. [CrossRef] [PubMed]
27. Pollack, S.J.; Kawalek, M.D.; Williams-Hill, D.M.; Hellberg, R.S. Evaluation of DNA barcoding methodologies for the identification of fish species in cooked products. *Food Control* **2018**, *84*, 297–304. [CrossRef]
28. Armani, A.; Tinacci, L.; Lorenzetti, R.; Benvenuti, A.; Susini, F.; Gasperetti, L.; Ricci, E.; Guarducci, M.; Guidi, A. Is raw better? A multiple DNA barcoding approach (full and mini) based on mitochondrial and nuclear markers reveals low rates of misdescription in sushi products sold on the Italian market. *Food Control* **2017**, *79*, 126–133. [CrossRef]
29. Feng, Y.; Li, Q.; Kong, L.; Zheng, X. DNA barcoding and phylogenetic analysis of Pectinidae (Mollusca: Bivalvia) based on mitochondrial COI and 16S rRNA genes. *Mol. Biol. Rep.* **2011**, *38*, 291–299. [CrossRef]
30. Saavedra, C.; Peña, J.B. Phylogenetics of American scallops (Bivalvia: Pectinidae) based on partial 16S and 12S ribosomal RNA gene sequences. *Mar. Biol.* **2006**, *150*, 111–119. [CrossRef]
31. Fernandez, A.; García, T.; Gonzalez, I.; Asensio, L.; Rodriguez, M.Á.; Hernández, P.E.; Martin, R. Polymerase chain reaction-restriction fragment length polymorphism analysis of a 16S rRNA gene fragment for authentication of four clam species. *J. Food Prot.* **2002**, *65*, 692–695. [CrossRef]
32. Bendezu, I.F.; Slater, J.W.; Carney, B.F. Identification of Mytilus spp. and Pecten maximus in Irish waters by standard PCR of the 18S rDNA gene and multiplex PCR of the 16S rDNA gene. *Mar. Biotechnol. (NY)* **2005**, *7*, 687–696. [CrossRef]
33. Colombo, F.; Trezzi, I.; Bernardi, C.; Cantoni, C.; Renon, P. A case of identification of pectinid scallop (Pecten jacobaeus, Pecten maximus) in a frozen and seasoned food product with PCR technique. *Food Control* **2004**, *15*, 527–529. [CrossRef]
34. Fernandes, T.J.R.; Amaral, J.S.; Mafra, I. DNA barcode markers applied to seafood authentication: An updated review. *Crit. Rev. Food Sci. Nutr.* **2020**, *61*, 1–32. [CrossRef]
35. Mafra, I.; Ferreira, I.M.P.L.V.O.; Oliveira, M.B.P.P. Food authentication by PCR-based methods. *Eur. Food Res. Technol.* **2008**, *227*, 649–665. [CrossRef]
36. Marín, A.; Fujimoto, T.; Arai, K. Rapid species identification of fresh and processed scallops by multiplex PCR. *Food Control* **2013**, *32*, 472–476. [CrossRef]
37. Marshall, H.D.; Johnstone, K.A.; Carr, S.M. Species-specific oligonucleotides and multiplex PCR for forensic discrimination of two species of scallops, *Placopecten magellanicus* and *Chlamys islandica. Forensic Sci. Int.* **2007**, *167*, 1–7. [CrossRef] [PubMed]
38. Zhan, A.; Hu, J.; Hu, X.; Lu, W.; Wang, M.; Peng, W.; Hui, M.; Bao, Z. Fast identification of scallop adductor muscles using species-specific microsatellite markers. *Eur. Food Res. Technol.* **2008**, *227*, 353–359. [CrossRef]
39. Meistertzheim, A.-L.; Héritier, L.; Lejart, M. High-Resolution Melting of 18S rDNA sequences (18S-HRM) for discrimination of bivalve's species at early juvenile stage: Application to a spat survey. *Mar. Biol.* **2017**, *164*, 133–141. [CrossRef]
40. Pereira, A.M.; Fernández-Tajes, J.; Gaspar, M.B.; Méndez, J. Identification of the wedge clam *Donax trunculus* by a simple PCR technique. *Food Control* **2012**, *23*, 268–270. [CrossRef]
41. Jilberto, F.; Araneda, C.; Larraín, M.A. High resolution melting analysis for identification of commercially-important *Mytilus* species. *Food Chem.* **2017**, *229*, 716–720. [CrossRef]
42. Rego, I.; Martínez, A.; González-Tizón, A.; Vieites, J.; Leira, F.; Méndez, J. PCR technique for identification of mussel species. *J. Agric. Food Chem.* **2002**, *50*, 1780–1784. [CrossRef]
43. Jin, Y.; Li, Q.; Kong, L.; Yu, H.; Zhong, X. High-resolution melting (HRM) analysis: A highly sensitive alternative for the identification of commercially important *Crassostrea* oysters. *J. Molluscan Stud.* **2015**, *81*, 167–170. [CrossRef]
44. Hellberg, R.S.; Pollack, S.J.; Hanner, R.H. Seafood species identification using DNA sequencing. In *Seafood Authenticity and Traceability*; Academic Press: London, UK, 2016; pp. 113–132. ISBN 9780128015926.
45. Galimberti, A.; Casiraghi, M.; Bruni, I.; Guzzetti, L.; Cortis, P.; Berterame, N.M.; Labra, M. From DNA barcoding to personalized nutrition: The evolution of food traceability. *Curr. Opin. Food Sci.* **2019**, *28*, 41–48. [CrossRef]
46. Günther, B.; Raupach, M.J.; Knebelsberger, T. Full-length and mini-length DNA barcoding for the identification of seafood commercially traded in Germany. *Food Control* **2017**, *73*, 922–929. [CrossRef]
47. Dobrovolny, S.; Blaschitz, M.; Weinmaier, T.; Pechatschek, J.; Cichna-Markl, M.; Indra, A.; Hufnagl, P.; Hochegger, R. Development of a DNA metabarcoding method for the identification of fifteen mammalian and six poultry species in food. *Food Chem.* **2019**, *272*, 354–361. [CrossRef]
48. Frigerio, J.; Agostinetto, G.; Sandionigi, A.; Mezzasalma, V.; Berterame, N.M.; Casiraghi, M.; Labra, M.; Galimberti, A. The hidden 'plant side' of insect novel foods: A DNA-based assessment. *Food Res. Int.* **2020**, *128*, 108751. [CrossRef] [PubMed]
49. Nicolè, S.; Negrisolo, E.; Eccher, G.; Mantovani, R.; Patarnello, T.; Erickson, D.L.; Kress, W.J.; Barcaccia, G. DNA Barcoding as a Reliable Method for the Authentication of Commercial Seafood Products. *Food Technol. Biotechnol.* **2012**, *50*, 387–398.
50. Chin Chin, T.; Adibah, A.B.; Danial Hariz, Z.A.; Siti Azizah, M.N. Detection of mislabelled seafood products in Malaysia by DNA barcoding: Improving transparency in food market. *Food Control* **2016**, *64*, 247–256. [CrossRef]
51. Giusti, A.; Armani, A.; Sotelo, C.G. Advances in the analysis of complex food matrices: Species identification in surimi-based products using Next Generation Sequencing technologies. *PLoS ONE* **2017**, *12*, e0185586. [CrossRef]

52. Giusti, A.; Tinacci, L.; Sotelo, C.G.; Marchetti, M.; Guidi, A.; Zheng, W.; Armani, A. Seafood Identification in Multispecies Products: Assessment of 16SrRNA, cytb, and COI Universal Primers' Efficiency as a Preliminary Analytical Step for Setting up Metabarcoding Next-Generation Sequencing Techniques. *J. Agric. Food Chem.* **2017**, *65*, 2902–2912. [CrossRef]
53. Arulandhu, A.J.; Staats, M.; Hagelaar, R.; Voorhuijzen, M.M.; Prins, T.W.; Scholtens, I.; Costessi, A.; Duijsings, D.; Rechenmann, F.; Gaspar, F.B.; et al. Development and validation of a multi-locus DNA metabarcoding method to identify endangered species in complex samples. *Gigascience* **2017**, *6*, 1–18. [CrossRef] [PubMed]
54. Carvalho, D.C.; Palhares, R.M.; Drummond, M.G.; Gadanho, M. Food metagenomics: Next generation sequencing identifies species mixtures and mislabeling within highly processed cod products. *Food Control* **2017**, *80*, 183–186. [CrossRef]
55. Schirmer, M.; D'Amore, R.; Ijaz, U.Z.; Hall, N.; Quince, C. Illumina error profiles: Resolving fine-scale variation in metagenomic sequencing data. *BMC Bioinform.* **2016**, *17*, 125. [CrossRef] [PubMed]
56. Bundesministerium für Arbeit, Soziales, Gesundheit und Konsumentenschutz. *Codexkapitel/B 35/Fische, Krebse, Weichtiere und dar-aus Hergestellte Erzeugnisse: BMGF-75210/0026-II/B/13/2017*; Bundesministerium für Arbeit, Soziales, Gesundheit und Konsumentenschutz: Vienna, Austria, 2007.
57. Martin, M. Cutadapt removes adapter sequences from high-throughput sequencing reads. *EMBnet J.* **2011**, *17*, 10. [CrossRef]
58. Edgar, R.C. Search and clustering orders of magnitude faster than BLAST. *Bioinformatics* **2010**, *26*, 2460–2461. [CrossRef] [PubMed]
59. European Commission. Directorate General for Maritime Affairs and Fisheries. In *The EU Fish Market*; Publications Office: Brussels, Belgium, 2019.
60. Del Rio-Lavín, A.; Jiménez, E.; Pardo, M.Á. SYBR-Green real-time PCR assay with melting curve analysis for the rapid identification of Mytilus species in food samples. *Food Control* **2021**, *130*, 108257. [CrossRef]

Article

Real-Time PCR Assay for the Detection and Quantification of Roe Deer to Detect Food Adulteration—Interlaboratory Validation Involving Laboratories in Austria, Germany, and Switzerland

Barbara Druml [1,2], Steffen Uhlig [3], Kirsten Simon [3], Kirstin Frost [3], Karina Hettwer [3], Margit Cichna-Markl [2,*] and Rupert Hochegger [1,*]

1   Department of Molecular Biology and Microbiology, Institute for Food Safety Vienna, Austrian Agency for Health and Food Safety (AGES), Spargelfeldstraße 191, 1220 Vienna, Austria; barbara.druml@univie.ac.at
2   Department of Analytical Chemistry, Faculty of Chemistry, University of Vienna, Währinger Straße 38, 1090 Vienna, Austria
3   QuoData GmbH, Prellerstraße 14, 01309 Dresden, Germany; steffen.uhlig@quodata.de (S.U.); kirsten.simon@quodata.de (K.S.); kirstin.frost@quodata.de (K.F.); karina.hettwer@quodata.de (K.H.)
*   Correspondence: margit.cichna@univie.ac.at (M.C.-M.); rupert.hochegger@ages.at (R.H.)

Abstract: Game meat products are particularly prone to be adulterated by replacing game meat with cheaper meat species. Recently, we have presented a real-time polymerase chain reaction (PCR) assay for the identification and quantification of roe deer in food. Quantification of the roe deer content in % (w/w) was achieved relatively by subjecting the DNA isolates to a reference real-time PCR assay in addition to the real-time PCR assay for roe deer. Aiming at harmonizing analytical methods for food authentication across EU Member States, the real-time PCR assay for roe deer has been tested in an interlaboratory ring trial including 14 laboratories from Austria, Germany, and Switzerland. Participating laboratories obtained aliquots of DNA isolates from a meat mixture containing 24.8% (w/w) roe deer in pork, roe deer meat, and 12 meat samples whose roe deer content was not disclosed. Performance characteristics included amplification efficiency, level of detection ($LOD_{95\%}$), repeatability, reproducibility, and accuracy of quantitative results. With a relative reproducibility standard deviation ranging from 13.35 to 25.08% (after outlier removal) and recoveries ranging from 84.4 to 114.3%, the real-time PCR assay was found to be applicable for the detection and quantification of roe deer in raw meat samples to detect food adulteration.

Keywords: real-time PCR; roe deer; game meat; detection; quantification; food authentication; validation; interlaboratory ring trial; probability of detection

Citation: Druml, B.; Uhlig, S.; Simon, K.; Frost, K.; Hettwer, K.; Cichna-Markl, M.; Hochegger, R. Real-Time PCR Assay for the Detection and Quantification of Roe Deer to Detect Food Adulteration—Interlaboratory Validation Involving Laboratories in Austria, Germany, and Switzerland. *Foods* **2021**, *10*, 2645. https://doi.org/10.3390/foods10112645

Academic Editor: Saskia Van Ruth

Received: 18 September 2021
Accepted: 26 October 2021
Published: 1 November 2021

1. Introduction

Game meat is appreciated because of its characteristic sensory properties, especially its distinct flavor and tenderness. In general, game meat is regarded as healthier than meat from domestic species due to its lower intramuscular fat and cholesterol content and its high content of polyunsaturated fatty acids [1]. Like other commercial food products, game meat products must comply with national and international food legal regulations. Hence, game meat products have to be not only safe but also authentic. Adulteration of meat products by complete or partial replacing of more expensive meat with cheaper meat species is, however, known to be a global issue [2–5]. Due to its high price and seasonal availability, game meat is particularly prone to this kind of adulteration.

Fraudulent labeling of game meat products can only be detected by applying specific and sensitive analytical methods. Both conventional polymerase chain reaction (PCR) and real-time PCR assays have been developed for the detection of a variety of game meat species in food [6–10]. According to the Codex Alimentarius Austriacus, a collection of standards and product descriptions serving as guidelines for food inspectors, the declaration "game sausage" may only be used if $\geq$38% of the total meat content in the sausage originates from game species [11]. Thus, methods providing quantitative information are required in addition to qualitative methods for game meat authentication.

Recently, we have developed real-time PCR assays for the detection and quantification of red deer, sika deer, fallow deer, roe deer, and wild boar in food [12–20]. Quantification of meat species in food by real-time PCR is a challenging task [2,21]. The main difficulty arises from the necessity to correlate the DNA concentration determined by real-time PCR to the meat content given in weight/weight (w/w). Factors such as tissue type, the number of cells per unit of mass, genome size, and DNA extractability may affect the accuracy of quantitative results [21]. Since the number of mitochondrial DNA copies varies between different animal species and tissue types, we have designed primers targeting single copy genes [12,14,18]. In order to compensate for differences in tissue composition, we pursued a relative quantification approach. In addition to the game species-specific real-time PCR assay, DNA isolates were subjected to a reference real-time PCR assay. The reference real-time PCR assay allows amplification of a conserved 97 bp fragment of the myostatin gene in mammalian and poultry species [22]. Relative quantification is less labor intensive than quantification by using matrix-specific calibrators, another quantification strategy applied in meat species authentication [23,24].

Roe deer (*Capreolus capreolus*) is one of the most frequently consumed deer species in Europe. The real-time PCR assay for roe deer developed recently targets a 62 bp sequence of the roe deer lactoferrin gene [14]. The real-time PCR assay did not show cross-reactivity with 23 animal and 43 plant species tested. An increase in the fluorescence signal was only observed for fallow deer. Since the difference of Ct values between roe deer and fallow deer was >13, low cross-reactivity was considered negligible. In order to investigate whether the real-time PCR is fit for its intended purpose [25], it was subjected to in-house validation, including determination of amplification efficiency, level of detection (LOD), limit of quantification (LOQ), repeatability, and robustness. In-house validation data suggested that the real-time PCR assay for roe deer is suitable for routine analysis. However, for method standardization, evaluation of interlaboratory variability is a prerequisite [25].

Aiming at harmonizing analytical methods for food authentication across EU Member States, in 2017 the real-time PCR assay for roe deer was tested in an interlaboratory ring trial on behalf of the Federal Office for Consumer Protection and Food Safety, Berlin, Germany for the Official Collection of Methods ASU § 64 LFGB. Performance characteristics included amplification efficiency, $LOD_{95\%}$, repeatability, reproducibility, and accuracy of quantitative results. Results of interlaboratory validation of the real-time PCR assay for roe deer are summarized in this paper.

2. Materials and Methods

2.1. Participating Laboratories

The interlaboratory ring trial was organized by the Austrian Agency for Health and Food Safety (AGES) on behalf of the Federal Office for Consumer Protection and Food Safety in Germany. The following laboratories participated in the ring trial (in alphabetical order): Austrian Agency for Health and Food Safety (AGES), Vienna, Austria; Cantonal Office of Consumer Protection Aargau, Aarau, Switzerland; Chemical and Veterinary Investigation Office Freiburg, Freiburg, Germany; Chemical and Veterinary Analytical Institute Muensterland-Emscher-Lippe, Muenster, Germany; German Federal Institute for Risk Assessment (BfR), Berlin, Germany; Impetus GmbH & Co. Bioscience KG, Bremerhaven, Germany; Institute of Hygiene and Environment, Hamburg, Germany; Max Rubner-Institut, Kulmbach, Germany; Official Food Control Authority of the Canton Zurich, Zurich, Switzerland; Saxon State Institute of Health and Veterinary Affairs, Dresden, Germany; Saxony-Anhalt State Office for Consumer Protection, Halle, Germany; State Laboratory Berlin-Brandenburg, Berlin, Germany; State Office Laboratory Hessen, Gießen,

Germany; State Office of Agriculture, Food Safety and Fisheries Mecklenburg-Vorpommern, Rostock, Germany.

2.2. Design of the Interlaboratory Ring Trial

The design of the interlaboratory ring trial is outlined in Figure 1. Participants obtained an aliquot of a DNA isolate from a meat mixture containing 24.8% (w/w) roe deer in pork, an aliquot of a DNA isolate from roe deer meat, and coded aliquots of DNA isolates from 12 meat samples.

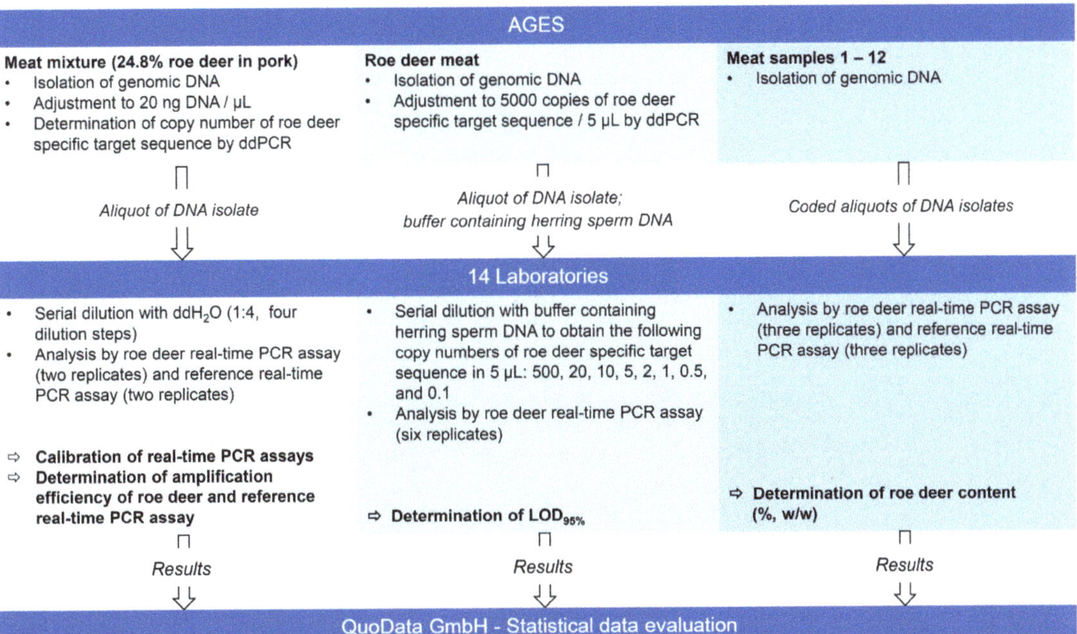

Figure 1. Design of the interlaboratory ring trial.

The DNA isolate from the meat mixture containing 24.8% (w/w) roe deer in pork was used for calibration of the roe deer real-time PCR assay and the reference real-time PCR assay. From the slope of the calibration curves, the amplification efficiency was calculated. The isolate had a DNA concentration of 20 ng/µL and contained 1440 copies of the roe deer specific target sequence per µL. The DNA isolate was serially diluted with bidistilled water (ddH₂O) to obtain DNA isolates with a concentration of 5, 1.25, 0.3125, and 0.078 ng/µL, corresponding to 360, 90, 22.5, and 5.625 copies of the roe deer-specific target sequence per µL, respectively. The diluted DNA isolates were analyzed by the roe deer real-time PCR assay and the reference real-time PCR assay in two PCR replicates each.

The DNA isolate from roe deer meat served for determination of LOD₉₅% of the roe deer real-time PCR assay. The DNA isolate containing 5000 copies of the roe deer-specific target sequence per 5 µL was serially diluted with a buffer containing herring sperm DNA (20 ng/µL; also provided by the organizer of the ring trial). DNA isolates containing 500, 20, 10, 5, 2, 1, 0.5, or 0.1 copies of the roe deer-specific target sequence per 5 µL were prepared. DNA isolates containing ≤ 20 copies of the roe deer specific target sequence per 5 µL were analyzed by the roe deer real-time PCR assay in six PCR replicates. Herring sperm DNA was used as no template control (NTC, two PCR replicates).

DNA isolates from 12 meat samples (Table 1) served for determination of the applicability of the roe deer real-time PCR assay for providing quantitative results. Participants directly analyzed the DNA isolates by the roe deer real-time PCR assay and the reference real-time PCR assay in three PCR replicates each.

Table 1. Meat samples.

Meat Sample	Sample Name	Proportion of Roe Deer (%, *w/w*)
1	meat mixture 1	0
2	meat mixture 2	1
3	meat mixture 3	4.9
4	meat mixture 4	9.5
5	meat mixture 5	24.8
6	meat mixture 6	37.2
7	meat mixture 7	49.4
8	meat mixture 8	25.1
9	meat mixture 9, boiled	24.9
10	model sausage, raw	21.0
11	sausage, brewed	unknown [1]
12	sausage, raw	unknown [1]

[1] declared to contain 5–10% (*w/w*) roe deer.

2.3. Meat Samples

Meat samples included nine meat mixtures and three sausages (Table 1). Meat mixtures were prepared at the AGES. Fresh roe deer and pork meat was taken in a slaughterhouse by a food inspector. After cutting and homogenizing roe deer and pork meat in a cutter (robot coupe R5 plus, Toperczer, Schwechat-Rannersdorf, Austria) for 5–10 min, at least 2 kg of mixtures were prepared by weighing out the respective amounts of meat and homogenizing the mixture in the cutter. With the exception of meat mixture 1 (meat sample 1), which was free of roe deer, meat mixtures contained roe deer in the range from 1 to 49.4% (*w/w*). Meat mixture 9 (meat sample 9) was boiled at 100 °C for 20 min. Immediately after preparation, meat mixtures were subjected to DNA isolation.

Sausage 1 (meat sample 10) was a model sausage, containing 21.0% roe deer. The other two sausages (meat sample 11 and 12) were purchased from a supermarket, with meat sample 11 being brewed and meat sample 12 being a raw sausage. Both sausages were declared to contain roe deer in the range from 5 to 10% (*w/w*). After homogenization, sausages were stored at −20 °C until DNA isolation.

Participants directly subjected DNA isolates to real-time PCR analysis.

2.4. Isolation of Genomic DNA

Isolation of genomic DNA from meat mixtures, sausages, and roe deer meat was carried out at the AGES by applying the official method L 00.00–119 [26]. After isolating genomic DNA twice, the undiluted DNA isolates were combined.

DNA concentration of the (combined) DNA isolate from the meat mixture containing 24.8% (*w/w*) roe deer in pork (used for calibration of the roe deer real-time PCR assay and the reference real-time PCR assay) was adjusted to 20 ng/μL. The copy number of the roe deer specific target sequence per 5 μL, determined by droplet digital PCR (ddPCR, QX200 Droplet Generator, QX200 Droplet Reader (Bio-Rad, Hercules, CA, USA)), was 1440 copies/μL.

After determining the copy number of the roe deer-specific target sequence in the DNA isolate from roe deer meat (serving for determination of $LOD_{95\%}$) by ddPCR, the DNA isolate was diluted to obtain 1000 copies/μL.

2.5. Real-Time PCR

Sequences and concentrations of primers and probes for the roe deer real-time PCR assay and the reference real-time PCR assay are given in Table 2. Primers and probes were provided by the AGES. All participants used the QuantiTect Multiplex PCR-Kit (NoROX, Qiagen, Hilden, Germany).

Table 2. Primers and probes.

Assay	Primer/Probe	Sequence (5′-3′) [1]	Final Concentration [nM]	Reference
roe deer	primer f	TGGCTGCTGCGTGCAGAA	200	[14]
	primer r	TCTAAAATGCTTGGGAACCAGATAT	200	
	probe	FAM-GAAGGGTCTCCGTCTGC-MGBNFQ	100	
myostatin	primer f	TTGTGCARATCCTGAGACTCAT	200	[22]
	primer r	ATACCAGTGCCTGGGTTCAT	200	
	probe	FAM-CCCATGAAAGACGGTACAAGRTATACTG-BHQ1	100	

[1] FAM: 6-carboxyfluorescein, MGBNFQ: minor groove binding non-fluorescent quencher, BHQ1: black hole quencher 1, R: A + G.

Real-time PCR reactions were performed in a total volume of 25 µL, consisting of 20 µL of reaction mix and 5 µL of DNA isolate. The following temperature program was applied for both the roe deer real-time PCR assay and the reference real-time PCR assay: 15 min at 95 °C, 45 cycles of 60 s at 94 °C and 60 s at 60 °C.

2.6. Data Evaluation and Statistical Analysis

For each laboratory, the amplification efficiency, E, was calculated from the slope of the standard curve: $E(\%) = (10^{\frac{-1}{slope}} - 1) \cdot 100$.

The probability of detection across laboratories, POD, was calculated as follows: $POD\,(x) = 1 - \exp(-\lambda_0 \cdot x^b)$ with λ_0 being the average amplification probability and b being the slope across laboratories. Here, both parameters λ_0 and b were estimated based on a generalized linear mixed model as described in Uhlig et al. [27].

$LOD_{95\%}$ based on the POD curve was calculated as

$$LOD_{95\%} = (-\ln(0.05)/\lambda_0)^{1/b}$$

The content of roe deer meat in relation to the total meat content of the meat sample was calculated as follows:

$$\text{concentration of roe deer DNA (ng/µL)} = 10^{\frac{Ct_{spec} - d_{spec}}{slope_{spec}}}$$

$$\text{concentration of total meat DNA (ng/µL)} = 10^{\frac{Ct_{ref} - d_{ref}}{slope_{ref}}}$$

with $_{spec}$ and $_{ref}$ referring to the roe deer real-time PCR assay and the reference real-time PCR assay, respectively

Ct: Ct value
d: intercept of the standard curve
slope: slope of the standard curve

$$\text{roe deer meat content (\%)} = \frac{\text{concentration of roe deer DNA (ng/µL)}}{\text{concentration of total meat DNA (ng/µL)}} \cdot 100$$

Repeatability, reproducibility, and accuracy of the roe deer meat content were determined using the statistical approaches according to ISO 5725-2 [28] as well as according to the specifications of the ASU § 64 LFGB [29].

Statistical analyses were performed by QuoData GmbH using the software package PROLab Plus [30]. Results were subjected to several outlier tests to check for outliers. The presence of outliers within the laboratories was tested as well as whether the variances of the laboratories were approximately the same and whether systematic errors affected the mean values.

3. Results and Discussion

3.1. Amplification Efficiency

Table 3 summarizes the slope of the laboratory-specific standard curve, coefficient of determination (R^2), and amplification efficiency (E) for both the roe deer real-time PCR assay and the reference real-time PCR assay, obtained by analyzing serially diluted DNA isolates from a meat mixture containing 24.8% (w/w) roe deer in pork in two PCR replicates each.

Table 3. Slope of the laboratory-specific standard curve, coefficient of determination (R^2), and amplification efficiency (E) for the roe deer real-time PCR assay and the reference real-time PCR assay.

Laboratory	Roe Deer Real-Time PCR			Reference Real-Time PCR		
	Slope	R^2	E (%)	Slope	R^2	E (%)
1	−3.5759	0.9966	90.39	−3.3792	0.9992	97.66
2	−3.4180	0.9963	96.14	−3.3263	0.9968	99.82
3	−3.5573	0.9969	91.03	−3.5396	0.9985	91.65
4	−3.6037	0.9961	89.45	−3.6082	0.9970	89.30
5	−3.5877	0.9942	89.99	−3.3749	0.9999	97.83
6	−3.4349	0.9987	95.49	−3.3983	0.9937	96.91
7	−3.4698	0.9973	94.18	−3.4947	0.9989	93.26
8	−3.5276	0.9970	92.08	−3.3273	0.9978	99.78
9	−3.2937	0.9981	101.19	−3.3817	0.9996	97.56
10	−3.5797	0.9989	90.26	−3.4860	0.9996	93.58
11	−3.0728 [1]	0.9986	111.56 [2]	−3.5304	0.9615 [2]	91.98
12	−3.4141	0.9980	96.29	−3.3037	0.9994	100.77
13	−3.4531	0.9975	94.80	−3.2621	0.9991	102.56
14	−3.4324	0.9986	95.59	−3.3850	0.9981	97.43

[1] outlier (Grubbs test, $\alpha = 0.05$), [2] outlier (Grubbs test, $\alpha = 0.01$).

In case of laboratory 11, the slope of the standard curve (Grubbs test, $\alpha = 0.05$) and the amplification efficiency (Grubbs test, $\alpha = 0.01$) obtained for the roe deer real-time PCR assay as well as the coefficient of determination obtained for the reference real-time PCR assay (Grubbs test, $\alpha = 0.01$) were identified as outliers.

According to the guidelines recommended by the European Network of GMO (Genetically Modified Organisms) Laboratories (ENGL) [31], the slope should be between −3.1 and −3.6, corresponding to an amplification efficiency of ~90 to 110%. In almost all cases, slope and amplification efficiency were within the recommended range, with the exception of some values from laboratory 4 and 11. The coefficient of determination, R^2, is recommended to be >0.98 [31]. All laboratories fulfilled this criterion, with the exception of laboratory 11.

3.2. Level of Detection ($LOD_{95\%}$)

Table 4 summarizes the laboratory-specific number of positive results obtained by repeated analysis (six PCR replicates) of a DNA isolate from roe deer meat, diluted to 20 to 0.1 copies of the roe deer specific target sequence per 5 µL. Table 5 gives the number of positive results obtained for each dilution step in relation to the total number of tests ($n = 84$). Down to a copy number of five copies per 5 µL, all tests resulted in an increase in the fluorescence signal within 45 cycles. For 2, 1, 0.5, and 0.1 copies per 5 µL, the percentage of positive results was decreased to 86.9, 75.0, 50.0, and 10.7%, respectively.

Table 4. Laboratory-specific number of positive results obtained for the DNA isolate from roe deer meat, diluted to 20 to 0.1 copies of the roe deer-specific target sequence per 5 µL. A result was considered positive in cases in which the Ct value was <45 and the copy number, calculated based on the standard curve, was >0.

Laboratory	Copy Number/5 µL						
	0.1	0.5	1	2	5	10	20
1	0	2	5	4	6	6	6
2	2	4	5	5	6	6	6
3	0	5	4	6	6	6	6
4	0	3	4	4	6	6	6
5	2	4	3	6	6	6	6
6	0	1	3	4	6	6	6
7	1	3	6	6	6	6	6
8	1	2	5	5	6	6	6
9	0	5	6	5	6	6	6
10	0	4	5	5	6	6	6
11	1	1	4	6	6	6	6
12	1	2	6	5	6	6	6
13	0	4	3	6	6	6	6
14	1	2	4	6	6	6	6

Table 5. Summary of results obtained for determination of $LOD_{95\%}$ of the roe deer real-time PCR assay.

Theoretical Copy Number of the Roe Deer-Specific Target Sequence Per 5 µL	Roe Deer Real-Time PCR			
	Number of Positive Tests/Total Number of Tests	Percentage of Positive Tests (%)	p_U (%) [1]	p_O (%) [2]
20	84/84	100.0	96.5	100.0
10	84/84	100.0	96.5	100.0
5	84/84	100.0	96.5	100.0
2	73/84	86.9	79.3	92.5
1	63/84	75.0	66.0	82.6
0.5	42/84	50.0	40.5	59.5
0.1	9/84	10.7	5.7	18.0

[1,2] Clopper–Pearson confidence intervals. p_U: lower limit of the 90% confidence interval for the detection probability p, p_O: upper limit of the 90% confidence interval for the detection probability p.

3.2.1. $LOD_{95\%}$ According to Simplified Calculation Approaches

In GMO analysis, $LOD_{95\%}$ of a real-time PCR assay is defined as the lowest copy number of a target DNA sequence in a sample, for which a positive result is obtained with a detection probability, p, of 95% ($LOD_{95\%}$). We used three simplified calculation approaches for the determination of $LOD_{95\%}$. In the first approach, $LOD_{95\%}$ was regarded as the lowest copy number for which all replicates in all laboratories were positive. In the second approach, $LOD_{95\%}$ was considered the lowest copy number of the roe deer-specific target sequence, for which the lower limit of the 90% confidence interval for the detection probability p, p_u, was achieved with a probability ≥95%. In the third approach, $LOD_{95\%}$ was defined as the lowest copy number of the target sequence, for which ≥95% of the tests yielded a positive result. With all three calculation approaches, $LOD_{95\%}$ of the real-time PCR assay for roe deer was determined to be five copies of the roe deer specific target sequence per 5 µL (Table 5).

3.2.2. $LOD_{95\%}$ Derived from the Mixed Model for the POD Curve

In addition, we determined $LOD_{95\%}$ by applying a statistical model for calculating the probability of detection (POD) across laboratories. Since its introduction by Uhlig et al. [27], this model has already been used several times to determine the sensitivity of real-time PCR assays [32–37]. Qualitative results obtained for the seven dilution steps of the DNA isolate from roe deer meat were used to determine the laboratory standard deviation σ_L, and the $LOD_{95\%}$ for the median laboratory, as described previously [27]. Table 6 summarizes the model parameters, including the estimated values for the average amplification probability λ_o, and the slope b for describing the POD curve across laboratories in dependence of the copy number of the target sequence (Figure 2). σ_L was determined to be 0.15, and the $LOD_{95\%}$ for the median laboratory 2.4 copies of the target sequence per 5 µL.

Table 6. Summary of the POD statistics for the real-time PCR assay for roe deer.

Parameter	Value
number of participating laboratories	14
number of PCR replicates per dilution level	6
model parameters of the POD curve:	
average amplification probability λ_o	1.25
95% confidence interval for the estimated value of λ_o	1.05–1.49
estimated value for slope b	1
laboratory standard deviation σ_L	0.15
$LOD_{95\%}$ for median laboratory (copy number of the target sequence per 5 µL)	2.4

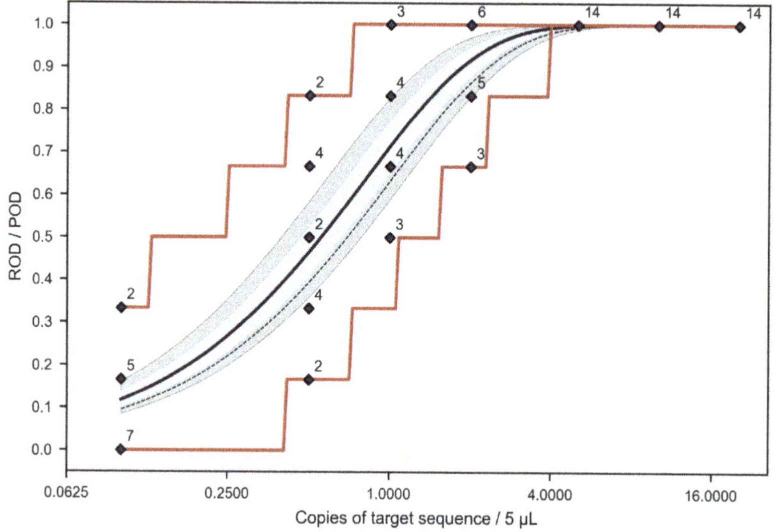

Figure 2. POD curve across laboratories (dark blue) with associated 95% confidence range (light blue) and 95% prediction range (light gray), laboratory-specific rate of detection (ROD) (blue diamonds, numerical values give the numbers of laboratories with the respective ROD) with associated 90% prediction interval (red). The ideal POD curve obtained under optimal conditions is given as dashed line.

Figure 2 shows the POD curve across laboratories together with the 95% confidence and prediction range as well as the laboratory-specific rates of detection (ROD) with the respective 90% confidence range. In addition, the ideal POD curve obtained under optimal conditions is given.

The POD curve across laboratories (dark blue) was found to lie above the ideal curve obtained under optimal conditions (dashed), which would mean that the obtained $LOD_{95\%}$ is better than theoretically achievable. The difference between both curves was statistically significant ($p < 0.05$), suggesting that the actual copy numbers were at least 1.05-fold higher than the nominal copy numbers in the diluted DNA isolates. However, by taking a standard measuring uncertainty of 10% of the DNA isolate from roe deer meat into account, the difference can be considered statistically insignificant.

3.3. Analysis of Meat Samples

DNA isolates from meat samples (meat samples 1–12, Table 1) were analyzed by the roe deer real-time PCR assay and the reference real-time PCR assay in three PCR replicates each.

3.3.1. False Positive and False Negative Results

Results obtained with the roe deer real-time PCR assay for the meat mixture that did not contain roe deer (meat sample 1) and samples containing roe deer (meat samples 2–12) were used to determine the rate of false positive and false negative results, respectively. Analysis of 12 meat samples in PCR triplicates in 14 laboratories resulted in a total of 504 results; 42 thereof were obtained for meat sample 1 and 462 for meat samples 2–12, containing roe deer. The qualitative result was correct for all meat samples, there were neither false positive nor false negative results.

3.3.2. Quantitative Results

Evaluation of quantitative results was based on results obtained for meat samples 2–12. Meat sample 1 was not taken into account since it did not contain roe deer. The roe deer content in % (w/w) was calculated by relating the DNA concentration (ng/μL) determined by the roe deer real-time PCR assay to the DNA concentration (ng/μL) determined by the reference real-time PCR assay.

Single outliers within one laboratory, detected for four samples in four different laboratories, were removed first. Furthermore, results for three meat samples (6, 8, and 12) obtained by one laboratory each show a statistically significantly excessive variance of the triplicates. Statistical evaluation according to ASU § 64 LFGB (based on ISO 5725-2) [28,29] was based on the data after outlier elimination. Table 7 gives the statistical parameters for the determination of the roe deer content (%) in the 11 meat samples containing roe deer. Reproducibility standard deviation, s_R, is a measure for the variability between laboratories, whereas the repeatability standard deviation, sr, characterizes the variability within a laboratory under repeatable conditions. Based on reproducibility and repeatability standard deviation, the reproducibility limit, R, and repeatability limit, r, were calculated. Reproducibility and repeatability limits are a measure of the maximally expected deviation between two values obtained for a specific sample in different laboratories and in the same laboratory, respectively.

Relative repeatability standard deviation ranged from 6.60% (sample 8) to 17.71% (sample 2), and relative reproducibility standard deviation from 13.35% (sample 8) to 30.22% (sample 5). The rather high relative reproducibility standard deviation obtained for sample 5 decreased to 21.42% when results obtained by laboratory 11 were not taken into account. According to the ENGL guidelines, relative repeatability and relative reproducibility standard deviations should be <25 and <35%, respectively [31]. The roe deer real-time PCR assay fulfilled these criteria and can therefore be considered suitable for achieving reproducible results.

The aim of the ring trial was to validate the real-time PCR assay for roe deer. The suitability of the DNA isolation method (official method L 00.00–119) has been demonstrated before. Thus, the participants did not have to isolate DNA from the samples. DNA isolates, prepared at the AGES, were provided by the organizer of the ring trial. The relative reproducibility standard deviation given above therefore does not include variability caused by DNA isolation. Furthermore, all participants of the ring trial used the same PCR kit (QuantiTect Multiplex PCR-Kit). In principle, the use of different PCR kits might result in higher relative reproducibility standard deviation than the value given above. However, in a preliminary experiment, PCR kits from different providers did not lead to significantly different results.

Table 7. Summary of the statistical parameters for determining the roe deer content in meat samples. The distribution of readings for sample 8 deviates statistically significantly from the normal distribution. Therefore, the results for this sample have limited applicability.

Parameter [1]	Sample 2	Sample 3	Sample 4	Sample 5	Sample 6	Sample 7	Sample 8	Sample 9	Sample 10	Sample 11	Sample 12
participating labs	14	14	14	14	14	14	14	14	14	14	14
labs with quantitative results	14	14	14	14	14	14	14	14	14	14	14
outlier labs	0	0	0	0	1	0	1	0	0	0	1
labs for determining parameters	14	14	14	14	13	14	13	14	14	14	13
theoretical value (%)	1.0	4.9	9.5	24.8	37.2	49.4	25.1	24.9	21.0	-	-
mean $\pm$ confidence level (%)	1.05 ± 0.07	5.22 ± 0.42	9.90 ± 0.76	24.02 ± 3.66	37.50 ± 3.19	47.40 ± 4.39	28.46 ± 1.93	132.57 ± 12.91	17.72 ± 1.57	7.63 ± 0.96	8.99 ± 0.86
s_R (%)	0.20	0.90	1.63	7.26	6.34	9.44	3.80	26.85	3.51	1.91	1.68
$s_{R\ rel}$	19.50%	17.26%	16.46%	30.22%	16.91%	19.91%	13.35%	20.25%	19.83%	25.08%	18.71%
R (%)	0.57	2.52	4.56	20.33	17.76	26.42	10.64	75.18	9.84	5.36	4.71
R_{rel}	54.59%	48.33%	46.09%	84.63%	47.36%	55.75%	37.38%	56.71%	55.52%	70.23%	52.40%
s_r (%)	0.19	0.56	0.98	2.93	3.29	5.71	1.88	14.36	2.35	0.83	0.82
$s_{r\ rel}$	17.71%	10.63%	9.87%	12.20%	8.76%	12.05%	6.60%	10.83%	13.24%	10.83%	9.09%
r (%)	0.52	1.56	2.73	8.21	9.20	16.00	5.26	40.21	6.57	2.32	2.29
r_{rel}	49.60%	29.77%	27.62%	34.17%	24.54%	33.75%	18.49%	30.33%	37.08%	30.34%	25.46%
recovery (%)	105.1	106.6	104.2	96.9	100.8	95.9	113.4	532.4	84.4	-[2]	-[2]

[1] s_R: reproducibility standard deviation; $s_{R\ rel}$: relative reproducibility standard deviation; R: reproducibility limit, R_{rel}: relative reproducibility limit; s_r: repeatability standard deviation; $s_{r\ rel}$: relative repeatability standard deviation; r (%): repeatability limit; r_{rel}: relative repeatability limit. [2] Recovery could not be determined because the exact roe deer content was unknown.

In addition to data on the repeatability and reproducibility, Table 7 contains recoveries obtained for meat samples 2 to 10. For meat samples 2 to 8 and 10, recovery ranged from 84.4 to 114.3%. With 528.2%, recovery obtained for sample 9 was drastically too high. Sample 9 was the only meat mixture that had been heat-treated (boiled at 100 °C for 20 min). Quantification of the meat content in heat-treated foods by real-time PCR is known to be challenging [38–42]. In several studies, DNA isolates from heat-treated food products were found to yield higher Ct values than DNA isolates from untreated ones [40–42]. The five-fold overestimation of the roe deer content for sample 9 can be explained by differences in the amplifiability of the reference sequence compared to the roe deer-specific sequence. With the referenced real-time PCR assay, higher Δ Ct values (difference in the Ct values obtained for DNA isolates from raw and heat-treated samples) were obtained than with the roe deer real-time PCR assay. This result suggests that the reference real-time PCR assay is not applicable for heat-treated meat mixtures. With an amplicon length of 97 bp, the amplicon was substantially longer than that obtained with the roe deer real-time PCR assay (62 bp). We assume that an alternative reference real-time PCR assay published recently [43] is more suitable for heat-treated meat mixtures since it results in a 70 bp amplicon.

For interlaboratory evaluation, combination scores of systematic deviations, RSZ (rescaled sum of z_U scores), and relative laboratory performance, RLP, [44] across all samples were used. RSZ is based on a standardized sum of all z_U scores (corrected z scores), measuring the deviations of the mean value of a laboratory from the total mean value. If the RSZ is within −2 and +2, the respective laboratory does not show a significant systematic deviation. RLP is ideally 1 or <1. An RLP of 1 indicates that deviations of the respective laboratory are on average. Figure 3 shows z_U scores and the respective combination scores.

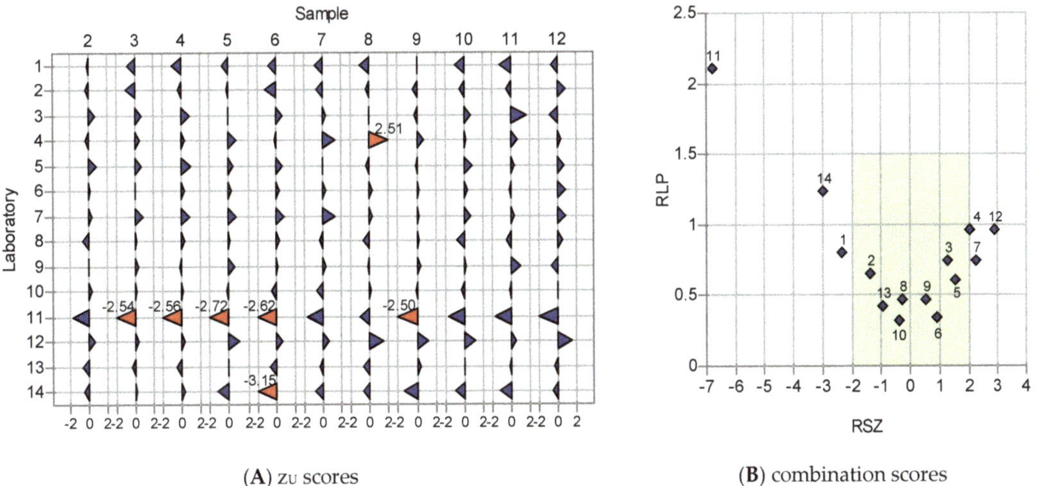

(**A**) z_U scores

(**B**) combination scores

Figure 3. z_U scores (**A**) and combination scores (**B**) for the determination of the roe deer content in meat samples. In (**A**), z_U scores within −2 and +2 are shown in blue, z_U scores <−2 or >+2 in red. The figure includes laboratories which were identified as outliers.

RSZ values of laboratories 4, 7, and 12 indicate a systematic positive bias (RSZ > +2) for the determination of the roe deer content, and RSZ values of the laboratories 1, 11, and 14 indicate a systematic negative bias (RSZ < −2). In fact, the vast majority of z_U scores were positive for the laboratories 4, 7, and 12, and negative for laboratories 1, 11, and 14. The z_U scores of laboratory 11 are particularly noticeable as results that were significantly too low (z_U score < −2) were obtained for five samples (3, 4, 5, 6, and 9). Interestingly, the lowest roe deer contents for all meat samples were determined by laboratory 11. With 111.56%, the amplification efficiency of the roe deer real-time PCR assay was considerably higher than 100%. However, with 91.98%, the amplification efficiency of the reference real-time PCR assay was much lower. These differences in the amplification efficiency explain why the roe deer content of the meat samples was systematically underestimated by laboratory 11.

4. Conclusions

Results obtained in the interlaboratory ring trial demonstrate the applicability of the real-time PCR assay for the detection and quantification of roe deer in meat samples to detect food adulteration. For none of the meat samples, false negative or false positive results were obtained. In ten out of eleven meat samples, the roe deer content was determined with satisfactory reproducibility and accuracy. Only for a heat-treated meat mixture, the roe deer content was ~five-fold overestimated. Overestimation of the roe deer content can be explained by differences in the amplifiability of the reference sequence compared to the roe deer specific sequence. A reference system published recently [43], amplifying a 70 bp fragment, is most probably more suitable for heat-treated products. This method has been successfully validated for the detection of animal components in vegan products by the Federal Office of Consumer Protection and Food Safety [45]. However, the applicability of the reference real-time PCR assay targeting a 70 bp fragment remains to be investigated in a further ring trial. Since heat-treatment procedures are known to affect DNA differently, the ring trial should include a variety of heat-treated model food products, e.g., brewed, cooked, and microwave treated ones.

Author Contributions: Conceptualization, R.H.; methodology, R.H., B.D. and M.C.-M.; validation, S.U., K.S., K.F. and K.H.; writing—original draft preparation, M.C.-M.; writing—review and editing, R.H., B.D., S.U., K.S., K.F. and K.H.; visualization, B.D. and M.C.-M.; supervision, R.H. All authors have read and agreed to the published version of the manuscript.

Funding: The ring trial was supported by the Federal Office of Consumer Protection and Food Safety, Germany (BVL). BVL funded all reagents for performing the tests and the statistical evaluation by QuoData GmbH.

Institutional Review Board Statement: Ethical review and approval was waived for this study as no live animals were used or slaughtered to achieve the aims of the study.

Data Availability Statement: The datasets generated during the current study are available from the corresponding authors on reasonable request.

Acknowledgments: We thank the members of the working group "Molecular biological methods for plant and animal species differentiation" within the scope of the official method collection according to §64 of the German Food and Feed Code (LFGB) for their participation in the study.

Conflicts of Interest: Evaluation of data obtained in an interlaboratory ring trial makes it necessary to apply a lot of specific statistical tests and this was done by the co-authors from QuoData. The authors declare no conflict of interest.

References

1. Hoffman, L.C.; Wiklund, E. Game and venison—Meat for the modern consumer. *Meat Sci.* **2006**, *74*, 197–208. [CrossRef]
2. Ballin, N.Z. Authentication of meat and meat products. *Meat Sci.* **2010**, *86*, 577–587. [CrossRef]
3. Premanandh, J. Horse meat scandal—A wake-up call for regulatory authorities. *Food Control* **2013**, *34*, 568–569. [CrossRef]
4. D'Amato, M.E.; Alechine, E.; Cloete, K.W.; Davison, S.; Corach, D. Where is the game? Wild meat products authentication in South Africa: A case study. *Investig. Genet.* **2013**, *4*, 6. [CrossRef] [PubMed]
5. Stamatis, C.; Sarri, C.A.; Moutou, K.A.; Argyrakoulis, N.; Galara, I.; Godosopoulos, V.; Kolovos, M.; Liakou, C.; Stasinou, V.; Mamuris, Z. What do we think we eat? Single tracing method across foodstuff of animal origin found in Greek market. *Food Res. Int.* **2015**, *69*, 151–155. [CrossRef]
6. Fajardo, V.; González, I.; López-Calleja, I.; Martín, I.; Rojas, M.; Hernández, P.E.; García, T.; Martín, R. Identification of meats from red deer (*Cervus elaphus*), fallow deer (*Dama dama*), and roe deer (*Capreolus capreolus*) using polymerase chain reaction targeting specific sequences from the mitochondrial 12S rRNA gene. *Meat Sci.* **2007**, *76*, 234–240. [CrossRef] [PubMed]

7. Fajardo, V.; González, I.; Martín, I.; Rojas, M.; Hernández, P.E.; García, T.; Martín, R. Differentiation of European wild boar (*Sus scrofa scrofa*) and domestic swine (*Sus scrofa domestica*) meats by PCR analysis targeting the mitochondrial D-loop and the nuclear melanocortin receptor 1 (*MC1R*) genes. *Meat Sci.* **2008**, *78*, 314–322. [CrossRef]

8. Rojas, M.; González, I.; Pavón, M.Á.; Pegels, N.; Lago, A.; Hernández, P.E.; García, T.; Martín, R. Novel TaqMan real-time polymerase chain reaction assay for verifying the authenticity of meat and commercial meat products from game birds. *Food Addit. Contam.* **2010**, *27*, 749–763. [CrossRef] [PubMed]

9. Amaral, J.S.; Santos, C.G.; Melo, V.S.; Oliveira, M.B.P.P.; Mafra, I. Authentication of a traditional game meat sausage (*Alheira*) by species-specific PCR assays to detect hare, rabbit, red deer, pork and cow meats. *Food Res. Int.* **2014**, *60*, 140–145. [CrossRef]

10. Rak, L.; Knapik, K.; Bania, J.; Sujkowski, J.; Gadzinowski, A. Detection of roe deer, red deer, and hare meat in raw materials and processed products available in Poland. *Eur. Food Res. Technol.* **2014**, *239*, 189–194. [CrossRef]

11. Codex Alimentarius Austriacus. *Österreichisches Lebensmittelbuch, Codexkapitel/B14/Fleisch und Fleischerzeugnisse*; Springer: Vienna, Austria, 2005.

12. Druml, B.; Grandits, S.; Mayer, W.; Hochegger, R.; Cichna-Markl, M. Authenticity control of game meat products—A single method to detect and quantify adulteration of fallow deer (*Dama dama*), red deer (*Cervus elaphus*) and sika deer (*Cervus nippon*) by real-time PCR. *Food Chem.* **2015**, *170*, 508–517. [CrossRef]

13. Druml, B.; Hochegger, R.; Cichna-Markl, M. Duplex real-time PCR assay for the simultaneous determination of the roe deer (*Capreolus capreolus*) and deer (sum of fallow deer, red deer and sika deer) content in game meat products. *Food Control* **2015**, *57*, 370–376. [CrossRef]

14. Druml, B.; Mayer, W.; Cichna-Markl, M.; Hochegger, R. Development and validation of a TaqMan real-time PCR assay for the identification and quantification of roe deer (*Capreolus capreolus*) in food to detect food adulteration. *Food Chem.* **2015**, *178*, 319–326. [CrossRef]

15. Kaltenbrunner, M.; Hochegger, R.; Cichna-Markl, M. Red deer (*Cervus elaphus*)-specific real-time PCR assay for the detection of food adulteration. *Food Control* **2018**, *89*, 157–166. [CrossRef]

16. Kaltenbrunner, M.; Hochegger, R.; Cichna-Markl, M. Sika deer (*Cervus nippon*)-specific real-time PCR method to detect fraudulent labelling of meat and meat products. *Sci. Rep.* **2018**, *8*, 7236. [CrossRef] [PubMed]

17. Kaltenbrunner, M.; Hochegger, R.; Cichna-Markl, M. Tetraplex real-time PCR assay for the simultaneous identification and quantification of roe deer, red deer, fallow deer and sika deer for deer meat authentication. *Food Chem.* **2018**, *269*, 486–494. [CrossRef]

18. Kaltenbrunner, M.; Hochegger, R.; Cichna-Markl, M. Development and validation of a fallow deer (*Dama dama*)-specific TaqMan real-time PCR assay for the detection of food adulteration. *Food Chem.* **2018**, *243*, 82–90. [CrossRef]

19. Kaltenbrunner, M.; Mayer, W.; Kerkhoff, K.; Epp, R.; Rüggeberg, H.; Hochegger, R.; Cichna-Markl, M. Differentiation between wild boar and domestic pig in food by targeting two gene loci by real-time PCR. *Sci. Rep.* **2019**, *9*, 9221. [CrossRef]

20. Kaltenbrunner, M.; Mayer, W.; Kerkhoff, K.; Epp, R.; Rüggeberg, H.; Hochegger, R.; Cichna-Markl, M. Applicability of a duplex and four singleplex real-time PCR assays for the qualitative and quantitative determination of wild boar and domestic pig meat in processed food products. *Sci. Rep.* **2020**, *10*, 17243. [CrossRef]

21. Ballin, N.Z.; Vogensen, F.K.; Karlsson, A.H. Species determination—Can we detect and quantify meat adulteration? *Meat Sci.* **2009**, *83*, 165–174. [CrossRef]

22. Laube, I.; Zagon, J.; Spiegelberg, A.; Butschke, A.; Kroh, L.W.; Broll, H. Development and design of a 'ready-to-use' reaction plate for a PCR-based simultaneous detection of animal species used in foods. *Int. J. Food Sci. Tech.* **2007**, *42*, 9–17. [CrossRef]

23. Eugster, A.; Ruf, J.; Rentsch, J.; Hübner, P.; Köppel, R. Quantification of beef and pork fraction in sausages by real-time PCR analysis: Results of an interlaboratory trial. *Eur. Food Res. Technol.* **2008**, *227*, 17–20. [CrossRef]

24. Eugster, A.; Ruf, J.; Rentsch, J.; Köppel, R. Quantification of beef, pork, chicken and turkey proportions in sausages: Use of matrix-adapted standards and comparison of single versus multiplex PCR in an interlaboratory trial. *Eur. Food Res. Technol.* **2009**, *230*, 55–61. [CrossRef]

25. Taverniers, I.; De Loose, M.; Van Bockstaele, E. Trends in quality in the analytical laboratory. II. Analytical method validation and quality assurance. *Trends Anal. Chem.* **2004**, *23*, 535–552. [CrossRef]

26. ISO 21571. *Foodstuffs – Methods of Analysis for the Detection of Genetically Modified Organisms and Derived Products: Nucleic Acid Extraction. Annex A3*; International Organization for Standardization: Geneva, Switzerland, 2005.

27. Uhlig, S.; Frost, K.; Colson, B.; Simon, K.; Mäde, D.; Reiting, R.; Gowik, P.; Grohmann, L. Validation of qualitative PCR methods on the basis of mathematical–statistical modelling of the probability of detection. *Accredit. Qual. Assur.* **2015**, *20*, 75–83. [CrossRef]

28. ISO 5725-2. *Accuracy (Trueness and Precision) of Measurement Methods and Results—Part 2: Basic Method for the Determination of Repeatability and Reproducibility of a Standard Measurement Method (ISO 5725-2:1994 Including Technical Corrigendum 1:2002)*; International Organization for Standardization: Geneva, Switzerland, 2002.

29. Official Collection of Test Methods, ASU. *Statistik—Planung und Statistische Auswertung von Ringversuchen zur Methodenvalidierung*; Federal Office of Consumer Protection and Food Safety (BVL), Beuth Verlag: Berlin, Germany, 2006.

30. PROLab Plus. Software for Planning, Organizing, Performing and Analyzing Interlaboratory Studies. Available online: www.quodata.de (accessed on 25 October 2021).

31. ENGL. European Network of GMO Laboratories. Definition of Minimum Performance Requirements for Analytical Methods of GMO Testing. Available online: http://gmo-crl.jrc.ec.europa.eu/guidancedocs.htm (accessed on 31 October 2016).

32. Grohmann, L.; Belter, A.; Speck, B.; Goerlich, O.; Guertler, P.; Angers-Loustau, A.; Patak, A. Screening for six genetically modified soybean lines by an event-specific multiplex PCR method: Collaborative trial validation of a novel approach for GMO detection. *J. Consum. Prot. Food Saf.* **2017**, *12*, 23–36. [CrossRef]

33. Cai, Y.; Wang, Q.; He, Y.; Pan, L. Interlaboratory validation of a real-time PCR detection method for bovine- and ovine-derived material. *Meat Sci.* **2017**, *134*, 119–123. [CrossRef]

34. Gondim, C.D.S.; Junqueira, R.G.; de Souza, S.V.C.; Callao, M.P.; Ruisánchez, I. Determining performance parameters in qualitative multivariate methods using probability of detection (POD) curves. Case study: Two common milk adulterants. *Talanta* **2017**, *168*, 23–30. [CrossRef]

35. Wang, Q.; Cai, Y.; He, Y.; Yang, L.; Pan, L. Collaborative ring trial of two real-time PCR assays for the detection of porcine- and chicken-derived material in meat products. *PLoS ONE* **2018**, *13*, e0206609. [CrossRef]

36. Fraiture, M.-A.; Papazova, N.; Roosens, N.H.C. DNA walking strategy to identify unauthorized genetically modified bacteria in microbial fermentation products. *Int. J. Food Microbiol.* **2021**, *337*, 108913. [CrossRef]

37. Zarske, M.; Zagon, J.; Schmolke, S.; Seidler, T.; Braeuning, A. Detection of silkworm (*Bombyx mori*) and Lepidoptera DNA in feeding stuff by real-time PCR. *Food Control* **2021**, *126*, 108059. [CrossRef]

38. Ali, M.E.; Rahman, M.M.; Hamid, S.B.A.; Mustafa, S.; Bhassu, S.; Hashim, U. Canine-specific PCR assay targeting cytochrome b gene for the detection of dog meat adulteration in commercial frankfurters. *Food Anal. Methods* **2014**, *7*, 234–241. [CrossRef]

39. Cammà, C.; Di Domenico, M.; Monaco, F. Development and validation of fast Real-Time PCR assays for species identification in raw and cooked meat mixtures. *Food Control* **2012**, *23*, 400–404. [CrossRef]

40. Fumière, O.; Dubois, M.; Baeten, V.; Von Holst, C.; Berben, G. Effective PCR detection of animal species in highly processed animal byproducts and compound feeds. *Anal. Bioanal. Chem.* **2006**, *385*, 1045–1054. [CrossRef]

41. Hird, H.; Chisholm, J.; Sanchez, A.; Hernandez, M.; Goodier, R.; Schneede, K.; Boltz, C.; Popping, B. Effect of heat and pressure processing on DNA fragmentation and implications for the detection of meat using a real-time polymerase chain reaction. *Food Addit. Contam.* **2006**, *23*, 645–650. [CrossRef] [PubMed]

42. Rojas, M.; González, I.; Pavón, M.Á.; Pegels, N.; Hernández, P.E.; García, T.; Martín, R. Application of a real-time PCR assay for the detection of ostrich (*Struthio camelus*) mislabelling in meat products from the retail market. *Food Control* **2011**, *22*, 523–531. [CrossRef]

43. Druml, B.; Kaltenbrunner, M.; Hochegger, R.; Cichna-Markl, M. A novel reference real-time PCR assay for the relative quantification of (game) meat species in raw and heat-processed food. *Food Control* **2016**, *70*, 392–400. [CrossRef]

44. Uhlig, S.; Lischer, P. Statistically-based performance characteristics in laboratory performance studies. *Analyst* **1998**, *123*, 167–172. [CrossRef]

45. Official Test Method L 00.00-170. *Untersuchung von Lebensmitteln. Nachweis von DNA aus Säugetieren und Geflügel in Lebensmitteln Mittels Real-Time PCR auf Basis des Myostatin-Gens*; Federal Office of Consumer Protection and Food Safety (BVL), Beuth Verlag: Berlin, Germany, 2020.

Article

Identification of Mammalian and Poultry Species in Food and Pet Food Samples Using 16S rDNA Metabarcoding

Laura Preckel [1], Claudia Brünen-Nieweler [1,*], Grégoire Denay [2], Henning Petersen [3], Margit Cichna-Markl [4], Stefanie Dobrovolny [5] and Rupert Hochegger [5]

[1] Chemical and Veterinary Analytical Institute Muensterland-Emscher-Lippe (CVUA-MEL), Joseph-Koenig-Strasse 40, 48147 Muenster, Germany; laura.preckel@cvua-mel.de

[2] Chemical and Veterinary Analytical Institute Rhein-Ruhr-Wupper (CVUA-RRW), Deutscher Ring 100, 47798 Krefeld, Germany; gregoire.denay@cvua-rrw.de

[3] Chemical and Veterinary Analytical Institute Ostwestfalen-Lippe (CVUA-OWL), Westerfeldstrasse 1, 32758 Detmold, Germany; henning.petersen@cvua-owl.de

[4] Department of Analytical Chemistry, Faculty of Chemistry, University of Vienna, Währinger Strasse 38, 1090 Vienna, Austria; margit.cichna@univie.ac.at

[5] Austrian Agency for Health and Food Safety (AGES), Institute for Food Safety Vienna, Department for Molecular Biology and Microbiology, Spargelfeldstrasse 191, 1220 Vienna, Austria; stefanie.dobrovolny@ages.at (S.D.); rupert.hochegger@ages.at (R.H.)

* Correspondence: claudia.bruenen-nieweler@cvua-mel.de

Abstract: The substitution of more appreciated animal species by animal species of lower commercial value is a common type of meat product adulteration. DNA metabarcoding, the combination of DNA barcoding with next-generation sequencing (NGS), plays an increasing role in food authentication. In the present study, we investigated the applicability of a DNA metabarcoding method for routine analysis of mammalian and poultry species in food and pet food products. We analyzed a total of 104 samples (25 reference samples, 56 food products and 23 pet food products) by DNA metabarcoding and by using a commercial DNA array and/or by real-time PCR. The qualitative and quantitative results obtained by the DNA metabarcoding method were in line with those obtained by PCR. Results from the independent analysis of a subset of seven reference samples in two laboratories demonstrate the robustness and reproducibility of the DNA metabarcoding method. DNA metabarcoding is particularly suitable for detecting unexpected species ignored by targeted methods such as real-time PCR and can also be an attractive alternative with respect to the expenses as indicated by current data from the cost accounting of the AGES laboratory. Our results for the commercial samples show that in addition to food products, DNA metabarcoding is particularly applicable to pet food products, which frequently contain multiple animal species and are also highly prone to adulteration as indicated by the high portion of analyzed pet food products containing undeclared species.

Keywords: DNA metabarcoding; 16S rDNA; meat species identification; authentication; food; pet food; feed; real-time PCR; PCR array

Citation: Preckel, L.; Brünen-Nieweler, C.; Denay, G.; Petersen, H.; Cichna-Markl, M.; Dobrovolny, S.; Hochegger, R. Identification of Mammalian and Poultry Species in Food and Pet Food Samples Using 16S rDNA Metabarcoding. *Foods* **2021**, *10*, 2875. https://doi.org/10.3390/foods10112875

Academic Editor: Bianca Castiglioni

Received: 23 September 2021
Accepted: 15 November 2021
Published: 20 November 2021

Publisher's Note: MDPI stays neutral with regard to jurisdictional claims in published maps and institutional affiliations.

1. Introduction

Commercial food and feed products must meet the requirements of national and international regulations. Manufacturers have to ensure that their products are both safe and authentic. However, food fraud has become a global issue, with meat products being particularly vulnerable to adulteration [1]. The term food fraud encompasses a variety of activities that are committed intentionally and aimed at deceiving consumers with respect to food quality. Meat products are frequently found to be adulterated by substitution of animal species given on the label by animal species of lower commercial value [2].

Food controls play a crucial role in the mitigation of food fraud. For the differentiation of animal species in food products, various molecular methodologies have been developed, including protein- and DNA-based ones [2–6]. DNA-based methodologies make use of genetic variations between species, e.g., single nucleotide polymorphisms (SNPs), insertions and deletions. They target either species-specific fragments in nuclear DNA or conserved regions in the mitochondrial genome. At present, DNA arrays and real-time PCR assays are mainly used for the authentication of meat products in official food laboratories.

DNA arrays are based on DNA hybridization [7]. In a first step, the target region, e.g., a conserved region of 16S rDNA, is amplified using biotinylated primers, resulting in the formation of biotinylated PCR products. The labeled PCR products are hybridized to species-specific oligonucleotide probes prespotted on a chip. After removing unbound PCR products by washing, hybridized PCR products are detected enzymatically. Commercial DNA arrays for animal species differentiation are fast, robust and cost-efficient [7]. They allow the simultaneous detection of the most relevant mammalian and poultry species for human consumption. Depending on sample matrix and processing grade, the limit of detection (LOD) ranges from 0.1% to 1%. A disadvantage of DNA arrays is that they do not yield quantitative information.

This limitation can be overcome by performing real-time PCR. However, quantification of animal species in meat products by real-time PCR is known to be a challenging task [1,3]. The main problem is to evaluate the meat content (w/w) one is actually interested in from the DNA concentration (e.g., ng/µL) determined by real-time PCR. Differences in tissue type, the number of cells per unit of mass, genome size, processing grade, and DNA extractability may impair the accuracy of quantitative results [8]. Various strategies have been proposed to compensate for these differences, e.g., the use of matrix-specific calibrators [9–11]. However, this strategy is very labor and time consuming. Thus, normalization with DNA extracts from material of defined composition [12] and relative quantification by using a reference real-time PCR assay [13–15] are widely applied in food control laboratories. With both approaches, the DNA ratios of the respective animal species in samples are obtained. Multiplex real-time PCR assays allow the identification of multiple species simultaneously, e.g., cattle, pig, turkey and chicken [16]; cattle, pig, equids and sheep [11]; roe deer, red deer, fallow deer and sika deer [17]; chicken, guinea fowl and pheasant or quail and turkey [18]. However, the number of species that can be targeted simultaneously is limited by the number of optical channels of the real-time PCR instrument.

In recent years, remarkable progress has been made towards developing DNA barcoding and DNA metabarcoding methods for food authentication [19–23]. DNA barcoding is based on amplification of short DNA barcode regions, followed by either high resolution melting (HRM) analysis [24,25] or Sanger sequencing [26,27]. DNA metabarcoding is the processing of multiple DNA templates using next-generation sequencing (NGS) technologies. While DNA barcoding via Sanger sequencing can only be applied for single species products, DNA metabarcoding also enables the identification of species in complex food and feed products containing multiple species. After amplifying the DNA barcode region, all amplicons, even those obtained for different samples, are sequenced in parallel. Finally, reads are analyzed using a bioinformatic workflow and compared to DNA reference sequences from well-characterized species for taxonomic assignment.

We have recently developed a DNA metabarcoding method allowing the identification of 15 mammalian and six poultry species [28]. The applicability of the method targeting a region of 16S rDNA was investigated by analyzing DNA extract mixtures and model sausages. The species of interest could be identified, differentiated and detected down to a proportion of 0.1%.

In the present study, we aimed at investigating the applicability of the DNA metabarcoding method for routine analysis in more detail. The design parameters and objectives of our study were as follows:

- The study included 25 reference samples with known composition, 56 commercial food and 23 pet food products.
- All samples were analyzed by the DNA metabarcoding method published previously [28] as well as by a commercial DNA array and/or by real-time PCR.

- Qualitative and quantitative results obtained by DNA metabarcoding were compared to those obtained by the two PCR methodologies currently playing the most important role in meat species authentication in official food laboratories.
- A subset of seven reference samples was analyzed by using the DNA metabarcoding method in two independent laboratories, yielding information on the robustness and reproducibility of the method.
- We evaluated whether the results obtained by DNA metabarcoding were in line with sample composition (reference samples) or declaration (commercial food and pet food products).

2. Materials and Methods

2.1. Samples

For this study, a collection of various samples was compiled. Reference samples, comprising eight meat mixtures (LGC7240-49), four dairy products (DLA45 1–4) and 13 boiled sausages (DLA44, DLAptAUS2, Lippold A–C 2019–2021), were supplied by regulatory authorities (LGC Standards Ltd., Teddington, UK; DLA—Proficiency Tests GmbH, Sievershütten, Germany; LVU Lippold, Herbolzheim, Germany). Food and pet food products were obtained from official food control agencies and supermarkets. The study mainly focused on sausages and pet food containing game species because these products are known to be vulnerable to the substitution of high-value game ingredients by lower-quality, cheaper meat species.

Reference samples were analyzed in "laboratory 1" (Chemical and Veterinary Analytical Institute Muensterland-Emscher-Lippe (CVUA-MEL) in cooperation with Chemical and Veterinary Analytical Institute Ostwestfalen-Lippe (CVUA-OWL), where sequencing was performed. A subset of seven reference samples was also analyzed in "laboratory 2" (Austrian Agency for Health and Food Safety (AGES)). Commercial food and pet food samples were analyzed independently either in laboratory 1 or laboratory 2.

2.2. DNA Extraction and Quantification

After homogenization and prior to DNA isolation, all samples were lysed in the presence of a lysis buffer and proteinase K solution at elevated temperature under constant shaking. Afterwards, DNA extraction was performed using commercially available kits. DNA from reference samples was isolated with either the Wizard Genomic DNA Purification Kit, the Wizard DNA Clean-Up Kit or the Maxwell 16 FFS Nucleic Acid Extraction Kit from Promega (Madison, WI, USA) according to the respective manufacturer's instruction sheet. DNA from food and pet food samples was extracted with either the DNeasy mericon Food Kit (Qiagen, Hilden, Germany) or the Maxwell 16 FFS Nucleic Acid Extraction Kit (Promega, Madison, WI, USA), following the instructions of the manufacturers. DNA isolates were stored at −20 °C. Before DNA library preparation, the concentration of individual DNA extracts was determined either with a spectrophotometer (Eppendorf, Hamburg, Germany) or a Qubit 2.0 fluorometer (Thermo Fisher Scientific, Waltham, MA, USA) by using the dsDNA BR assay kit (Thermo Fisher Scientific, Waltham, MA, USA).

2.3. DNA-Library Preparation and NGS

A ~120 base pair fragment of the mitochondrial 16S rDNA gene was used as barcode region for species identification. Library preparation was carried out as described previously [28] with minor modifications. PCR products were indexed using the Illumina Nextera XT Index Kit v2 set A-D or the IDT-Illumina Nextera DNA UD Indexes Kit (Illumina, San Diego, CA, USA). Paired-end sequencing (2 × 150 bp) was performed with either the MiSeq Reagent Kit v2 or the MiSeq Reagent Kit v2 Micro (Illumina, San Diego, CA, USA) at a final loading concentration between 8–10 pM, depending on the instrument and the reagent kit, using the MiSeq system. PhiX DNA, added at a concentration of ~5%, served as sequencing control.

2.4. NGS Data Analysis Using Galaxy

After paired-end sequencing and FastQ file generation via on-board MiSeq Control software (version 2.6.2.1, Illumina, San Diego, CA, USA) and MiSeq Reporter software (version 2.6.2.3, Illumina, San Diego, CA, USA), the resulting FastQ files were used as input for data analysis. Afterwards, the previously uploaded files were processed according to the analysis pipeline as described previously [28] by using the Galaxy platform with the following modifications: the target-specific primer sequences were trimmed off with Cutadapt, Galaxy Version 1.16.6 [29] instead of using the tool Trim (Galaxy Version 0.0.1). Moreover, NGS reads were not clustered into operational taxonomic units (OTUs). After completely identical reads were collapsed into a representative sequence with the tool Dereplicate, Galaxy Version 1.0.0 [30], these sequences were directly matched against a customized database including 51 mitochondrial genomes from animals using BLASTn.

2.5. DNA Array and Real-Time PCR Assays

The LCD Array Kit MEAT 5.0 (Chipron GmbH, Berlin, Germany), allowing the simultaneous detection of 17 mammalian and seven bird species, was performed following the manufacturer's instruction. Data analysis was done with the SlideReader Software (version 12, 2012-01, Chipron GmbH, Berlin, Germany).

Real-time PCR assays for the detection and quantification of meat species were performed following protocols published previously [11,14,16,18,31–35]. Quantification was carried out either by normalization with DNA extract from material of defined composition or relatively by using a reference real-time PCR assay [13].

3. Results and Discussion

In order to investigate the applicability of the DNA metabarcoding method for routine analysis, a total of 104 samples were analyzed. The samples consisted of 25 reference samples, 56 food products, and 23 pet food products. In addition to DNA metabarcoding, each sample was analyzed by real-time PCR and/or a commercial DNA array to evaluate the reliability of the DNA metabarcoding method. Results obtained by DNA metabarcoding are expressed as the ratio of the number of reads that were assigned to the respective meat species and the total number of reads that passed the amplicon analysis pipeline. The results obtained by the commercial DNA array are given as "positive" or "negative", results obtained by real-time PCR as a ratio of DNA (%).

3.1. Reference Samples

Twenty-five reference samples were analyzed, comprising eight meat mixtures, four dairy products and thirteen boiled sausages. Reference samples contained from two to 14 meat species in a ratio from 1.0 to 99.0% (*w/w*) (Table 1). In total, 20 different animal species, including 14 mammalian species (moose, kangaroo, sheep, buffalo, horse, cattle, hare, goat, red deer, pork, rabbit, roe deer, reindeer and fallow deer) and six poultry species (ostrich, pheasant, Muscovy duck, turkey, goose, and chicken) were present in the reference samples. Results obtained by DNA metabarcoding, DNA array and real-time PCR assays are summarized in Table 1.

Table 1. Results obtained for reference samples. DNA array and real-time PCR results were obtained in laboratory 1. DNA metabarcoding results were obtained in laboratory 1, except those marked by footnote 5.

Reference Sample	Composition		DNA Metabarcoding Ratio of Reads (%) [4]	Real-Time PCR (Ratio of DNA (%)) or DNA Array (Positive/Negative)
	Species	**Ratio (%, *w/w*)**		
LGC7242	cattle	99.0	98.2	98.9 [1]
	pork	1.0	1.8	1.1 [1]
LGC7240	cattle	99.0	98.8	95.9 [1]
	horse	1.0	1.2	1.3 (Equidae) [1]
LGC7249	sheep	95.0	90.1	90.3 [1]
	cattle	5.0	9.9	9.7 [1]
LGC7248	sheep	99.0	97.7	97.0 [1]
	cattle	1.0	2.3	3.0 [1]
LGC7245	sheep	95.0	98.4	93.0 [1]
	chicken	5.0	1.6	7.0 [1]
LGC7244	sheep	99.0	100.0	99.9 [1]
	chicken	1.0	<0.1	0.1 [1]
LGC7247	sheep	95.0	96.3	94.9 [1]
	turkey	5.0	3.8	5.1 [1]
LGC7246	sheep	99.0	98.8	98.8 [1]
	turkey	1.0	1.2	1.2 [1]
DLA44-1, 2019	pork	93.4	89.6	88.5 [1]
	horse	6.6	10.4	11.5 (Equidae) [1]
DLA44-3, 2019	pork	87.3	87.4	85.1 [1]
	turkey	7.0	7.6	11.3 [1]
	cattle	5.6	5.1	3.6 [1]
DLA45-1, 2019	cattle	92.0	91.8/94.2 [5]	90.7 [1]
	buffalo	8.0	8.0/5.7 [5]	9.3 [1]
DLA45-2, 2019	buffalo	81.0	72.5/72.3 [5]	71.5 [1]
	cattle	10.0	10.5/11.6 [5]	7.6 [1]
	sheep	9.0	16.7/15.7 [5]	20.9 [1]
	goat	not added [6]	0.3/0.3 [5]	negative [3]
DLA45-3, 2019	cattle	89.0	65.5 / 73.0 [5]	56.2 [1]
	goat	11.0	34.5/27.0 [5]	43.8 [1]
DLA45-4, 2019	goat	90.0	95.2/94.2 [5]	96.9 [1]
	sheep	10.0	4.7/5.6 [5]	3.4 [1]
DLAptAUS2-3.1, 2020	pork	90.9	98.7	99.7 [1]
	donkey	9.1	1.1	positive [3]
	horse	not added [6]	0.2	0.3 (Equidae) [1]
Lippold-A, 2013	cattle	27.8	18.5	14.7 [2]
	sheep	16.7	14.0	6.6 [2]
	chicken	22.2	10.8	15.3 [2]
	goose	11.1	15.7	positive [3]
	Muscovy duck	11.1	12.8	positive [3]
	roe deer	11.1	28.2	18.1 [2]

Table 1. *Cont.*

Reference Sample	Composition		Results	
	Species	Ratio (%, *w/w*)	DNA Metabarcoding Ratio of Reads (%) [4]	Real-Time PCR (Ratio of DNA (%)) or DNA Array (Positive/Negative)
Lippold-A, 2019	red deer	16.0	22.8/24.4 [5]	13.2 [2]
	cattle	15.6	9.1/11.2 [5]	22.2 [2]
	ostrich	15.3	17.6/19.9 [5]	positive [3]
	hare	14.4	8.6/7.6 [5]	positive [3]
	kangaroo	14.2	16.8/9.1 [5]	positive [3]
	sheep	12.6	12.5/13.9 [5]	10.3 [2]
	pheasant	12.0	12.5/14.0 [5]	10.5 [2]
Lippold-B, 2019	goose	16.4	23.2/23.0 [5]	positive [3]
	rabbit	15.5	3.7/2.6 [5]	positive [3]
	chicken	14.9	7.6/6.8 [5]	16.6 [2]
	pork	13.6	21.4/21.7 [5]	2.9 [2]
	moose	13.6	13.0/13.3 [5]	positive [3]
	roe deer	13.5	24.5/26.4 [5]	23.8 [2]
	turkey	12.4	6.6/6.2 [5]	8.7 [2]
Lippold-C, 2019	pork	28.9	9.6/8.8 [5]	8.2 [2]
	horse	17.8	19.4/17.2 [5]	10.6 (Equidae) [2]
	Muscovy duck	16.4	19.9/22.5 [5]	positive [3]
	reindeer	13.8	32.0/32.4 [5]	positive [3]
	goat	12.0	6.7/6.8 [5]	2.8 [2]
	fallow deer	11.1	-	12.6 [2]
	cattle	traces [7]	1.1/1.2 [5]	1.8 [2]
Lippold-A, 2020	goose	38.8	49.9	positive [3]
	horse	25.0	28.5	12.9 (Equidae) [2]
	pork	12.5	3.7	9.1 [2]
	hare	11.2	6.8	positive [3]
	Muscovy duck	10.0	9.6	positive [3]
	turkey	2.5	1.5	2.3 [2]
Lippold-B, 2020	pork	31.3	12.2	10.2 [2]
	fallow deer	24.1	-	12.9 [2]
	reindeer	17.9	45.0	positive [3]
	chicken	12.5	9.4	15.9 [2]
	goat	11.7	7.5	3.7 [2]
	turkey	2.4	1.8	1.6 [2]
Lippold-C, 2020	goose	8.1	14.5	positive [3]
	red deer	8.1	10.5	10.8 [2]
	cattle	7.9	3.9	21.2 [2]
	rabbit	7.7	4.0	positive [3]
	chicken	7.4	4.2	13.0 [2]
	hare	7.3	2.2	positive [3]
	kangaroo	7.2	6.5	positive [3]
	pork	6.7	11.3	2.5 [2]
	moose	6.7	7.1	positive [3]
	roe deer	6.7	14.4	22.4 [2]
	sheep	6.3	5.2	2.8 [2]
	turkey	6.1	3.5	5.4 [2]
	pheasant	6.0	5.0	positive [3]
	ostrich	7.7	7.7	positive [3]

Table 1. *Cont.*

Reference Sample	Composition		Results	
	Species	Ratio (%, *w/w*)	DNA Metabarcoding Ratio of Reads (%) [4]	Real-Time PCR (Ratio of DNA (%)) or DNA Array (Positive/Negative)
Lippold-A, 2021	cattle	8.5	8.0	4.1 [2]
	pork	6.3	10.6	3.1 [2]
	sheep	7.8	4.7	6.2 [2]
	horse	6.3	3.8	3.5 (Equidae) [2]
	red deer	7.8	14.1	7.4 [2]
	fallow deer	6.3	-	3.8 [2]
	roe deer	6.3	11.6	11.3 [2]
	moose	6.3	6.4	positive [3]
	kangaroo	7.4	8.1	positive [3]
	rabbit	7.1	1.7	positive [3]
	reindeer	6.1	9.8	positive [3]
	chicken	9.8	4.6	12.2 [2]
	turkey	6.3	2.6	5.7 [2]
	ostrich	7.8	7.8	positive [3]
Lippold-B, 2021	cattle	traces [7]	2.8	1.8 [2]
	pork	32.6	10.6	14.2 [2]
	horse	4.3	4.0	2.0 (Equidae) [2]
	roe deer	14.4	27.4	27.4 [2]
	moose	10.9	19.7	positive [3]
	kangaroo	13.9	12.7	positive [3]
	hare	10.9	8.4	positive [3]
	pheasant	13.1	14.4	positive [3]
Lippold-C, 2021	cattle	25.0	14.9	6.2 [2]
	pork	13.9	14.5	2.3 [2]
	sheep	14.3	12.9	3.9 [2]
	goat	16.4	7.3	2.2 [2]
	red deer	12.1	20.2	6.6 [2]
	goose	7.8	15.9	positive [3]
	Muscovy duck	10.4	14.5	positive [3]

-: Not detected. [1] Relative quantification based on normalization. [2] Relative quantification by using a reference real-time PCR assay. [3] Obtained by the DNA array. [4] For samples containing fallow deer, ratios of reads refer to 100% minus ratio (%, *w/w*) of fallow deer. [5] Obtained in laboratory 2 (AGES). [6] Proficiency test results were inconsistent, some were positive, some negative. [7] Species not added intentionally, but identified by 86% (Lippold-C, 2019) and 97% (Lippold-B, 2021) of the participants of the proficiency test.

3.1.1. Qualitative Results

The DNA metabarcoding method allowed the detection of 19 out of the 20 animal species covered by the reference samples. Fallow deer could not be detected because the DNA barcode region of fallow deer is not amplified due to two mismatches in the reverse primer (unpublished data). The DNA metabarcoding method allowed accurate identification of animal species in meat mixtures, dairy products, and boiled sausages. Species could be identified correctly down to a ratio of 1% (*w/w*). Goat DNA was detected at low concentration (0.3%) in one dairy sample (DLA45-2), although goat was not added intentionally. Notably, for this sample, proficiency test results were inconsistent (some were positive, some negative) [36].

The commercial DNA array and real-time PCR assays also allowed correct identification of all species contained. In contrast to the DNA metabarcoding method, goat was not detected in the dairy sample DLA45-2.

A subset of seven reference samples, including four dairy products (DLA45 1–4) and three boiled sausages (Lippold A–C, 2019), was independently subjected to DNA metabarcoding analysis at the AGES (laboratory 2, Table 1). In spite of small differences in the workflow, including a different sequencing chemistry, the species identified were identical, demonstrating the robustness of the DNA metabarcoding method. In line with laboratory 1, goat DNA was detected in dairy sample DLA45-2.

3.1.2. Quantitative Results

In order to investigate the applicability of the DNA metabarcoding method for obtaining quantitative results, we calculated the relative quantification error (RQE, absolute difference between the expected and experimentally determined ratio of the species contained in the sample, normalized by the expected value). RQE of the DNA metabarcoding method depended on the ratio of the species in the reference sample (Figure 1A). For species being present at a concentration ratio ≤5%, the median of RQE was 33%. For concentration ratios ranging from 5% to 20%, the median RQE was slightly higher (42%). As expected, the lowest RQE (7%) was obtained for concentration ratios >20%.

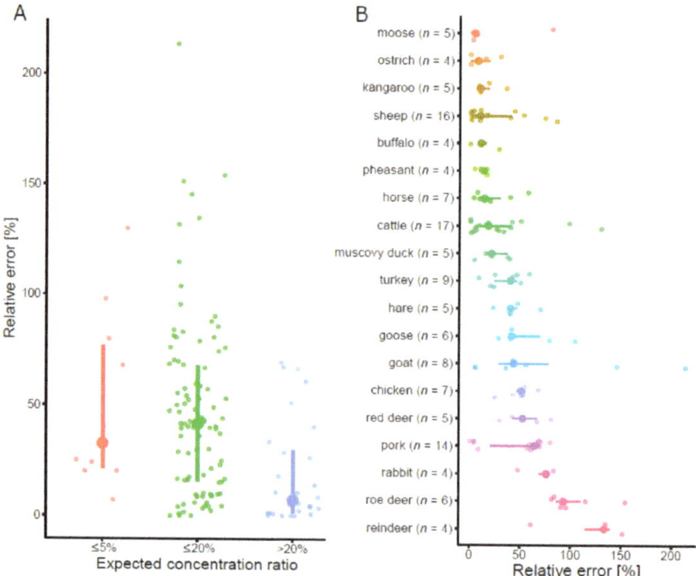

Figure 1. Relative quantification error (RQE) of the DNA-metabarcoding method on reference samples. RQE was calculated as the difference between the expected concentration ratio of a species and the proportion of reads assigned to that species, normalized by the expected concentration ratio. (**A**) RQE for different concentration ratio ranges. Small points represent a single measurement, large points and lines represent the median and inter-quantile range, respectively. Red: expected concentration <5%, green: expected concentration between 5% and 20%, blue: expected concentration >20%. (**B**) RQE by species. RQE calculated as for (**A**) is represented for each species, the number of data points (including those obtained in laboratory 2 (AGES)) is indicated in parenthesis. Species are sorted according to their median RQE from top (lowest) to bottom (highest). Small points represent a single measurement, large points and lines represent the median and inter-quantile range, respectively.

In Figure 1B, the RQE is shown for each of the 19 species detected by DNA metabarcoding. For eight mammalian (moose, kangaroo, sheep, buffalo, horse, cattle, hare, and goat) and five poultry species (ostrich, pheasant, Muscovy duck, turkey, and goose), the median RQE was <50%. For four mammalian species (red deer, pork, rabbit, and roe deer) and chicken, the median RQE was between 50% and 100%. The highest median RQE was obtained for reindeer (133%).

RQE was also calculated for real-time PCR (difference between the ratio of the species contained in the reference sample (Table 1, column 3) and the ratio of DNA (%) determined by real-time PCR (Table 1, column 5), divided by the ratio of the species contained in the reference sample (Table 1, column 3)). The boxplot in Figure 2A shows the distributions of RQE determined by DNA metabarcoding and real-time PCR. Median and interquartile ranges for NGS and PCR errors are 39.7% (7.8%–59.9%) and 36.9% (11.4%–67.9%), respectively, indicating that the two distributions largely overlap.

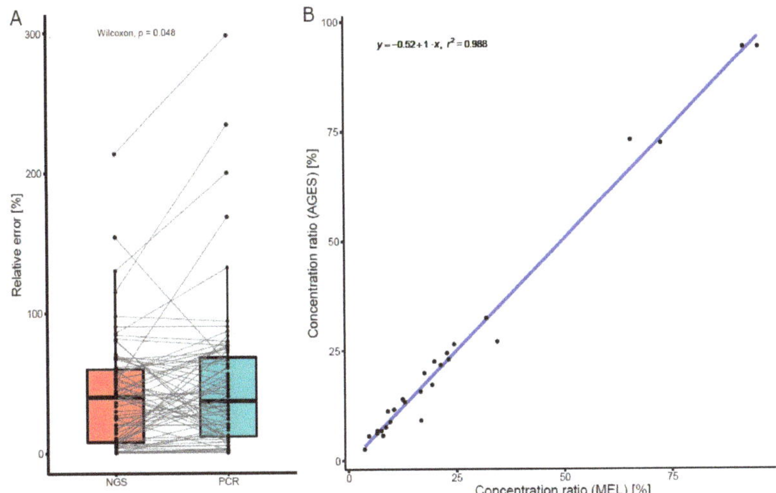

Figure 2. Precision and reproducibility of the DNA-metabarcoding method. (**A**) Comparison of RQE of the DNA metabarcoding method (red) compared to that of real-time PCR (blue). Only species for which quantitative PCR was performed are represented. Black points represent single measurement and grey lines connect paired values. Colored boxes represent the interquartile range with the horizontal line at the median and whiskers represent the Tukey-corrected minimum and maximum. Although a significant difference between the two distributions was calculated (paired Wilcoxon rank test $p = 0.048$), the quantitative difference is too small to be biologically relevant. (**B**) Reproducibility of DNA metabarcoding quantification in two different laboratories. A subset of the samples was quantified with the DNA metabarcoding method in laboratory 1 (CVUA-MEL, x-axis) and laboratory 2 (AGES, y-axis), with highly similar results. A linear regression (blue) of both datasets showed a slope of 1 and a Pearson correlation coefficient $r^2 = 0.988$. Each point represents a single observation.

For all major components (cattle, sheep; 95% or 99%) in meat mixtures, the RQE of the DNA barcoding method and real-time PCR was <6%. For the minor component (horse, turkey; 1%, 5%) in samples LGC7240, LGC7247, and LGC7246, the RQE of both methods was <30%. Both DNA metabarcoding and real-time PCR led to substantially too high ratios (RQE 94%—200%) for cattle as minor component (LGC7249, 5%; LGC7248, 1%). The content of pork (1%) in sample LGC7242 was substantially overestimated (RQE 80%) by DNA metabarcoding, but not by real-time PCR.

Each of the four dairy products contained one major component (cattle, buffalo, or goat) and one, two, or three minor components (buffalo, cattle, sheep, or goat). The major components could be quantified with the RQE <30% with both methods. Only in sample DLA45-3, cattle was substantially underestimated by real-time PCR (RQE 37%). Due to high lipid content and harsh processing procedures, DNA isolated from dairy products is frequently not amplified efficiently [37]. Underestimation of cow milk compared to goat milk by real-time PCR has already been reported by Rentsch et al. and was explained by the relatively low number of somatic cell counts in cow milk compared to goat milk [31]. In the case of minor components, for buffalo (8%) and cattle (10%) in samples DLA45-1 and DLA45-2, respectively, the RQE of DNA metabarcoding and real-time PCR was ≤24%. Goat (11%) was substantially overestimated in sample DLA45-3 (RQE 214% and 298%), and sheep (10%) substantially underestimated in DLA45-4 (RQE 53% and 66%) by DNA barcoding and real-time PCR.

The number of species in 13 boiled sausages ranged from two (DLA44-1) to 14 (Lippold-C, 2020 and Lippold-A, 2021). For major components at a ratio >85% (pork in samples DLA44-1, DLA44-3, and DLAptAUS2-3.1), the RQE of DNA metabarcoding and real-time PCR was <10%. The major components at a ratio of between 85% and 20% (Lippold-A, 2013: cattle, chicken; Lippold-C, 2019: pork; Lippold-A, 2020: horse; Lippold-B, 2020: pork; Lippold-B, 2021: pork; Lippold-C, 2021: cattle) were underestimated by DNA metabarcoding and real-time PCR, with the RQE ranging from 33% to 67% and 31% to 75%, respectively. A number of minor components at a ratio of between 20% and 5% could be quantified with RQE <30% by either DNA metabarcoding (e.g., Lippold-A, 2013: sheep, Muscovy duck; Lippold-A, 2019; Lippold-C, 2020: sheep), or real-time PCR (e.g., Lippold-A, 2019: red deer; Lippold-B, 2019: chicken, turkey) or both methods (e.g., Lippold-A, 2019: sheep, pheasant; Lippold-B, 2020: chicken).

For cattle in samples Lippold-C, 2019 and Lippold-B, 2021 ratios of 1.1/1.2% (NGS) and 1.8% (PCR) or 2.8% (NGS) and 1.8% (PCR) were determined, respectively. Cattle was not added intentionally to these samples, but was contained as traces probably due to production-related carryover. Results of both proficiency tests showed that most participants (86% and 97%) also identified cattle in these samples.

Quantitative data sets obtained for the subset of seven reference samples analyzed in laboratory 1 and laboratory 2 by DNA metabarcoding showed a very good correlation (r^2 = 0.988) (Figure 2B), indicating the high reproducibility of the method. In conclusion, we found that the RQE was quite variable and depended on both the concentration and the identity of the analyte. Additionally, the error was comparable to that of PCR, the current gold-standard method.

Overall our data confirm the limitations known for DNA quantification in meat products [23]. Due to the differences in tissue type, the number of cells per unit of mass, genome size, processing grade and DNA extractability, quantitative results derived from DNA-based methods should serve only as rough estimates for weight ratios of different species in food and feed [8]. During manual and industrial production of meat products production-related carryover of undeclared animal species regularly occurs. In routine analysis of samples in public laboratories, mass concentrations below 1% (*w/w*) are generally reported as possible process contaminants and do not constitute a violation of declaration. Considering the high quantitation errors of DNA-based methods, in most cases a factor of five might be appropriate to discriminate between production-related carryover of undeclared species and mislabeling.

3.2. Commercial Food Products

Table 2 summarizes the results obtained by DNA metabarcoding, real-time PCR and DNA array for 56 commercial food products obtained from food control agencies or purchased at local supermarkets. The samples comprised 34 sausages, including seven wild boar sausages, 20 deer sausages and seven further sausages, six vertical rotating meat spits, seven pâtés, two minced meat products, one steak, two convenience foods, and four milk products.

Table 2 indicates that DNA metabarcoding and real-time PCR and/or the commercial DNA array led to identical qualitative results for the 56 commercial food products. However, for discrimination of meat from wild boar (*Sus scrofa scrofa*) and meat from domestic pig (*Sus scrofa domesticus*), results of two singleplex real-time PCR assays and/or a duplex real-time PCR assay developed recently had to be taken into account [38]. Neither the DNA metabarcoding method nor common real-time PCR assays for pork allow distinguishing between wild boar and pork, yielding only information on the total ratio of wild boar and pork DNA. This is due to the fact that the genomes of the two subspecies are highly homologous and hybridization and back-crossings increased sequence homologies and intra-subspecies variability [39,40].

Table 2. Results for commercial food products.

Sample	Animal Species Declared	Result			Comment
		Animal Species Detected	DNA Metabarcoding Ratio of Reads (%)	Real-Time PCR (Ratio of DNA (%)) or DNA Array (Positive/Negative)	
wild boar sausage 1	wild boar, pork, pork bacon	pork	83.0 [4]	52.5 [1]	
		wild boar		35.2 [1]	
		red deer	15.1	3.9 [1]	not declared, >5%
		cattle	1.7	8.4 [1]	not declared, 1%–5%
wild boar sausage 2	wild boar, pork, pork bacon	wild boar	86.9 [4]	23.7 [1]	
		pork		64.1 [1]	
		red deer	13.1	12.3 [1]	not declared, >5%
wild boar sausage 3	55% wild boar, 36% roe deer	roe deer	60.7	40.8 [1]	
		pork	25.1 [4]	50.5 [1]	
		wild boar		<1.0 [1]	declared and detected [3]
		cattle	14.0	8.7 [1]	not declared, >5%
wild boar sausage 4	74% red deer, 22% wild boar bacon	cattle	30.2	46.4 [1]	not declared, >5%
		pork	28.8 [4]	43.5 [1]	not declared, >5%
		wild boar		<1.0 [1]	declared and detected, r.s.
		red deer	22.8	10.1 [1]	declared and detected, r.s.
		chamois	18.2	-	not declared, >5%
wild boar sausage 5	chamois, wild boar, roe deer, pork bacon	pork	48.8 [4]	8.9 [1]	
		wild boar		38.2 [1]	
		red deer	36.8	42.0 [1]	not declared, >5%
		roe deer	14.0	10.9 [1]	
		chamois	0.0	-	declared, but not detected
wild boar sausage 6	no declaration	pork	62.2 [4]	28.5 [1]	
		wild boar		26.1 [1]	
		roe deer	24.4	16.0 [1]	
		cattle	13.4	29.4 [1]	

Table 2. *Cont.*

Sample	Animal Species Declared	Result			Comment
		Animal Species Detected	DNA Metabarcoding Ratio of Reads (%)	Real-Time PCR (Ratio of DNA (%)) or DNA Array (Positive/Negative)	
wild boar sausage 7	game, cattle, pork bacon	pork	70.2 [4]	60.3 [1]	
		wild boar		16.0 [1]	
		cattle	28.7	23.7 [1]	
		roe deer	<1.0	<1.0 [1]	
		sheep	<1.0	<1.0 [1]	
deer sausage 1	deer, pork, pork bacon	red deer	72.0	52.6 [1]	
		pork	19.5	41.6 [1]	
		cattle	8.5	5.8 [1]	not declared, >5%
deer sausage 2	roe deer, pork, pork bacon	roe deer	52.0	28.3 [1]	
		pork		54.3 [1]	
		wild boar	22.8 [4]	<1.0	
		cattle	10.8	7.9 [1]	not declared, >5%
		red deer	14.3	9.5 [1]	not declared, >5%
deer sausage 3	roe deer, pork	roe deer	89.9	75.1 [1]	
		pork	5.9	23.0 [1]	
		cattle	4.2	1.9 [1]	not declared, 1%–5%
deer sausage 4	deer, pork	red deer	67.0	52.3 [1]	
		pork		47.7 [1]	
		wild boar	33.0 [4]	<1.0 [1]	
deer sausage 5	roe deer, pork, pork bacon	roe deer	81.5	78.5 [1]	
		pork	9.3	15.8 [1]	
		cattle	9.0	5.7 [1]	
		red deer	< 1.0	<1.0 [1]	not declared, >5%
deer sausage 6	game, pork	red deer	83.8	66.9 [1]	
		cattle	9.3	14.9 [1]	not declared, >5%
		pork	5.2 [4]	15.3 [1]	
		wild boar		<1.0	
		roe deer	1.7	3.0 [1]	

Table 2. *Cont.*

Sample	Animal Species Declared	Result			Comment
		Animal Species Detected	DNA Metabarcoding Ratio of Reads (%)	Real-Time PCR (Ratio of DNA (%)) or DNA Array (Positive/Negative)	
deer sausage 7	70% red deer, 30% pork	red deer	70.4	72.2 [1]	
		pork		<1.0 [1]	
		wild boar	29.6 [4]	27.8 [1]	
deer sausage 8	deer, pork, pork bacon	red deer	68.0	46.7 [1]	
		pork	31.7	53.3 [1]	
		sika deer	<1.0	-	
deer sausage 9	deer, pork, pork bacon	roe deer	79.0	58.7 [1]	
		pork	20.9	41.3 [1]	
deer sausage 10	deer, pork, pork bacon	red deer	74.1	38.5 [1]	
		pork	25.3	61.5 [1]	
deer sausage 11	deer, pork, pork bacon	red deer	72.0	36.4 [1]	
		pork	27.8	63.6 [1]	
deer sausage 12	pork, red deer	pork	66.6	51.9 [2]	
		red deer	33.4	48.1 [2]	
deer sausage 13	roe deer, pork, pork bacon	roe deer	81.6	67.4 [1]	
		pork		32.6 [1]	
		wild boar	18.3 [4]	<1.0 [1]	
deer sausage 14	deer, pork, pork bacon, cattle casing [5]	red deer	70.6	48.7 [1]	
		pork	25.7	50.0 [1]	
		sika deer	2.6	1.3 [1]	
		roe deer	<1.0	<1.0 [1]	
deer sausage 15	pork, deer, cattle	red deer	51.4	39.9 [1]	
		pork	31.1	45.3 [1]	
		cattle	16.9	14.8 [1]	
		roe deer	< 1.0	<1.0 [1]	

Table 2. *Cont.*

Sample	Animal Species Declared	Result		Comment
		Animal Species Detected	DNA Metabarcoding Ratio of Reads (%)	Real-Time PCR (Ratio of DNA (%)) or DNA Array (Positive/Negative)	
deer sausage 16	deer, pork, pork bacon, cattle casing [5]	red deer	53.6	27.2 [1]	not declared, >5%
		pork	24.2	61.3 [1]	
		sheep	21.6	11.4 [1]	
		roe deer	<1.0	<1.0 [1]	
		fallow deer	-	<1.0 [1]	
deer sausage 17	deer, pork, cattle	roe deer	55.8	59.0 [1]	
		red deer	24.5	24.6 [1]	
		pork		8.9 [1]	
		wild boar	10.2 [4]	<1.0 [1]	
		cattle	9.4	7.5 [1]	
deer sausage 18	deer, cattle	red deer	66.1	53.8 [1]	
		cattle	32.2	46.2 [1]	
		sika deer	1.3	<1.0 [1]	
deer sausage 19	deer, cattle	red deer	76.9	43.9 [1]	
		cattle	20.1	56.1 [1]	
		sika deer	2.9	<1.0 [1]	
deer sausage 20	game, cattle, pork bacon	red deer	78.3	80.9 [1]	
		roe deer	14.5	7.0 [1]	
		pork	5.4	10.4 [1]	
		cattle	1.8	1.7 [1]	
sausage 1	chamois, cattle, pork bacon	red deer	35.8	11.9 [1]	not declared, >5%
		pork	29.2	70.0 [1]	declared and detected, r.s.
		cattle	19.8	8.9 [1]	
		roe deer	13.4	9.2 [1]	not declared, >5%
		chamois	1.6	-	declared and detected, r.s.

Table 2. *Cont.*

Sample	Animal Species Declared	Result			Comment
		Animal Species Detected	DNA Metabarcoding Ratio of Reads (%)	Real-Time PCR (Ratio of DNA (%)) or DNA Array (Positive/Negative)	
sausage 2	60% sheep, 35% pork, 5% goat	sheep	44.5	35.7 [1]	
		pork	27.0	49.7 [1]	
		red deer	12.5	3.9 [1]	not declared, >5%
		cattle	8.5	9.4 [1]	not declared, >5%
		goat	7.4	1.4 [1]	
sausage 3	cattle	water buffalo	67.0	-	not declared, >5%
		cattle	33.0	22.9 [2]	
sausage 4	42% cattle, 35% chicken	chicken	86.0	96.4 [1]	
		cattle	13.5	3.6 [1]	declared and detected, r.s.
sausage 5	40% poultry, 15% cattle, cattle fat	turkey	44.4	36.4 [1]	
		chicken	30.1	32.9 [1]	
		cattle	25.0	30.7 [1]	
sausage 6	pork, cattle or lamb	cattle	53.6	59.5 [1]	
		pork	46.1	40.5 [1]	
sausage 7	lamb, pork	sheep	80.1	71.7 [1]	
		pork	19.8	28.3 [1]	
vertical rotating meat spit 1	95% beef	cattle	64.9	85.4 [1]	
		turkey	35.1	14.6 [1]	not declared, >5%
vertical rotating meat spit 2	75% veal, 20% turkey	cattle	57.5	76.0 [1]	
		turkey	35.4	21.1 [1]	
		chicken	7.1	2.9 [1]	not declared, >5%
vertical rotating meat spit 3	70% veal, 20% turkey	cattle	59.2	74.1 [1]	
		turkey	33.7	21.5 [1]	
		chicken	7.2	4.4 [1]	not declared, >5%
vertical rotating meat spit 4	turkey	turkey	98.2	94.8 [1]	
		cattle	1.7	5.2 [1]	not declared, 1%–5%

Table 2. *Cont.*

Sample	Animal Species Declared	Result			Comment
		Animal Species Detected	DNA Metabarcoding Ratio of Reads (%)	Real-Time PCR (Ratio of DNA (%)) or DNA Array (Positive/Negative)	
vertical rotating meat spit 5	55% cattle, 10% turkey, 25% chicken	cattle chicken turkey	58.8 23.0 18.1	29.1 [1] 35.0 [1] 35.9 [1]	
vertical rotating meat spit 6	55% cattle, 35% poultry	cattle chicken turkey	56.0 43.5 < 1.0	41.9 [1] 58.1 [1] <1.0 [1]	
pâté 1	wild boar, pork	pork wild boar	99.8 [4]	100.0 [1] <1.0 [1]	declared and detected, r.s.
pâté 2	game, pork	pork red deer	57.6 42.1	77.5 [1] 22.5 [1]	
pâté 3	49% pork, lamb liver	pork sheep	66.6 33.4	71.9 [1] 28.1 [1]	
pâté 4	pork neck and liver, rabbit meat	pork rabbit	96.2 3.8	46.5 [2] positive [3]	
pâté 5	duck meat and breast, poultry liver	turkey mallard Muscovy duck	49.1 28.1 22.8	21.3 [2] positive [3] positive [3]	
pâté 6	50% pork meat, 20% red deer meat	pork red deer	57.9 42.0	84.4 [2] 15.6 [2]	
pâté 7	33% pork meat, 20% roe deer meat	roe deer pork	59.7 40.3	59.9 [2] 40.1 [2]	
minced meat product 1	chicken, cattle	chicken cattle buffalo, kangaroo, fish	76.2 23.0 <1.0	81.9 [2] 18.1 [2] positive [3]	

Table 2. *Cont.*

Sample	Animal Species Declared	Result		Comment
		Animal Species Detected	DNA Metabarcoding Ratio of Reads (%)	Real-Time PCR (Ratio of DNA (%)) or DNA Array (Positive/Negative)	

Sample	Animal Species Declared	Animal Species Detected	DNA Metabarcoding Ratio of Reads (%)	Real-Time PCR (Ratio of DNA (%)) or DNA Array (Positive/Negative)	Comment
minced meat product 2	lamb, cattle	cattle	70.3	51.6 [1]	
		sheep	29.4	48.4 [1]	
steak	reindeer	reindeer	100.0	positive [3]	
convenience food 1	37% pork and cattle, cattle soup	pork	67.9	82.0 [1]	
		cattle	32.1	18.0 [1]	
convenience food 2	25% pork, cattle soup	pork	87.2	93.9 [1]	
		cattle	12.5	6.1 [1]	
milk product 1	goat	goat	97.4	positive [3]	
		cattle	2.6	positive [3]	not declared, 1%–5%
milk product 2	goat milk	goat	62.8	positive [3]	
		sheep	36.2	positive [3]	
		ibex	<1.0	-	not declared, >5%
		cattle	<1.0	positive [3]	
milk product 3	goat milk	goat	62.9	positive [3]	
		sheep	36.0	positive [3]	
		ibex	<1.0	-	not declared, >5%
		cattle	<1.0	negative [3]	
milk product 4	sheep milk	sheep	95.4	positive [3]	
		goat	4.5	positive [3]	not declared, 1%–5%

-: Not detected, r.s.: ratio suspicious. [1] Relative quantification based on normalization. [2] Relative quantification by using a reference real-time PCR assay. [3] Obtained by the DNA array. [4] Sum of pork and wild boar. [5] In most cases species of casings are not detectable by DNA-based methods.

The ingredient list of 14 out of 20 deer sausages did not contain any information on the deer species (red deer, sika deer, fallow deer). Red deer, roe deer, red deer and roe deer, and red deer and sika deer were detected with DNA ratios >1% in eight, one, three and two of these sausages, respectively. Four and two out of the 20 deer sausages were declared to contain roe deer and red deer, respectively. Our results confirmed the presence of these deer species in the respective food products.

For all species detected in deer sausage 17, sausage 5 and 6, pâté 7 and minced meat product 1 (Table 2), the ratios obtained by DNA metabarcoding and real-time PCR differed by less than 30%. However, in the cases of the other food products, differences >30% were observed for at least one of the species identified.

Comparison of our results, obtained by DNA metabarcoding and real-time PCR and/or the DNA array, with the food ingredient lists revealed multiple discrepancies (Table 2). In a number of commercial food products, species that were not given on the food label were detected by both DNA metabarcoding and real-time PCR and/or the DNA array. Most frequently, the DNA of undeclared species was found in high ratios >5%, indicating that the replacement of meat species by cheaper alternatives is an ongoing food fraud issue. For some products, the species detected were declared but the DNA ratios determined did not correspond with declaration ("declared and detected, ratio suspicious"). In further products, the DNA of undeclared species was detected in traces between 1% and 5%, which were possibly contained due to production-related carry-over. In only one product (wild boar sausage 5), a species declared (chamois) was not detected. Figure 3A summarizes the number of mislabeled species by type of fraud in commercial foodstuffs, Figure 3B the number of mislabeled species by type of food product.

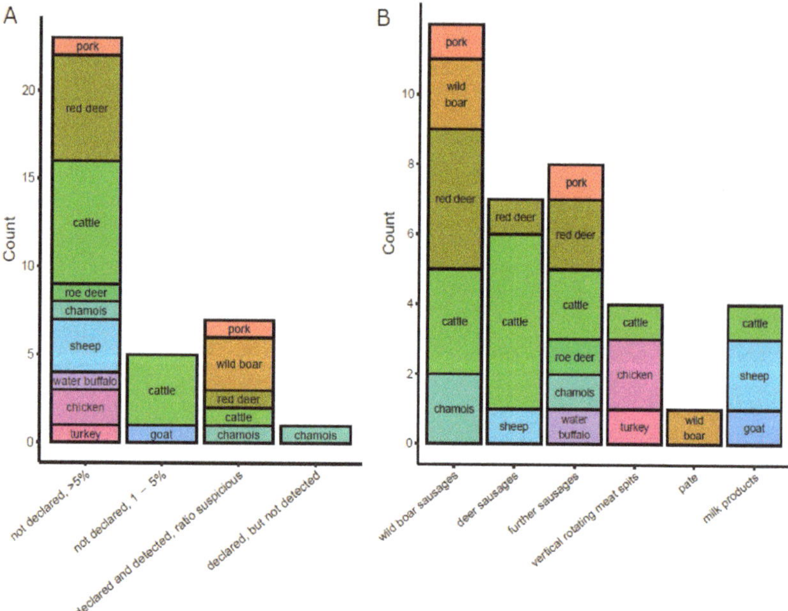

Figure 3. Wrong declarations in foodstuffs (**A**) Breakdown of wrongly labeled species by type of fraud in foodstuffs. Each box represents a single species, the size of the box indicates the number of times that this species appeared for each type of fraud in the dataset. (**B**) Breakdown of wrongly labelled species by type of food product. Each box represents a single species, the size of the box indicates the number of times that this species appeared for each type of food product in the dataset.

3.3. Commercial Pet Food Products

The applicability of the DNA metabarcoding method was also investigated by analyzing 23 pet food products. The following species were given on the food label: deer, roe deer, cattle, sheep, rabbit, chicken, turkey, duck, Muscovy duck, and ostrich. Table 3 indicates that qualitative results obtained by DNA metabarcoding were in line with those obtained by real-time PCR and/or the commercial DNA array. For some animal species, e.g., red deer in samples 1, 3, 12; pork in sample 2, 10, 19, 21; and chicken in samples 19, 22; the ratios determined by DNA metabarcoding and real-time PCR differed by less than 30%. However, in other cases, differences in the ratios >30% were obtained (Table 3).

Fifteen out of the 23 pet food products were declared to contain deer, without disclosing the deer species. DNA metabarcoding and real-time PCR and/or the commercial DNA array detected red deer in six, red deer and roe deer in four and reindeer in one out of these 15 pet food products. In four pet food products (samples 5, 8, 11, and 21), deer was neither detected by DNA metabarcoding nor by real-time PCR and/or the commercial DNA array. Identical qualitative results were also obtained for three pet food products declared to contain roe deer (samples 12, 16, and 18). Each of the methodologies applied yielded a negative result for roe deer, but a positive result for red deer.

In sample 18, sika deer (16.6%) was detected by DNA metabarcoding. Since sika deer is rarely used in pet food products, sample 18 was not analyzed by a real-time PCR assay for sika deer and the DNA array used does not detect sika deer. This example illustrates one of the main limitations of using PCR for meat species authentication: animal species that are not expected will not be detected [41].

In a high number of commercial pet food products, undeclared species were detected by each of the methodologies applied. Most frequently, undeclared species, were present at a ratio >5%, e.g., pork, chicken, cattle, mallard, and turkey (Figure 4). These animal species of lower commercial value mainly replaced deer, either totally or in part. The results show that inspection of pet food for authenticity has high relevance.

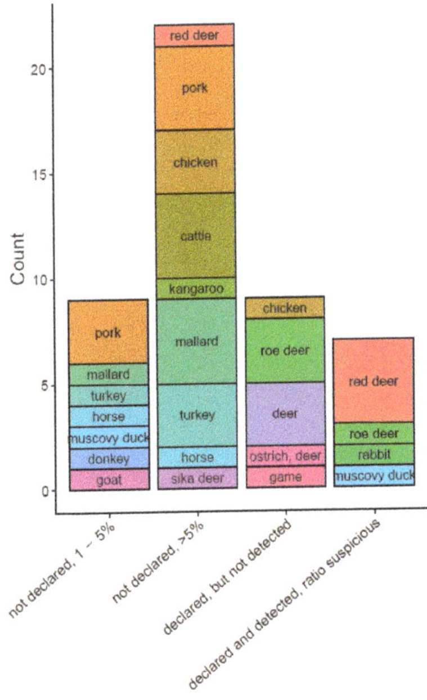

Figure 4. Breakdown of wrongly labelled species by type of fraud in pet food products. Each box represents a single species, the size of the box indicated the number of times that this species appeared for each type of fraud in the dataset.

Table 3. Results for pet food products.

Sample	Animal Species Declared	Result			Comment
		Animal Species Detected	DNA Metabarcoding Ratio of Reads (%)	Real-Time PCR (Ratio of DNA (%)) or DNA Array (Positive/Negative)	
1	65% deer (heart, liver, lung, rumen)	red deer	96.3	92.9 [1]	
		pork	1.7	<1.0 [1]	not declared, 1%–5%
		fallow deer	-	6.0 [1]	
		sheep	<1.0	<1.0 [1]	
		chicken	<1.0	<1.0 [1]	
		cattle	<1.0	<1.0 [1]	
		kangaroo	<1.0	positive [2]	
2	60% deer meat	pork	47.1	32.6 [1]	not declared, >5%
		roe deer	36.0	55.0 [1]	
		red deer	16.9	12.4 [1]	
3	51% deer meat, <2.5% chicken liver	red deer	96.2	95.9 [1]	
		roe deer	2.4	3.1 [1]	
		pork	1.0	<1.0 [1]	not declared, 1%–5%
		rabbit	<1.0	positive [2]	
		chicken	negative	negative [1]	declared, but not detected
4	59% fresh meat from deer and roe deer, 1.2% eggshell powder	red deer	62.4	53.0 [1]	
		mallard	29.8	positive [2]	not declared, >5%
		chicken	6.5	16.9 [1]	
		fallow deer	-	2.3 [1]	
		roe deer	<1.0	<1.0 [1]	declared and detected, r.s.
		pork, sheep, cattle	<1.0	<1.0 [1]	
5	10% deer meat (dried and ground)	chicken	38.1	25.7 [1]	not declared, >5%
		turkey	12.3	7.3 [1]	not declared, >5%
		mallard	10.7	positive [2]	not declared, >5%
		horse	33.0	15.8 (Equidae) [1]	not declared, >5%
		Muscovy duck	4.6	positive [2]	not declared, 1%–5%
		donkey	1.1	positive [2]	not declared, 1%–5%
		cattle	<1.0	<1.0 [1]	
		deer	negative	negative [1,2]	declared, but not detected

Table 3. *Cont.*

Sample	Animal Species Declared	Result			Comment
		Animal Species Detected	DNA Metabarcoding Ratio of Reads (%)	Real-Time PCR (Ratio of DNA (%)) or DNA Array (Positive/Negative)	
6	28% fresh and 26% dried deer meat, 9% chicken fat, 2% dried eggs, 2% fresh and 2% dried herrings, 1% fish oil	pork	92.2	39.4 [1]	not declared, >5%
		fish	-	positive [2]	
		chicken	3.9	36.0 [1]	
		red deer	2.7	2.3 [1]	declared and detected, r.s.
		turkey	<1.0	22.0 [1]	
		sheep	<1.0	<1.0 [1]	
7	18% dried Muscovy duck meat, 9.4% dried and ground deer meat, 6.3% dried whiting, 6.3% ground wild bones, egg yolk powder	cattle	67.8	59.9 [1]	not declared, >5%
		chicken	9.7	13.3 [1]	not declared, >5%
		mallard	7.1	positive [2]	not declared, >5%
		red deer	7.5	2.8 [1]	
		turkey	5.0	2.3 [1]	not declared, >5%
		Muscovy duck	1.8	positive [2]	declared and detected, r.s.
		sheep	<1.0	<1.0 [1]	
		sika deer	<1.0	-	
		goat	<1.0	positive [2]	
		fish	-	positive[2]	
8	50% meat and animal byproducts, 4% ostrich and deer	pork	52.4	3.8 [1]	
		cattle	30.5	69.6 [1]	
		chicken	16.8	26.5 [1]	
		turkey	<1.0	<1.0 [1]	
		mallard	<1.0	negative [2]	
		ostrich, deer	negative	negative [2]	declared, but not detected
9	35% cattle, 31% poultry, 4% deer	cattle	71.4	43.6 [1]	
		turkey	9.0	8.3 [1]	
		reindeer	12.3	positive [2]	
		chicken	6.0	6.4 [1]	
		mallard	<1.0	positive [2]	
		pork, sheep, horse	<1.0	<1.0 [1]	
		red deer	negative	negative [1]	

Table 3. *Cont.*

Sample	Animal Species Declared	Result			Comment
		Animal Species Detected	DNA Metabarcoding Ratio of Reads (%)	Real-Time PCR (Ratio of DNA (%)) or DNA Array (Positive/Negative)	
10	lung, meat, kidney, liver, udder, 5% deer	pork	89.3	88.8 [1]	
		cattle	9.7	10.5 [1]	
		red deer	<1.0	<1.0 [1]	declared and detected, r.s.
		chicken	<1.0	<1.0 [1]	
11	48% fresh deer meat, 4% entrails of deer	mallard	96.3	31.0 [1]	not declared, >5%
		goat	1.7	<1.0 [1]	not declared, 1%–5%
		turkey, chicken	<1.0	<1.0 [1]	
		pork, sheep	<1.0	<1.0 [1]	
		deer	negative	negative [1]	declared, but not detected
12	50% roe deer (60% meat, 25% heart, 10% lung, 5% liver)	red deer	98.1	97.9 [1]	not declared, >5%
		horse	1.6	<1.0 (Equidae) [1]	not declared, 1%–5%
		cattle	<1.0	1.7 [1]	
		fallow deer	-	<1.0 [1]	
		roe deer	negative	negative [1]	declared, but not detected
13	99% deer meat	chicken	71.4	51.8 [1]	not declared, >5%
		kangaroo	17.6	positive [2]	not declared, >5%
		red deer	10.3	3.8 [1]	declared and detected, r.s.
		rabbit	<1.0	positive [2]	
		pork, cattle	<1.0	<1.0 [1]	
14	75% deer (meat, heart, lung)	pork	84.3	45.1 [1]	not declared, >5%
		cattle	6.4	10.0 [1]	not declared, >5%
		roe deer	4.2	38.9 [1]	
		mallard	2.2	1.5 [1]	not declared, 1%–5%
		turkey	1.4	<1.0 [1]	not declared, 1%–5%
		red deer	1.5	1.7 [1]	

Table 3. *Cont.*

Sample	Animal Species Declared	Result			Comment
		Animal Species Detected	DNA Metabarcoding Ratio of Reads (%)	Real-Time PCR (Ratio of DNA (%)) or DNA Array (Positive/Negative)	
15	100% deer meat	roe deer	65.8	12.6 [1]	
		cattle	31.2	87.0 [1]	not declared, >5%
		chicken	<1.0	<1.0 [1]	
		pork	1.9	<1.0 [1]	not declared, 1%–5%
		red deer	1.0	<1.0 [1]	
16	50% roe deer	turkey	98.2	99.6 [1]	not declared, >5%
		red deer, horse	<1.0	<1.0 [1]	
		pork	<1.0	<1.0 [1]	
		roe deer	negative	negative [1]	declared, but not detected
17	60% deer	red deer	40.4	25.3 [1]	
		cattle	36.3	73.0 [1]	not declared, >5%
		pork	22.9	1.7 [1]	not declared, >5%
		chicken	<1.0	<0.1 [1]	
18	46% poultry meat, 8% roe deer	chicken	55.7	83.8 [1]	
		turkey	26.7	13.0 [1]	
		sika deer	16.6	-	not declared, >5%
		cattle	<1.0	<1.0 [1]	
		red deer, pork	<1.0	<1.0 [1]	
		fallow deer	-	3.0 [1]	
		roe deer	negative	negative [1]	declared, but not detected
19	51% meat and animal byproducts, 12% chicken, turkey, duck	pork	58.5	55.0 [1]	
		chicken	26.1	28.4 [1]	
		turkey	9.0	10.1 [1]	
		cattle	5.3	4.6 [1]	
		mallard	<1.0	<1.0 [1]	
		guinea fowl	<1.0	<1.0 [1]	

Table 3. *Cont.*

Sample	Animal Species Declared	Result			Comment
		Animal Species Detected	DNA Metabarcoding Ratio of Reads (%)	Real-Time PCR (Ratio of DNA (%) or DNA Array (Positive/Negative)	
20	51% meat and animal byproducts, 12% cattle, sheep, chicken	chicken	45.2	53.6 [1]	
		pork	40.2	15.3 [1]	
		cattle	10.9	26.8 [1]	
		sheep	3.5	4.3 [1]	
		turkey	<1.0	positive [2]	
21	33% meat and animal byproducts, 4% poultry, 4% deer	pork	94.4	83.6 [1]	
		chicken	3.9	15.3 [1]	
		guinea fowl	<1.0	<1.0 [1]	
		turkey	<1.0	1.0 [1]	
		deer	negative	negative [2]	declared, but not detected
22	meat and animal byproducts (4% turkey, 4% duck, 4% game)	chicken	49.4	57.6 [1]	
		pork	25.1	13.4 [1]	
		cattle	12.6	7.3 [1]	
		turkey	6.3	19.5 [1]	
		duck	4.1	<1.0 [1]	
		sheep	1.9	<1.0 [1]	
		horse	<1.0	<1.0 (Equidae) [1]	
		fish	-	positive [2]	
		game	negative	negative [2]	declared, but not detected
23	40% chicken (heart, meat, liver, stomachs, necks), 28.7% broth, 28% rabbit	chicken	99.1	positive [2]	
		cattle	<1.0	positive [2]	
		rabbit	<1.0	positive [2]	declared and detected, r.s.

-: Not detected, r.s.: ratio suspicious. [1] Relative quantification by using a reference real-time PCR assay. [2] Obtained with the DNA array.

In some products, undeclared species were detected in a ratio between 1% and 5%. Most probably, these species were present due to production-related carry-over. Chicken, roe deer, deer, ostrich and game could not be identified in several pet food products although they were declared to contain these species. In some products, the declared species was detected but the DNA ratio determined drastically differed from the content given on the label (Figure 4). These results are probably caused by total or partial degradation of DNA due to high processing grades of the respective raw materials.

3.4. Cost Analysis

Metabarcoding could be an attractive alternative to real-time PCR in species differentiation, especially due to the possibility of analyzing many samples simultaneously for many species. A detailed cost comparison with the standard real-time PCR method is not yet available. For the present publication a break-even analysis was performed, based on current data from AGES cost accounting, to show what effect the number of samples and the number of parameters (animal species) has on the choice of methodology used. The break-even point or volume (BEP) represents the number of tested samples/parameters where the real-time PCR-based cost equals the NGS-based cost. Above this threshold, an NGS-based approach generates savings. Figure 5A shows the BEP for NGS of a maximum of 21 animal species, corresponding to 21 real-time PCR methods for animal species available in the AGES laboratory. The analysis shows that the use of NGS is more cost-effective for the detection of 21 animal species from the tenth sample onwards. If no multiplex methods for real-time PCR are available in the laboratory, NGS is already profitable from the fifth sample onwards. If the scope of testing is limited to only up to seven animal species per sample, real-time PCR is always cheaper than NGS analysis. Figure 5B shows the BEP at full capacity of the sequencing kit. If the sequencing kit is fully utilized (Illumina MiSeq v2 chemistry, 75 samples, 200,000 reads per sample), the costs per sample are significantly reduced. In this case, NGS is already cheaper from the first sample onwards, if at least 15 parameters are analyzed. Below a parameter number of seven, however, real-time PCR always remains the cheaper method.

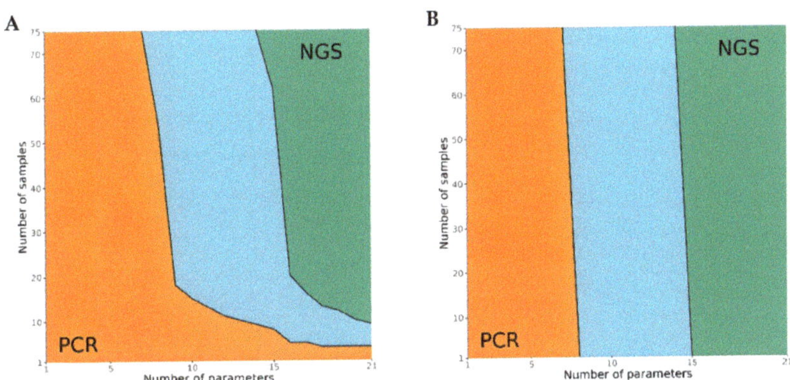

Figure 5. Break-even point analysis of NGS-metabarcoding and real-time PCR for the qualitative identification of bird and mammal species. The left-most orange area corresponds to combinations of sample number/parameter number for which PCR is always cheaper than NGS. The right-most green area corresponds to combinations for which NGS is always cheaper than PCR. The middle blue zone corresponds to combinations for which the cost difference largely depends on the degree of multiplexing of the PCRs. NGS costs were calculated for two exemplary laboratories: (**A**) a laboratory running exclusively meat-metabarcoding runs, and (**B**) a laboratory running full-capacity sequencing runs, for example, mixing samples with other type of assays.

4. Conclusions

By analyzing 25 reference samples, 56 commercial food and 23 pet food products using DNA metabarcoding and real-time PCR and/or a commercial DNA array, we demonstrated that the DNA metabarcoding method developed recently is a suitable screening method for meat species authentication. Qualitative and quantitative results of the DNA metabarcoding method were in line with those obtained by real-time PCR. The results from independent analyses in two laboratories indicate the robustness and reproducibility of the DNA metabarcoding method. Our data on reference samples confirm the limitations known for DNA quantification in meat products. Quantitative results derived from DNA-based methods should serve only as rough estimates for weight ratios of different species in food and feed.

A major advantage of metabarcoding is the parallel detection of a large number of animal species including species not tested routinely or for which no real-time PCR methods are available. Our results indicate that in addition to food products, DNA metabarcoding is particularly applicable to pet food products, which frequently contain multiple animal species and were shown to be also highly prone to adulteration.

For a large number of samples or parameters, metabarcoding is the more cost-effective analysis. By combining different applications (joint sequencing of plant and animal species, bacteria, etc.), an additional cost reduction is possible, as the sequencing kits, the biggest cost driver, can be better utilized.

Author Contributions: Conceptualization, C.B.-N. and R.H.; methodology, S.D. and R.H.; validation, C.B.-N., S.D., R.H. and H.P.; investigation, S.D., H.P. and L.P.; resources, C.B.-N., R.H. and H.P.; writing—original draft preparation, C.B.-N., M.C.-M. and S.D.; writing—review and editing, C.B.-N., M.C.-M., G.D., S.D., R.H., H.P. and L.P.; visualization, G.D.; supervision, C.B.-N. and R.H.; project administration, C.B.-N.; R.H. funding acquisition, C.B.-N. All authors have read and agreed to the published version of the manuscript.

Funding: We gratefully acknowledge funding for metabarcoding analyses of CVUA-MEL by the Ministry for Environment, Agriculture, Conservation and Consumer Protection of the State of North Rhine-Westphalia.

Institutional Review Board Statement: Not applicable, exclude statement. The study did not involve living animals.

Informed Consent Statement: Not applicable, exclude statement.

Data Availability Statement: The data presented in this study are available on request.

Conflicts of Interest: The authors declare no conflict of interest.

References

1. Ballin, N. Authentication of meat and meat products. *Meat Sci.* **2010**, *86*, 577–587. [CrossRef]
2. Montowska, M.; Pospiech, E. Authenticity Determination of Meat and Meat Products on the Protein and DNA Basis. *Food Rev. Int.* **2010**, *27*, 84–100. [CrossRef]
3. Ballin, N.Z.; Vogensen, F.; Karlsson, A.H. Species determination—Can we detect and quantify meat adulteration? *Meat Sci.* **2009**, *83*, 165–174. [CrossRef]
4. Kumar, A.; Kumar, R.R.; Sharma, B.D.; Gokulakrishnan, P.; Mendiratta, S.K.; Sharma, D. Identification of Species Origin of Meat and Meat Products on the DNA Basis: A Review. *Crit. Rev. Food Sci. Nutr.* **2013**, *55*, 1340–1351. [CrossRef]
5. Amaral, J.; Meira, L.; Oliveira, B.; Mafra, I. Advances in Authenticity Testing for Meat Speciation. In *Advances in Food Authenticity Testing*; Elsevier: Amsterdam, The Netherlands, 2016; pp. 369–414.
6. Lo, Y.-T.; Shaw, P.-C. DNA-based techniques for authentication of processed food and food supplements. *Food Chem.* **2018**, *240*, 767–774. [CrossRef] [PubMed]
7. Iwobi, A.N.; Huber, I.; Hauner, G.; Miller, A.; Busch, U. Biochip Technology for the Detection of Animal Species in Meat Products. *Food Anal. Methods* **2010**, *4*, 389–398. [CrossRef]
8. *ISO 20813: 2019—Molecular Biomarker Analysis—Methods of Analysis for the Detection and Identification of Animal Species in Foods and Food Products (Nucleic Acid-Based Methods)—General Requirements and Definitions*; International Organization for Standardization: Geneva, Switzerland, 2019.

9. Eugster, A.; Ruf, J.; Rentsch, J.; Hübner, P.; Köppel, R. Quantification of beef and pork fraction in sausages by real-time PCR analysis: Results of an interlaboratory trial. *Eur. Food Res. Technol.* **2007**, *227*, 17–20. [CrossRef]

10. Eugster, A.; Ruf, J.; Rentsch, J.; Köppel, R. Quantification of beef, pork, chicken and turkey proportions in sausages: Use of matrix-adapted standards and comparison of single versus multiplex PCR in an interlaboratory trial. *Eur. Food Res. Technol.* **2009**, *230*, 55–61. [CrossRef]

11. Köppel, R.; Ruf, J.; Rentsch, J. Multiplex real-time PCR for the detection and quantification of DNA from beef, pork, horse and sheep. *Eur. Food Res. Technol.* **2011**, *232*, 151–155. [CrossRef]

12. Köppel, R.; Eugster, A.; Ruf, J.; Rentsch, J. Quantification of Meat Proportions by Measuring DNA Contents in Raw and Boiled Sausages Using Matrix-Adapted Calibrators and Multiplex Real-Time PCR. *J. AOAC Int.* **2012**, *95*, 494–499. [CrossRef]

13. Laube, I.; Zagon, J.; Spiegelberg, A.; Butschke, A.; Kroh, L.W.; Broll, H. Development and design of a 'ready-to-use' reaction plate for a PCR-based simultaneous detection of animal species used in foods. *Int. J. Food Sci. Technol.* **2007**, *42*, 9–17. [CrossRef]

14. Druml, B.; Kaltenbrunner, M.; Hochegger, R.; Cichna-Markl, M. A novel reference real-time PCR assay for the relative quantification of (game) meat species in raw and heat-processed food. *Food Control* **2016**, *70*, 392–400. [CrossRef]

15. Iwobi, A.; Sebah, D.; Spielmann, G.; Maggipinto, M.; Schrempp, M.; Kraemer, I.; Gerdes, L.; Busch, U.; Huber, I. A multiplex real-time PCR method for the quantitative determination of equine (horse) fractions in meat products. *Food Control* **2017**, *74*, 89–97. [CrossRef]

16. Köppel, R.; Ruf, J.; Zimmerli, F.; Breitenmoser, A. Multiplex real-time PCR for the detection and quantification of DNA from beef, pork, chicken and turkey. *Eur. Food Res. Technol.* **2008**, *227*, 1199–1203. [CrossRef]

17. Kaltenbrunner, M.; Hochegger, R.; Cichna-Markl, M. Tetraplex real-time PCR assay for the simultaneous identification and quantification of roe deer, red deer, fallow deer and sika deer for deer meat authentication. *Food Chem.* **2018**, *269*, 486–494. [CrossRef] [PubMed]

18. Dolch, K.; Andrée, S.; Schwägele, F. Comparison of Real-Time PCR Quantification Methods in the Identification of Poultry Species in Meat Products. *Foods* **2020**, *9*, 1049. [CrossRef]

19. Staats, M.; Arulandhu, A.J.; Gravendeel, B.; Holst-Jensen, A.; Scholtens, I.; Peelen, T.; Prins, T.W.; Kok, E. Advances in DNA metabarcoding for food and wildlife forensic species identification. *Anal. Bioanal. Chem.* **2016**, *408*, 4615–4630. [CrossRef]

20. Fernandes, T.J.R.; Amaral, J.S.; Mafra, I. DNA barcode markers applied to seafood authentication: An updated review. *Crit. Rev. Food Sci. Nutr.* **2020**. [CrossRef]

21. Franco, C.M.; Ambrosio, R.L.; Cepeda, A.; Anastasio, A. Fish intended for human consumption: From DNA barcoding to a next-generation sequencing (NGS)-based approach. *Curr. Opin. Food Sci.* **2021**, *42*, 86–92. [CrossRef]

22. Nehal, N.; Choudhary, B.; Nagpure, A.; Gupta, R.K. DNA barcoding: A modern age tool for detection of adulteration in food. *Crit. Rev. Biotechnol.* **2021**, *41*, 767–791. [CrossRef]

23. Cottenet, G.; Blancpain, C.; Chuah, P.F.; Cavin, C. Evaluation and application of a next generation sequencing approach for meat species identification. *Food Control* **2020**, *110*, 107003. [CrossRef]

24. Druml, B.; Cichna-Markl, M. High resolution melting (HRM) analysis of DNA—Its role and potential in food analysis. *Food Chem.* **2014**, *158*, 245–254. [CrossRef] [PubMed]

25. López-Oceja, A.; Nuñez, C.; Baeta, M.; Gamarra, D.; de Pancorbo, M. Species identification in meat products: A new screening method based on high resolution melting analysis of cyt b gene. *Food Chem.* **2017**, *237*, 701–706. [CrossRef]

26. Parvathy, V.A.; Swetha, V.P.; Sheeja, T.E.; Leela, N.K.; Chempakam, B.; Sasikumar, B. DNA Barcoding to Detect Chilli Adulteration in Traded Black Pepper Powder. *Food Biotechnol.* **2014**, *28*, 25–40. [CrossRef]

27. Chin, T.C.; Adibah, A.; Hariz, Z.D.; Azizah, M.S. Detection of mislabelled seafood products in Malaysia by DNA barcoding: Improving transparency in food market. *Food Control* **2016**, *64*, 247–256. [CrossRef]

28. Dobrovolny, S.; Blaschitz, M.; Weinmaier, T.; Pechatschek, J.; Cichna-Markl, M.; Indra, A.; Hufnagl, P.; Hochegger, R. Development of a DNA metabarcoding method for the identification of fifteen mammalian and six poultry species in food. *Food Chem.* **2019**, *272*, 354–361. [CrossRef] [PubMed]

29. Martin, M. Cutadapt Removes Adapter Sequences from High-Throughput Sequencing Reads. *EMBnet J.* **2011**, *17*, 10. [CrossRef]

30. Edgar, R.C. Search and clustering orders of magnitude faster than BLAST. *Bioinformatics* **2010**, *26*, 2460–2461. [CrossRef]

31. Rentsch, J.; Weibel, S.; Ruf, J.; Eugster, A.; Beck, K.; Köppel, R. Interlaboratory validation of two multiplex quantitative real-time PCR methods to determine species DNA of cow, sheep and goat as a measure of milk proportions in cheese. *Eur. Food Res. Technol.* **2012**, *236*, 217–227. [CrossRef]

32. Köppel, R.; Daniels, M.; Felderer, N.; Brünen-Nieweler, C. Multiplex real-time PCR for the detection and quantification of DNA from duck, goose, chicken, turkey and pork. *Eur. Food Res. Technol.* **2013**, *236*, 1093–1098. [CrossRef]

33. Druml, B.; Mayer, W.; Cichna-Markl, M.; Hochegger, R. Development and validation of a TaqMan real-time PCR assay for the identification and quantification of roe deer (*Capreolus capreolus*) in food to detect food adulteration. *Food Chem.* **2015**, *178*, 319–326. [CrossRef]

34. Kaltenbrunner, M.; Hochegger, R.; Cichna-Markl, M. Red deer (*Cervus elaphus*)-specific real-time PCR assay for the detection of food adulteration. *Food Control* **2018**, *89*, 157–166. [CrossRef]

35. Kaltenbrunner, M.; Hochegger, R.; Cichna-Markl, M. Development and validation of a fallow deer (Dama dama)-specific TaqMan real-time PCR assay for the detection of food adulteration. *Food Chem.* **2018**, *243*, 82–90. [CrossRef] [PubMed]

36. Evaluation Report Proficiency TestDLA 45/2019 Animal Species-Screening III: Buffalo Milk, Cow's Milk, Sheep's Milk and Goat's Milk in Dairy Product (Herder Cheese) (2020) DLA—Proficiency Tests GmbH. Available online: http://www.dla-lvu.de/Auswerteberichte%202019/PT%20-%20DLA%2045-2019%20Final%20Report%20Animal%20Species-Screening%20III.pdf (accessed on 30 January 2020).
37. Pirondini, A.; Bonas, U.; Maestri, E.; Visioli, G.; Marmiroli, M.; Marmiroli, N. Yield and amplificability of different DNA extraction procedures for traceability in the dairy food chain. *Food Control* **2010**, *21*, 663–668. [CrossRef]
38. Kaltenbrunner, M.; Mayer, W.; Kerkhoff, K.; Epp, R.; Rüggeberg, H.; Hochegger, R.; Cichna-Markl, M. Applicability of a duplex and four singleplex real-time PCR assays for the qualitative and quantitative determination of wild boar and domestic pig meat in processed food products. *Sci. Rep.* **2020**, *10*, 117243. [CrossRef] [PubMed]
39. Goedbloed, D.J.; Megens, H.; Van Hooft, P.; Herrero-Medrano, J.M.; Lutz, W.; Alexandri, P.; Crooijmans, R.P.M.A.; Groenen, M.; Van Wieren, S.E.; Ydenberg, R.C.; et al. Genome-wide single nucleotide polymorphism analysis reveals recent genetic introgression from domestic pigs into Northwest European wild boar populations. *Mol. Ecol.* **2013**, *22*, 856–866. [CrossRef]
40. Dzialuk, A.; Zastempowska, E.; Skórzewski, R.; Twarużek, M.; Grajewski, J. High domestic pig contribution to the local gene pool of free-living European wild boar: A case study in Poland. *Mammal Res.* **2017**, *63*, 65–71. [CrossRef]
41. Böhme, K.; Calo-Mata, P.; Barros-Velázquez, J.; Ortea, I. Review of Recent DNA-Based Methods for Main Food-Authentication Topics. *J. Agric. Food Chem.* **2019**, *67*, 3854–3864. [CrossRef]

 *foods*

Article

Authenticity of Hay Milk vs. Milk from Maize or Grass Silage by Lipid Analysis

Sebastian Imperiale [1], Elke Kaneppele [2], Ksenia Morozova [1], Federico Fava [2], Demian Martini-Lösch [2], Peter Robatscher [2], Giovanni Peratoner [2], Elena Venir [2], Daniela Eisenstecken [2] and Matteo Scampicchio [1,*]

[1] Faculty of Science and Technology, Free University of Bozen-Bolzano, Piazza Università 5, 39100 Bolzano, Italy; Sebastian.imperiale@natec.unibz.it (S.I.); ksenia.morozova@unibz.it (K.M.)

[2] Laimburg Research Centre, Laimburg 6, Pfatten-Vadena, 39040 Auer-Ora, Italy; elke.kaneppele@student.uibk.ac.at (E.K.); federico.fava@laimburg.it (F.F.); demian.martini-loesch@laimburg.it (D.M.-L.); peter.robatscher@laimburg.it (P.R.); giovanni.peratoner@laimburg.it (G.P.); elena.venir@laimburg.it (E.V.); daniela.eisenstecken@laimburg.it (D.E.)

* Correspondence: matteo.scampicchio@unibz.it; Tel.: +39-0471-017210

Abstract: Hay milk is a traditional dairy product recently launched on the market. It is protected as "traditional specialty guaranteed" (TSG) and subjected to strict regulations. One of the most important restrictions is that the cow's feed ration must be free from silage. There is the need for analytical methods that can discriminate milk obtained from a feeding regime including silage. This study proposes two analytical approaches to assess the authenticity of hay milk. Hay milk and milk from cows fed either with maize or grass silage were analyzed by targeted GC-MS for cyclopropane fatty acid (dihydrosterculic acid, DHSA) detection, since this fatty acid is strictly related to the bacterial strains found in silage, and by HPLC-HRMS. The presence of DHSA was correlated to the presence of maize silage in the feed, whereas it was ambiguous with grass silage. HPLC-HRMS analysis resulted in the identification of 14 triacylglycerol biomarkers in milk. With the use of these biomarkers and multivariate statistical analysis, we were able to predict the use of maize and grass silage in the cow's diet with 100% recognition. Our findings suggest that the use of analytical approaches based on HRMS is a viable authentication method for hay milk.

Keywords: bovine feeding; LC-MS; milk; lipidomics; silage; hay milk; GC-MS; food authenticity; cyclopropane fatty acids; CPFAs

Citation: Imperiale, S.; Kaneppele, E.; Morozova, K.; Fava, F.; Martini-Lösch, D.; Robatscher, P.; Peratoner, G.; Venir, E.; Eisenstecken, D.; Scampicchio, M. Authenticity of Hay Milk vs. Milk from Maize or Grass Silage by Lipid Analysis. *Foods* **2021**, *10*, 2926. https://doi.org/10.3390/foods10122926

Academic Editor: Pedro Vitoriano de Oliveira

Received: 1 October 2021
Accepted: 23 November 2021
Published: 26 November 2021

Publisher's Note: MDPI stays neutral with regard to jurisdictional claims in published maps and institutional affiliations.

1. Introduction

In recent years, there has been an increasing interest to sustain and develop European mountain areas and decrease land abandonment [1]. Mountain dairy farming is more challenging due to harsher environmental and morphological conditions, which lead to higher workload and management costs [2]. To counteract the economical disadvantages derived from natural constraints in these areas, the European Union (EU) is applying new policies to promote the quality and authenticity of mountain products [1,3]. The EU scheme "traditional specialty guaranteed" (TSG) represents an important policy for the valorization of traditional products.

Hay milk is one of the dairy products that received the TSG label and is subjected to strict production regulations. This product is obtained with traditional methods [4], and is perceived by consumers as healthier and more natural [5]. Hay milk has been regaining popularity in recent years, especially in the alpine region [2,4,5], thanks to marketing, labelling, and certification strategies. This effort was made to valorize and differentiate local mountain production and to fully benefit from the TSG label [2].

In the specific case of hay milk, its TSG designation is controlled by the European Commission regulation 2016/304 [6]. It states that any form of fermented fodder, i.e., silage from maize and grass, moist hay, fermented hay, and any genetically modified feed is banned. Roughage, including fresh herbage and hay, but possibly also green rapeseed, green maize, green rye, and fodder beets, as well as hay, lucerne, maize pellets and similar types of feed, must make up at least 75% of the yearly ration of dry feed [6].

Even though the above regulation on the feeding is very clear, to the best of our knowledge, no clear markers or analytical methods for milk analysis are available to determine the presence of fermented fodder in the feed ration, especially for grass silage.

Some efforts to discriminate milk authenticity have been reported recently, especially through the analysis of the milk fat fraction, which is likely the most affected by the animal's diet [7]. For instance, it has been shown that different diets influence the fatty acids profile of milk fat [8]. Additionally, fatty acid characterization can provide information related to both diet composition and the ruminal fermentation pattern. Finally, many studies aimed to identify markers within this lipid class [9–14].

Recent literature reported the analysis of a group of specific fatty acids (cyclopropane fatty acids, CPFAs) used as markers for the authenticity of Parmigiano Reggiano cheese [9]. Indeed, similarly to hay milk, Parmigiano Reggiano cheese is produced with milk from cows fed without any silage [15]. In that work, CPFAs were found only in dairy products obtained with maize silage in the cow's diet [9]. Later, a method was proposed and validated [11], and CPFAs have been included in the Parmigiano Reggiano PDO regulations for the verification of its authenticity. Lactic acid bacteria (LAB) produce CPFAs as response to fermentative stress [16–20]. LAB convert soluble carbohydrates, present in matrices rich in starch like maize, into lactic acid [21]. However, it is not clear if the proposed method is also suitable to determine the presence of grass silage in the animal's diet.

For this reason, this study investigates the capacity of the CPFA method to assess the authenticity of hay milk with respect to milk obtained from cows fed with maize and, to the best of our knowledge, for the first time, with grass silage in the feed ration. Furthermore, this work proposes the use of an approach based on high-performance liquid chromatography (HPLC), for the detection and identification of new markers, or groups of markers. HPLC has advantages such as high analysis speed, resolution, and sensitivity [22]. When coupled to high-resolution mass spectrometry (HRMS), HPLC represents a powerful tool to analyze the composition of milk fat, especially for untargeted studies. Several lipidomic studies use HPLC and mass spectrometry to characterize milk lipids from cows fed with different diets [23,24]. Craige Trennery et al. performed LC-MS to study the effects of feeding on the milk lipid profile [25]. Because of the high amount of data obtained by HPLC-HRMS, the implementation of chemometrics and multivariate statistical methods is essential to elucidate the characteristics of milk.

For these purposes, this work analyzed three types of milk, namely hay milk (HM), milk from cows fed with maize silage in the ration (SM-M), and milk from cows fed with grass silage in the ration (SM-G), by GC-MS and HPLC-HRMS. The first part of the study followed a targeted approach to determine the amount of the CPFA dihydrosterculic acid (DHSA) in each milk sample, using a GC-MS method adapted from Marseglia et al. [9]. In the second part, HPLC-HRMS was used to characterize the lipid profile of each sample. Untargeted pattern recognition and correlation of the different feeding practices were conducted using multivariate statistical analysis. The combination of the presence of DHSA with multivariate analysis on the resulting HRMS data provided an analytical fingerprint that allows discrimination among the forages implemented in dairy farming.

2. Materials and Methods

2.1. Milk Samples

A total of 27 fresh and unpasteurized bulk milk samples were collected from 9 dairy farms located in the north of Italy (South Tirol, Italy), at altitudes ranging from 616 m a.s.l. to 1404 m a.s.l. Sample collection was dispersed over the winter feeding interval from October 2019 to March 2020. At each farm, bulk milk was sampled weekly over three consecutive weeks, in order to capture variability over time (Figure 1).

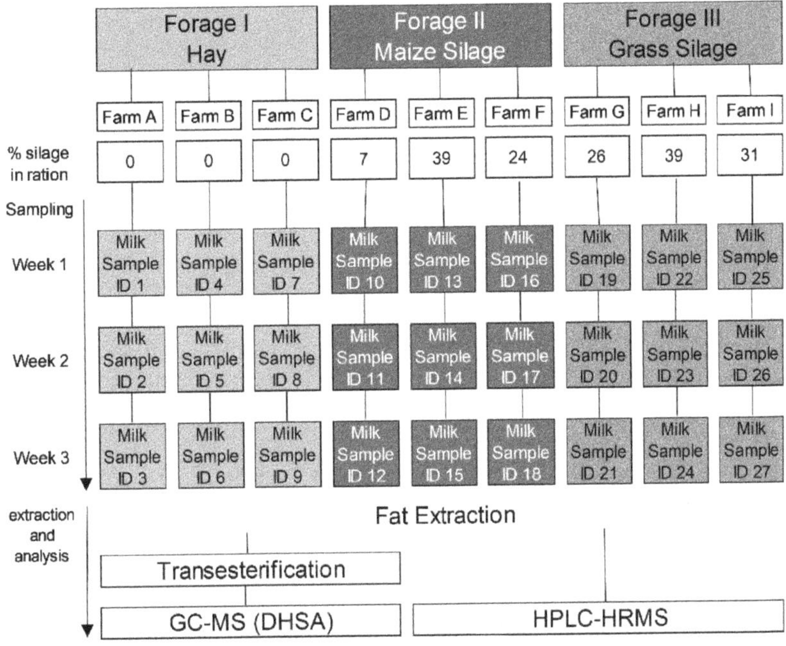

Figure 1. Sampling scheme for targeted GC-MS and HPLC-HR MS analysis of milk. Nine farms were selected based on the feeding regimen. Each sample is shown in the figure.

The farmers were interviewed by means of a structured survey concerning several aspects of the farm structure, characteristics, and the composition of the feed ration during the sampling period, and the proportion of silage in the diet was computed on a dry matter basis. At three farms, the animals were fed according to the EU-Regulation 2016/304 of hay milk production, i.e., without fermented fodder and roughage making up at least 75% dry matter of the yearly ration (hay milk, HM). The other six farms included silage in amounts ranging from 7% to 39% dry matter of the total feed ration: three included only maize silage (maize silage milk, SM-M), and three only grass silage (grass silage milk, SM-G). Each farm had between 16–35 cows. The milk samples were taken in the morning from the farm's own milk tank, being therefore a mix of the evening and morning milk. Before taking the samples, the milk was mixed by hand and then the samples were taken with a liquid sampler. At each sampling event, aliquots of 30 mL for each sample were taken from the same tank for the analysis of DHSA, for the lipid profile via HPLC-HRMS and for quality routine analysis, respectively. During transportation, the samples were kept refrigerated at a temperature of about 4 °C before being stored at −80 °C until analysis.

2.2. Chemicals and Reagents

Methanol was purchased from VWR (Radnor, PA, USA), and pentane was obtained from Fluka Analytical (Honeywell International Inc., Charlotte, NC, USA). Hexane and sodium methoxide solution 25 wt.% were purchased from Sigma Aldrich (St. Louise, MO, USA). Sodium sulfate was purchased from Titolchimica (Rovigo, Italy, IT) and the CPFA cis-9,10-methylene-octadecanoic acid (dihydrosterculic acid DHSA, as methyl ester, purity ≥98%) was obtained from Chem Cruz (TE Huissen, The Netherlands, NL). LC-MS grade formic acid and LC-MS grade ammonium formate were purchased from Sigma Aldrich (Steinheim, Germany). LC-MS grade methanol and LC-MS grade acetonitrile were purchased from Honeywell (Selze, Germany), and LC-MS grade 2-propanol and methyl tert-butyl ether (MTBE) were purchased from Merck KGaA (Darmstadt, Germany) and not purified further. If not otherwise stated, Milli-Q water was employed.

2.3. Sample Preparation and Analysis via GC-MS

2.3.1. Milk Fat Extraction

The milk samples were thawed in a water bath at 40 °C for 2 h. The fat was separated following a modified method based on Feng et al. [26]. A volume of 20 mL of milk was added in a 50 mL conical plastic tube and centrifuged at 12,000 rpm (17,800× g) for 30 min at 4 °C. After centrifugation, the fat cake (top layer) was transferred into a 15 mL conical plastic tube and stored over night at −80 °C. The fat was resuspended in a volume of 10 mL of a 9:1 (v/v) n-pentane/methanol, then vortexed for 2 min at room temperature and kept in a room temperature ultrasound bath (45 kHz) for 5 min, shaken for 5 min with a MultiRotator (PTR-60 Grant Intruments, Royston, UK), then vortexed again for 2 min with a final centrifugation at 4000 rpm (1900× g) for 2 min at room temperature. The organic phase was transferred into a dark glass vial and flushed with N_2 until dryness. The fat was stored at −80 °C until transesterification.

2.3.2. Transesterification

Transesterification was carried out according to Christie et al. [27]. Milk fat (100 mg ± 5 mg) was dissolved in 5 mL hexane. Then, 0.2 mL of sodium methoxide in dry methanol (1 mL sodium methoxide solution 25 wt% diluted with 1.25 mL methanol) was added, and the solution was briefly agitated (30 s) to ensure thorough mixing. The reaction was quenched by adding 0.5 g sodium sulfate, and after a brief agitation (30 s), it was centrifuged at 2000× g for 5 min at room temperature and the supernatant was used for analyses.

2.3.3. Analysis of the Cyclopropane Fatty Acid Dihydrosterculic Acid (DHSA)

The GC-MS analysis was carried out on a Shimadzu GC MS-QP2010 SE (Kyoto, Japan) equipped with an autosampler, a split/splitless injection port, a GC oven and a single quadrupole mass spectrometer. Each sample was measured in triplicate. For the analyses, 100 μL of transesterified mixture was taken, diluted with 900 μL hexane, and 1 μL was injected using a split ratio of 1:10. Helium was used as carrier gas with a flow rate of 1 mL/min and a low-polarity SLB-5 ms column (30 m × 0.25 mm i.d × 0.25 μm) (Supelco, Bellefonte, PA, USA) was used for chromatographic separation of the analyte. The run was conducted following a modified temperature program according to Marseglia et al. [9]. Temperature was kept at 40 °C for 5 min, increased at 280 °C with a rate of 10 °C/min and held for 10 min. The injector temperature and transfer line temperature were maintained at 280 °C and ion source temperature at 230 °C. The mass spectra were acquired in full scan mode (mass range 40–500 m/z) and in SIM Mode (using 55 m/z as quantifier, 69, 278, and 310 m/z qualifier). The quantification of DHSA in the samples was carried out by comparing the peak area of the samples with the peak area of known amounts of the DHSA standard, considering the matrix effect by spiked hay milk with DHSA following extraction and transesterification. The limit of detection (LOD) of the method was 7.5 mg DHSA/kg of fat and the limit of quantification (LOQ) was 25.0 mg DHSA/kg of fat. Linear range was from 25.0 mg/kg to 1500 mg/kg. Recovery of spiked fat was 101.5% (0.2 RSD%). Intraday repeatability was of 3.3, 5.4, 2.5 RSD% for 80, 400, and 1000 mg DHSA/kg of milk fat, respectively.

2.4. Sample Preparation and Analysis via HPLC-HRMS

2.4.1. Milk Fat Extraction

The milk samples were thawed at 8 °C overnight. Fat extraction from milk samples was carried out according to Breitkopf et al. based on the extraction method by Matyash et al. with modifications [28,29]. In short, 200 μL of milk was pipetted into a 15 mL centrifuge tube, 1.5 mL methanol was added and vortexed for 1 min. Then, 5 mL of MTBE was added and shaken at 200 rpm for 1 h at room temperature. After the incubation, 1.2 mL of water was added and vortexed for 1 min. The mixture was centrifuged for 10 min at $1000 \times g$ at room temperature. The upper phase was collected, and the bottom phase re-extracted with 2 volume parts of MTBE/methanol/water (10/3/2.5, $v/v/v$). The combined upper phases were dried under nitrogen flow at room temperature (MultiVap 8, LabTech S.r.l., Milano, Italy). Finally, the dried extract was dissolved in 5 mL methanol/2-propanol (50/50, v/v), diluted 1:100 with the same solvent mix and filtered with a 0.45 μm PTFE syringe filter prior to injection.

2.4.2. High-Performance Liquid Chromatography Coupled to High Resolution Mass Spectrometry (HPLC-HRMS)

The system consisted of a Q-Exactive hybrid quadrupole Orbitrap HRMS instrument (Thermo Fisher Scientific, Waltham, MA, USA) coupled to an Ultimate 3000 UHPLC instrument (Thermo Fisher Scientific, Waltham, MA, USA) with UV-vis detector. The separation of the compounds was done at a flow rate of 0.2 mL/min with a C18 column (Accucore RP-MS, 100 mm × 2.1 mm i.d., 2.6 μm particle size, Thermo Fisher Scientific, Waltham, MA, USA) with a security guard cartridge system (Thermo Fisher Scientific, Waltham, MA, USA). The mobile phase consisted of a combination of solvent A (water/acetonitrile 40/60 v/v with the addition of 0.1% formic acid and 10 mM ammonium formate) and B (acetonitrile/2-propanol 10/90 v/v with addition of 0.1% formic acid and 10 mM ammonium formate). The gradient was set as follows: 70 % B (v/v) for 2 min, then from 70 % B to 83 % B at 3 min, hold until 8 min then to 84 % B at 13 min and hold until 14 min. Sample injection volume was 5 μL using an autosampler with a 20 μL injection loop. After each sample, a wash step with a blank (2-propanol) was introduced with the same chromatographic set-up as before, but with a different gradient: from 84 % B at 0 min to 97 % B at 2 min, hold 97 % until 7 min, from 97% at 7 min to 70 % B at 8 min followed by a re-equilibration step (70% B) from 8 to 10 min. Blank injection volume was 20 μL. During this wash and re-equilibration step, the flow from the HPLC was diverted to waste using a Rheodyne switch valve, while a flow of 3 μL/min 2-propanol was delivered to the MS using an infusion syringe pump (Thermo Fisher Scientific, Waltham, MA, USA) to avoid clogging and minimize carry-over effects. The HRMS instrument was operated in positive ionization mode with a heated electrospray ionization ion source set as follows: sheath gas flow at 40 (arbitrary units), aux gas flow at 10 (arbitrary units), sweep gas flow at 0 (arbitrary units), spray voltage at 4.00 kV, capillary temperature at 300 °C, S-lens RF level at 50%, and aux gas temperature at 100 °C. Full-MS experiments were performed in a scan range from 150 to 1500 m/z with a resolution of 35,000 (at 200 m/z), an automatic gain control (AGC) target of 2×10^5 and a maximum injection time (IT) of 200 ms. Targeted SIM (t-SIM) experiments were performed with a resolution of 35,000, AGC target of 2×10^5, max IT of 125 ms, and an isolation window of 4 m/z. The MS^2 measurements of the selected ions were performed with a resolution of 17,500 and AGC target set at 1×10^5 and maximum IT of 50 ms, with a stepped normalized collision energy of 20, 30, and 60 eV.

2.5. Data Processing and Statistical Analysis

Correlation of chemical compounds relative abundances and integration of the area under each peak (HPLC-MS XIC integrations) was done using Compound Discoverer 3.1 and Xcalibur (Thermo Scientific, Milano, Italy) and by employing online (LIPIDMAPS) and local databases.

Multivariate statistical analysis was conducted using XLSTAT annual version 2021.1.1 1092 (Addinsoft 2021, New York, NY, USA).

3. Results and Discussion

3.1. GC-MS Analysis of DHSA

Table 1 shows the results of the determination of the CPFA dihydrosterculic acid (DHSA) in milk samples grouped into three categories: HM (hay milk, i.e., cows fed without the use of silage), SM-G (milk obtained from cows fed with grass silage in the ration), and SM-M (milk obtained from cows fed with maize silage in the ration).

Table 1. Quantification of dihydrosterculic acid (DHSA) (average ± standard deviation, $n = 3$) in milk samples using GC-MS.

Milk Sample	Farm	DHSA (mg/kg Fat)	RSD (%)
HM	A	<LOD	
	B	<LOD	
	C	<LOD	
SM-M	D	94 ± 13	13
	E	94 ± 68	72
	F	52 ± 10	19
SM-G	G	<LOQ	
	H	<LOD	
	I	30 ± 11	28

HM = hay milk, SM-G = milk obtained from cows fed with grass silage in the ration, SM-M = milk obtained from cows fed with maize silage in the ration. LOD (Limit of detection) = 7.5 mg/kg fat, LOQ (Limit of quantification) = 25.0 mg/kg fat, RSD, relative standard deviation.

As expected, and reported in a previous work [9], no CPFA, in this case DHSA, was detected in the HM samples, whereas it was detected in all SM-M samples. According to Caligiani et al., their GC-MS method was proposed as analytical tool for the detection of the marker in milk for the presence of silage in the feed ration of the cows [11]. The results obtained in the current study highlights how, for the SM-G samples, the determination of the DHSA is ambiguous. Indeed, in this study DHSA was not detected in four of the nine SM-G samples (three in farm H, and one in farm G).

These limitations might be due to a series of factors, one of them being the sensitivity of the method in detecting the marker. For the method employed for DHSA detection in milk fat, a limit of detection (LOD) of 7.5 mg/kg and a limit of quantification (LOQ) of 25.0 mg/kg were obtained. The concentrations of DHSA found in the SM-G samples were generally lower than in the SM-M samples. Supposedly, a lower amount of available carbohydrates, such as during grass fermentation might lead to a reduced LAB stress response and to lower or no content of DHSA in the milk samples produced from grass silage in the ration. Overall, the results obtained with GC-MS allow us to discriminate HM vs. SM-M samples, but they were unable to discriminate all SM-G samples from the HM samples. For this reason, an approach using HPLC-HRMS was further proposed for the discrimination of milk obtained with different types of forages during the winter feeding period for the 27 samples collected.

3.2. Milk Fat Analysis by HPLC-HRMS

For non-target milk fat analysis, the milk fat profiles were determined using HPLC-HRMS in full-scan mode (full-MS, 150 to 1500 m/z). Figure 2 shows a typical total ion current (TIC) chromatogram obtained from a milk fat sample in positive ionization mode. For every sample of each milk type, the lipid profile was obtained.

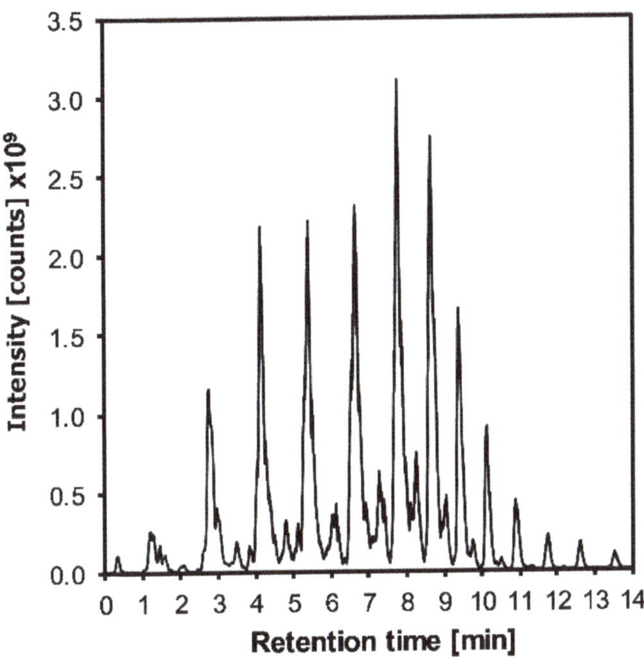

Figure 2. Total ion chromatogram acquired in full-MS showing the lipid profile of a milk fat extract obtained by HPLC-HRMS in positive ionization mode.

For each profile, distinct cluster peaks could be observed representing a unique fingerprint of each milk sample. With HRMS it is possible to obtain the whole lipidome of milk [28]. However, in order to find the target markers that could be used to distinguish milk samples with different feeding, the use of chemometric tools was needed to process the obtained data. This was achieved by using a software for untargeted MS analysis (Compound Discoverer 3.1). With Compound Discoverer it was possible to build up a peak table with the most abundant masses for all analyzed samples and match the compounds with online and custom databases.

The custom database in the software included a mass list with possible compounds of interest. In detail, we created a mass list comprising the most abundant lipid class in milk, the triacylglycerols (TAGs) which account for hundreds of different species (Figure 3) [30]. TAGs are composed of a glycerol molecule esterified with three fatty acids, which can be the same or different. When using the 16 fatty acids (FA) most common in milk [31], randomly distributed, it was possible to calculate 4096 (16^3) theoretical TAGs and their corresponding m/z values. Considering isomers with identical exact mass, the list was reduced to 253 groups of TAG molecular species, which contained the same number of carbons (CN) and the same number of double bonds (DB) in their FA residues (Supplementary Table S1).

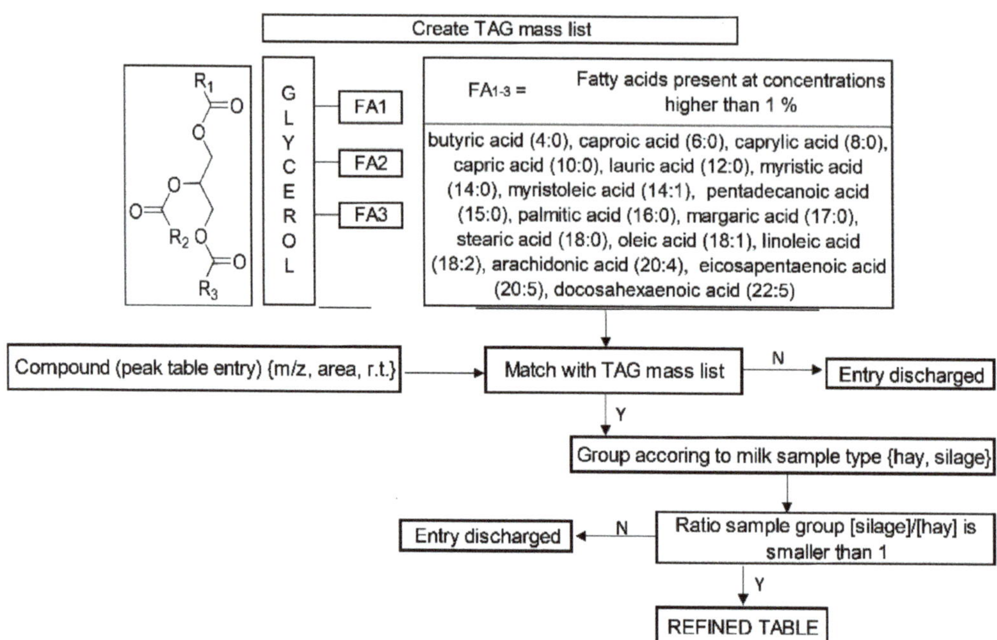

Figure 3. Creation of triacylglycerol (TAG) mass list and implemented algorithm for the identification of target molecules using HRMS. FA = Fatty Acid, N = no, Y = yes, r.t. = retention time.

The custom database with 253 masses of selected groups of TAG molecular species was used to match the masses in the peak table with the highest abundancy detected in the milk fat samples (Figure 3). All compounds that did not match the mass list were discarded. For the further data analysis, only these groups of TAG molecular species were selected (232 masses).

The scope of the study was to discriminate between hay milk and non-hay milk samples. For this reason, the relative intensities of matched groups of TAG molecular species were grouped in hay and silage samples. The extracted ion chromatograms of the selected TAG molecular species were generated with the following integration of the peaks. Variation in the TAG profiles between sample groups were reflected in the relative areas. Increase or decrease of matched TAG molecular species in one group could be used to differentiate between hay and silage samples. Therefore, we calculated the ratio of each area between the silage group and the hay group. All TAG molecular species that demonstrated a ratio inferior to one between groups were selected as potential markers to create the refined peak table (Figure 3). This allowed us to identify 14 groups of TAG molecular species that demonstrated the biggest differences between hay vs. silage sample groups (Table 2).

Table 2. Classification of the 14 groups of target TAG molecular species and tentative identification of their FA moieties.

m/z [M+NH$_4$]$^+$	Predicted Formula	Classif. (CN:DB *)	m/z Fragments	Tentatively Identified Fatty Acid Moieties
488.3946	C$_{27}$H$_{50}$O$_6$	TG 24:0	355.2843, 327.253, 299.2217, 155.143, 127.1117, 109.1012, 99.0804, 81.0699	butyric (4:0); caproic (6:0); caprylic (8:0)
678.5665	C$_{41}$H$_{73}$O$_6$	TG 38:3	573.4869, 405.3011, 383.3157, 261.2213, 239.2355, 71.04924	butyric (4:0); palmitic (16:0); linolenic (18:3)
696.6133	C$_{42}$H$_{78}$O$_6$	TG 39:1	591.5349, 437.3627, 409.3312, 409.3258, 397.3312, 397.3220, 265.2524, 99.0804, 71.0492	butyric (4:0); caproic (6:0); pentadecanoic (15:0); margaric (17:0); oleic (18:1)
698.6292	C$_{42}$H$_{80}$O$_6$	TG 39:0	593.5503, 565.5195, 509.4567, 453.3939, 439.3783, 425.3625, 411.3469, 397.3313, 313.2728, 267.2680, 253.2525, 239.2368, 235.2419, 225.2211, 221.2262, 211.2056, 207.2107, 193.1951, 173.1171, 155.1431, 137.1325, 127.1117, 99.0804, 81.0699, 71.0855, 53.0025	butyric (4:0); caproic (6:0); caprylic (8:0), capric (10:0); pentadecanoic (15:0); margaric (17:0); stearic (18:0)
706.5977	C$_{43}$H$_{76}$O$_6$	TG 40:3	409.3316, 407.3159, 99.0805, 411.3470, 601.5199, 265.2528, 145.0860, 263.2371, 261.2213, 405.3002, 433.3314, 119.0857, 127.1118, 573.4892, 247.2422, 245.2266, 243.2115, 239.2368, 173.1324, 313.2731, 339.2894, 155.1433, 53.0026, 99.1169, 71.0492	butyric (4:0); caproic (6:0); palmitic (16:0); stearic (18:0); oleic (18:1); linoleic (18:2); linolenic (18:3)
708.6133	C$_{43}$H$_{78}$O$_6$	TG 40:2	603.5757, 409.3601, 339.2887, 265.2522, 247.2463, 145.1018, 71.04978	butyric (4:0); oleic (18:1)
722.6289	C$_{44}$H$_{80}$O$_6$	TG 41:2	589.5209, 435.3455, 425.3624	caproic (6:0); margaric (17:0); linoleic (18:2);
730.5979	C$_{45}$H$_{76}$O$_6$	TG42:5	625.5178, 457.3311, 383.3153, 313.2517, 239.2368, 145.1012, 71.0492	docosapentaenoic (22:5); palmitic (16:0); butyric (4:0)
734.6289	C$_{45}$H$_{80}$O$_6$	TG 42:3	717.6027, 601.5206, 435.3476, 437.3621, 265.2532, 263.2367, 99.0805, 81.0699	caproic (6:0); oleic (18:1); linoleic (18:2)
758.6288	C$_{47}$H$_{80}$O$_6$	TG 44:5	411.3466, 285.0094, 239.0951, 201.1638, 109.1014	caprylic (8:0); capric (10:0); lauric (12:0); myristic (14:0); docosapentaenoic (22:5);
766.6917	C$_{47}$H$_{88}$O$_6$	TG 44:1	605.5507, 577.5189, 549.4877, 523.4722, 521.4564, 493.4251, 467.4093, 465.3938, 109.1011, 95.0854, 85.1011 81.0698, 71.0855, 57.0701	caprylic (8:0); capric (10:0); lauric (12:0); myristoleic (14:1); myristic (14:0); palmitic (16:0); oleic (18:1); stearic (18:0)
786.6596	C$_{49}$H$_{84}$O$_6$	TG 46:5	569.456, 313.278256, 467.4245, 239.2007, 211.2057	linoleic (12:0); myristic (14:0); docosapentaenoic (22:5); eicosapentaenoic (20:5)

Table 2. *Cont.*

m/z [M+NH$_4$]$^+$	Predicted Formula	Classif. (CN:DB *)	*m/z* Fragments	Tentatively Identified Fatty Acid Moieties
856.7379	C$_{54}$H$_{94}$O$_6$	TG 51:5	639.5344, 571.4151, 537.4889, 507.1079, 373,7151 239.2007, 299.2578, 191.1788	lauric (12:0); myristoleic (14:1); pentadecenoic (15:1); myristic (17:0); myristoleic (17:1); docosapentaenoic (22:5); docosatetraenoic (22:4); docosatrienoic (22:3)
876.8004	C$_{55}$H$_{102}$O$_6$	TG 52:2	221.2262, 239.2367, 245.2261, 263.2366, 267.2684, 313.2734, 575.5031, 579.5311, 603.5344	linoleic (18:2); stearic (18:0); palmitic (16:0)

* CN:DB = carbon number: total double bond number, of the 3 FA.

3.3. Tentative Identification of TAGs Marker

Tentative identification of the 14 groups of TAG molecular species was performed using a data-dependent HPLC-HRMS-MS^2 experiment (t-SIM-ddMS2). The exact mass of the molecular ions and their corresponding fragmentation spectra were compared with the entries in the lipidomic database LIPIDMAPS. For each group of TAG molecular species, the chemical formula of the neutral mass was calculated. Their classification was based on the number of carbons of the fatty acid residues (CN, TG x:−) and the number of double bonds in the fatty acid residues (DB, TG −:y), as shown in Table 2. The 14 groups of TAG molecular species were identified after fragmentation and determination of all FA moieties present in each group. The fragmentation spectra were compared with the theoretical spectra generated in LIPIDMAPS to characterize the groups of target molecules.

From Table 2, it can be derived that the fatty acid moieties identified in the TAG molecular species were characterized by a high abundance of unsaturated fatty acids, mainly oleic, linoleic, and linolenic acid. They were contained in the TAG molecular species present at higher concentrations in the hay milk samples. This was confirmed by the findings of the work of Bugaud et al., in which hay milk contained higher quantities of polyunsaturated FA [32]. Indeed, milk obtained from cows fed with diets rich in hay have an increased content of linolenic acid [33,34], whereas diets including maize silage lead to milk richer in short-chain FA, as well as myristic, palmitic, stearic, and oleic acid [35]. Diets rich in grass silage increase the content of myristic and palmitic acid at the expense of mono- and polyunsaturated FA [7]. It has been reported that the concentration of α-linolenic acid in milk obtained with silages generally decreases [7].

The higher relative abundancies of unsaturated fatty acid residues in the target TAGs reflects the cows' diet. In silage-based diets, their decrease could also correlate with the fermentative activity of LAB [36], but other factors influence the final FA composition of milk fat, and LAB activity could be only one of those. In order to consider all those factors, a group of markers, like the 14 groups of target TAGs, represents a promising approach.

3.4. Discrimination of Milk Samples Using Multivariate Statistical Analysis

We assessed the 14 groups of target TAGs for the discrimination of the type of milk (hay milk (HM), milk from silage (SM-G/M)). The 14 groups of target TAG molecular species were acquired in targeted single ion monitoring (SIM) mode. For each TAG group in all milk samples, the resulting peaks were integrated from an extracted ion chromatogram (XIC). The relative intensities were used for statistical analysis and discrimination of the samples.

Principal component analysis (PCA) was first performed using the areas of the target TAG groups. We evaluated whether the selected variables could be fitted to build discrimination models. From the PCA, the first and the third principal component explained 92.98% of the total variance and could display the data structure (Figure 4). The loading plot (Figure 4a) shows the relationship between the variables and how much they influence the system. It was possible to observe that the 14 target groups of TAGs form a group based on which the score plot can be built. The score plot shows a separation of the samples into distinct groups according to the type of milk for the samples considered in this experimental plan (Figure 4b: hay milk and milk from silage). The hay milk samples are located on the positive side of the PC1, which indicates higher amounts of the selected TAGs in these samples.

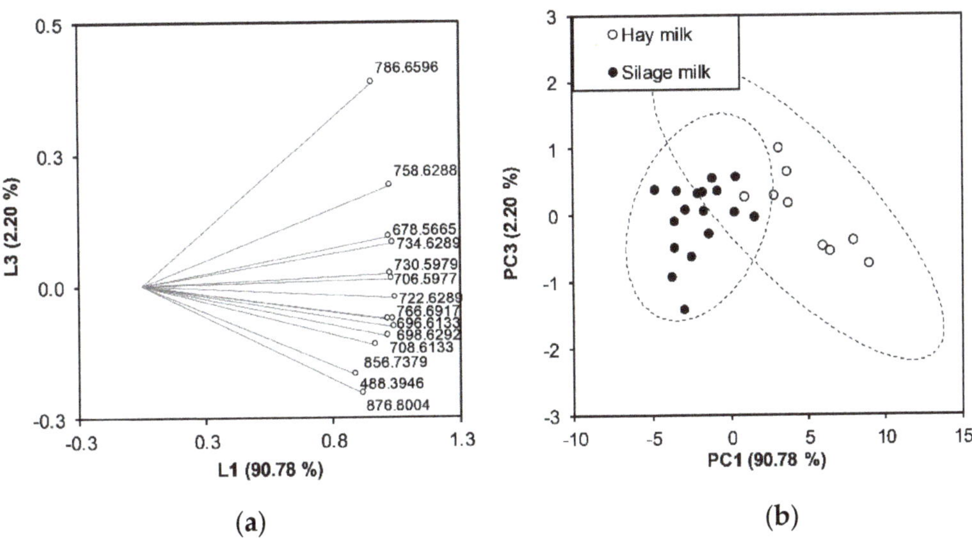

(a)   (b)

Figure 4. Principal component analysis of target TAGs obtained from the analysis of the hay milk and silage milk (milk obtained from cows fed with grass or maize silage in the ration) samples. Of the total variance, 92.98% is explained by the first and the third principal component. Loading of the 14 variables representing the target TAGs (**a**). Score plot showing the samples separated according to the type of milk produced (hay milk, silage milk) (**b**).

Next, the capacity of the target TAGs to predict the type of milk was assessed using the linear discriminant analysis (LDA) model based on two classes representing the type of milk: hay milk and milk from silage. The sample set was divided into a training sample and a validation sample, with final cross validation using the leave one out (LOO) algorithm. The LDA gave an overall recognition percentage of 100% (error rate 0%, the same for the LOO cross-validation). All milk samples were classified correctly according to their type based on silage and hay feeding during the winter-feeding period considered in our experimental plan (Table 3).

Table 3. Prediction of the type of feed used in the rations of cows based on target TAGs with LDA classification model and based on the presence of DHSA applied in milk. Rows represent the true class; columns represent the assigned class. Percentages of correct classified samples appear in brackets.

	Class	Hay	Silage	Sub-Class	Grass	Maize	Total
Fitting	Hay	9 (100%)	0 (%)		0 (%)	0 (%)	9
	Silage	0 (%)	18 (100%)	Grass	8 (89%)	1 (11%)	18
				Maize	1 (11%)	8 (89%)	
Cross validation, leave one out	Hay	9 (100%)	0 (%)		0 (%)	0 (%)	9
	Silage	0 (%)	18 (100%)	Grass	8 (89%)	1 (11%)	18
				Maize	1 (11%)	8 (89%)	
DHSA present	Hay	9 (100%)	0 (%)		0 (%)	0 (%)	9
	Silage	4 (22%)	14 (78%)	Grass	5 (56%)	-	18
				Maize	-	9 (100%)	

Then, we used the prediction model to determine whether the model could also predict the type of silage implemented in the ration for bovine feeding (grass or maize silage). Therefore, the three types of feed in the ration (hay, grass silage, and maize silage) were selected as classes. LDA showed that the first two canonical functions could classify the observations between groups. Figure 5 shows the corresponding canonical score plot in which the samples were grouped according to the class, i.e., hay milk, grass silage, and maize silage.

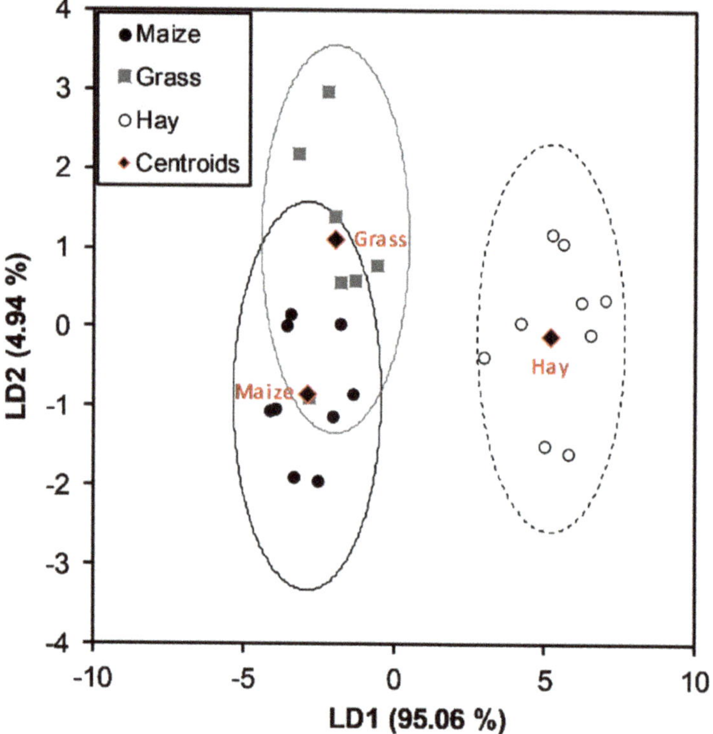

Figure 5. Linear discriminant analysis of the marker TAGs integrated areas of the milk samples according to the implemented feed in the ration (hay, grass silage, and maize silage).

The LDA classification model was also repeated, considering the silage sub-classes maize silage and grass silage (Table 3). The model had an overall recognition of 92% for the fitting and LOO cross-validation.

3.5. Comparison between DHSA and TAGs

Finally, the results of the GC-MS method and the HPLC-HRMS method were compared regarding their ability to discriminate the milk samples based on the type of feed. For the GC-MS method, the identity of the milk samples was assessed by the presence of DHSA. When this CPFA was present in the sample, it could directly be linked to the presence of silage in the ration. However, this was only the case with the SM-M samples, whereas not all SM-G samples were affirmative for DHSA (Table 1). As shown in Table 3, all HM samples were classified as such; all SM-M samples but not all SM-G samples were classified correctly with the GC-MS method. The absence of DHSA could therefore not be used as an indicator of hay milk when SM-G samples were also considered. When constructing a classification model using the presence of DHSA as an indicator of silage in the ration, in overall 84% of the milk samples were assigned correctly. In comparison, a higher recognition percentage was obtained with the LDA classification models using the HPLC-HRMS method. This was demonstrated by the 100% recognition obtained with the classification model based on the target TAGs (Table 3).

Overall, the HPLC-HRMS method resulted in better discrimination of the type of feed than the GC-MS method for the winter feeding period considered in the experimental plan. The untargeted approach benefited from the high resolution of HRMS, which could provide a detailed profile of each milk sample [37]. The variability between the profiles was better caught thanks to the non-targeted approach combined with multivariate analysis. Furthermore, building a prediction model based on a group of markers, rather than a single marker, was less susceptible to variations derived from the heterogeneity of the sample set.

4. Conclusions

This study proposed an HPLC-HRMS approach for the detection and identification of markers or groups of markers to assess the authenticity of hay milk in comparison to the targeted GC-MS method. This investigation included 27 samples collected during one winter season. HPLC-HRMS resulted in the identification of 14 groups of target TAG molecular species able to discriminate the type of implemented feed in the ration for milk production. Classification models based on LDA could predict the presence of silage in the ration with 100% recognition. Good comparability of the HPLC-HRMS method with the target GC-MS method using DHSA as marker was obtained when considering the HM samples vs. the SM-M samples. However, when also considering SM-G samples, a better recognition percentage was obtained with the target TAGs than with DHSA. The target TAGs might not account for the eventual presence or absence of DHSA, but on other dietary factors affecting the FA profile of the milk. Ultimately, by using a group of TAG markers, rather than a single marker, and with the aid of multivariate analysis, the variability in the milk sample set could be correlated to the presence of any silage (maize or grass) in the ration. To confirm the validity of the method, a bigger data set will be needed including samples from summer and winter seasons from at least two years.

Supplementary Materials: The following are available online at https://www.mdpi.com/article/10.3390/foods10122926/s1, Table S1: title, Mass list to match 253 groups of TAG molecular species with the same number of carbons (CN) and number of double bonds (DB) in the FA residues with the entries of the peak table generated in Compound Discoverer.

Author Contributions: Conceptualization, D.E., E.V., G.P. and M.S.; investigation, S.I., E.K.; formal analysis, S.I., K.M. and M.S.; methodology, S.I., E.K., F.F., E.V., D.E., M.S., K.M., G.P. and D.M.-L.; writing—original draft preparation, S.I., F.F. and D.M.-L.; writing—review and editing, D.E., M.S., K.M., G.P., E.V. and P.R.; visualization, S.I.; supervision, M.S., D.E., P.R., K.M., G.P. and E.V.; project administration, D.E., M.S., E.V., G.P. and P.R. All authors have read and agreed to the published version of the manuscript.

Funding: This research is funded by the FESR-EFRE 2014–2020 "Investitionen in Wachstum und Beschäftigung", project Heumilch, FESR1129 CUP: H36H19000000007. Laimburg Research Centre is funded by the Autonomous Province of Bolzano. This article was supported by the Open Access Publishing Fund of the Free University of Bozen-Bolzano.

Data Availability Statement: The data presented in this study are available on request from the corresponding author.

Acknowledgments: We would like to thank the Sennereiverband Südtirol and the BRING Beratungsring Berglandwirtschaft for their valuable support for the sampling.

Conflicts of Interest: All authors declare no conflict of interest. The funders had no role in the design of the study; in the collection, analyses, or interpretation of data; in the writing of the manuscript, or in the decision to publish the results.

References

1. MacDonald, D.; Crabtree, J.; Wiesinger, G.; Dax, T.; Stamou, N.; Fleury, P.; Gutierrez Lazpita, J.; Gibon, A. Agricultural abandonment in mountain areas of Europe: Environmental consequences and policy response. *J. Environ. Manag.* **2000**, *59*, 47–69. [CrossRef]
2. Busch, G.; Kühl, S.; Gauly, M. Consumer expectations regarding hay and pasture-raised milk in South Tyrol. *Austrian J. Agric. Econ. Rural. Stud.* **2018**, *27*, 79–86.
3. Tosato, A. The Protection of Traditional Foods in the EU: Traditional Specialities Guaranteed. *Eur. Law J.* **2013**, *19*, 545–576. [CrossRef]
4. Wölk, M.; Milkovska-Stamenova, S.; Schröter, T.; Hoffmann, R. Influence of seasonal variation and processing on protein glycation and oxidation in regular and hay milk. *Food Chem.* **2021**, *337*, 127690. [CrossRef]
5. Palmieri, N.; Pesce, A.; Verrascina, M.; Perito, M.A. Market Opportunities for Hay Milk: Factors Influencing Perceptions among Italian Consumers. *Animals* **2021**, *11*, 431. [CrossRef] [PubMed]
6. Commission Implementing Regulation (EU) 2016/304 of 2 March 2016. Available online: https://eur-lex.europa.eu/eli/reg_impl/2016/304/oj (accessed on 29 September 2021).
7. Chilliard, Y.; Ferlay, A.; Doreau, M. Effect of different types of forages, animal fat or marine oils in cow's diet on milk fat secretion and composition, especially conjugated linoleic acid (CLA) and polyunsaturated fatty acids. *Livest. Prod. Sci.* **2001**, *70*, 31–48. [CrossRef]
8. Cabiddu, A.; Peratoner, G.; Valenti, B.; Monteils, V.; Martin, B.; Coppa, M. A quantitative review of on-farm feeding practices to enhance the quality of grassland-based ruminant dairy and meat products. *Animal* **2021**, 100375. [CrossRef]
9. Marseglia, A.; Caligiani, A.; Comino, L.; Righi, F.; Quarantelli, A.; Palla, G. Cyclopropyl and ω-cyclohexyl fatty acids as quality markers of cow milk and cheese. *Food Chem.* **2013**, *140*, 711–716. [CrossRef]
10. Bergamaschi, M.; Cipolat-Gotet, C.; Cecchinato, A.; Schiavon, S.; Bittante, G. Chemometric authentication of farming systems of origin of food (milk and ripened cheese) using infrared spectra, fatty acid profiles, flavor fingerprints, and sensory descriptions. *Food Chem.* **2020**, *305*, 125480. [CrossRef] [PubMed]
11. Caligiani, A.; Nocetti, M.; Lolli, V.; Marseglia, A.; Palla, G. Development of a Quantitative GC-MS Method for the Detection of Cyclopropane Fatty Acids in Cheese as New Molecular Markers for Parmigiano Reggiano Authentication. *J. Agric. Food Chem.* **2016**, *64*, 4158–4164. [CrossRef]
12. Borreani, G.; Coppa, M.; Revello-Chion, A.; Comino, L.; Giaccone, D.; Ferlay, A.; Tabacco, E. Effect of different feeding strategies in intensive dairy farming systems on milk fatty acid profiles, and implications on feeding costs in Italy. *J. Dairy Sci.* **2013**, *96*, 6840–6855. [CrossRef]
13. Butler, G.; Collomb, M.; Rehberger, B.; Sanderson, R.; Eyre, M.; Leifert, C. Conjugated linoleic acid isomer concentrations in milk from high- and low-input management dairy systems. *J. Sci. Food Agric.* **2009**, *89*, 697–705. [CrossRef]
14. Collomb, M.; Bütikofer, U.; Sieber, R.; Jeangros, B.; Bosset, J.-O. Correlation between fatty acids in cows' milk fat produced in the Lowlands, Mountains and Highlands of Switzerland and botanical composition of the fodder. *Int. Dairy J.* **2002**, *12*, 661–666. [CrossRef]
15. Disciplinare e Normative del Consorzio-Parmigiano Reggiano. Available online: https://www.parmigianoreggiano.com/it/consorzio-disciplinare-normative/ (accessed on 10 June 2021).
16. Chen, Y.Y.; Gänzle, M.G. Influence of cyclopropane fatty acids on heat, high pressure, acid and oxidative resistance in Escherichia coli. *Int. J. Food Microbiol.* **2016**, *222*, 16–22. [CrossRef]
17. Russell, N.J.; Evans, R.I.; Steeg, P.F.; Hellemons, J.; Verheul, A.; Abee, T. Membranes as a target for stress adaptation. *Int. J. Food Microbiol.* **1995**, *28*, 255–261. [CrossRef]
18. Montanari, C.; Sado Kamdem, S.L.; Serrazanetti, D.I.; Etoa, F.-X.; Guerzoni, M.E. Synthesis of cyclopropane fatty acids in Lactobacillus helveticus and Lactobacillus sanfranciscensis and their cellular fatty acids changes following short term acid and cold stresses. *Food Microbiol.* **2010**, *27*, 493–502. [CrossRef]
19. Brown, J.L.; Ross, T.; McMeekin, T.A.; Nichols, P.D. Acid habituation of Escherichia coli and the potential role of cyclopropane fatty acids in low pH tolerance. *Int. J. Food Microbiol.* **1997**, *37*, 163–173. [CrossRef]
20. Chang, Y.Y.; Cronan, J.E. Membrane cyclopropane fatty acid content is a major factor in acid resistance of Escherichia coli. *Mol. Microbiol.* **1999**, *33*, 249–259. [CrossRef]
21. Woolford, M.K. *The Silage Fermentation*; Microbiological Series, No.14; Marcel Dekker: New York, NY, USA; Basle, Switzerland, 1984.
22. Lima, E.; Abdalla, D. High-performance liquid chromatography of fatty acids in biological samples. *Anal. Chim. Acta* **2002**, *465*, 81–91. [CrossRef]
23. Tzompa-Sosa, D.A.; Meurs, P.P.; van Valenberg, H.J.F. Triacylglycerol Profile of Summer and Winter Bovine Milk Fat and the Feasibility of Triacylglycerol Fragmentation. *Eur. J. Lipid Sci. Technol.* **2018**, *120*, 1700291. [CrossRef]
24. Lopez, C.; Briard-Bion, V.; Menard, O.; Rousseau, F.; Pradel, P.; Besle, J.-M. Phospholipid, sphingolipid, and fatty acid compositions of the milk fat globule membrane are modified by diet. *J. Agric. Food Chem.* **2008**, *56*, 5226–5236. [CrossRef]
25. Craige Trenerry, V.; Akbaridoust, G.; Plozza, T.; Rochfort, S.; Wales, W.J.; Auldist, M.; Ajlouni, S. Ultra-high-performance liquid chromatography–ion trap mass spectrometry characterisation of milk polar lipids from dairy cows fed different diets. *Food Chem.* **2013**, *141*, 1451–1460. [CrossRef] [PubMed]

26. Feng, S.; Lock, A.L.; Garnsworthy, P.C. Technical note: A rapid lipid separation method for determining fatty acid composition of milk. *J. Dairy Sci.* **2004**, *87*, 3785–3788. [CrossRef]

27. Christie, W.W. A simple procedure for rapid transmethylation of glycerolipids and cholesteryl esters. *J. Lipid Res.* **1982**, *23*, 1072–1075. [CrossRef]

28. Breitkopf, S.B.; Ricoult, S.J.H.; Yuan, M.; Xu, Y.; Peake, D.A.; Manning, B.D.; Asara, J.M. A relative quantitative positive/negative ion switching method for untargeted lipidomics via high resolution LC-MS/MS from any biological source. *Metabolomics* **2017**, *13*, 30. [CrossRef]

29. Matyash, V.; Liebisch, G.; Kurzchalia, T.V.; Shevchenko, A.; Schwudke, D. Lipid extraction by methyl-tert-butyl ether for high-throughput lipidomics. *J. Lipid Res.* **2008**, *49*, 1137–1146. [CrossRef] [PubMed]

30. Jensen, R.G. *Handbook of Milk Composition*; Academic Press: San Diego, CA, USA, 1995; ISBN 0123844304.

31. Jensen, R.G. The Composition of Bovine Milk Lipids: January 1995 to December 2000. *J. Dairy Sci.* **2002**, *85*, 295–350. [CrossRef]

32. Bugaud, C.; Buchin, S.; Coulon, J.-B.; Hauwuy, A.; Dupont, D. Influence of the nature of alpine pastures on plasmin activity, fatty acid and volatile compound composition of milk. *Lait* **2001**, *81*, 401–414. [CrossRef]

33. Decaen, C.; Adda, J. Changes in secretion of fatty acids in milk fat of cows during lactation. *Ann. Biol. Anim. Biochim. Biophys.* **1970**, *10*, 659–677. [CrossRef]

34. Bartsch, B.D.; Graham, E.R.; McLean, D.M. Protein and fat composition and some manufacturing properties of milk from dairy cows fed on hay and concentrate in various ratios. *Aust. J. Agric. Res.* **1979**, *30*, 191. [CrossRef]

35. Doreau, M.; Chilliard, Y. Effects of ruminal or postruminal fish oil supplementation on intake and digestion in dairy cows. *Reprod. Nutr. Dev.* **1997**, *37*, 113–124. [CrossRef] [PubMed]

36. Shabala, L.; Ross, T. Cyclopropane fatty acids improve Escherichia coli survival in acidified minimal media by reducing membrane permeability to H+ and enhanced ability to extrude H+. *Res. Microbiol.* **2008**, *159*, 458–461. [CrossRef] [PubMed]

37. Cossignani, L.; Pollini, L.; Blasi, F. Invited review: Authentication of milk by direct and indirect analysis of triacylglycerol molecular species. *J. Dairy Sci.* **2019**, *102*, 5871–5882. [CrossRef] [PubMed]

 *foods*

Article

Authentication of Argan (*Argania spinosa* L.) Oil Using Novel DNA-Based Approaches: Detection of Olive and Soybean Oils as Potential Adulterants

Joana S. Amaral [1,2,3,*], Fatima Z. Raja [1,4], Joana Costa [1], Liliana Grazina [1], Caterina Villa [1], Zoubida Charrouf [4] and Isabel Mafra [1,*]

[1] REQUIMTE-LAQV, Faculdade de Farmácia, Universidade do Porto, Rua de Jorge Viterbo Ferreira, 228, 4050-313 Porto, Portugal

[2] Centro de Investigação de Montanha (CIMO), Instituto Politécnico de Bragança, Campus de Santa Apolónia, 5300-253 Bragança, Portugal

[3] Laboratório Associado para a Sustentabilidade e Tecnologia em Regiões de Montanha (SusTEC), Instituto Politécnico de Bragança, Campus de Santa Apolónia, 5300-253 Bragança, Portugal

[4] Faculty of Science, University Mohammed V, Avenue Ibn Battouta, Rabat 10106, Morocco

* Correspondence: jamaral@ipb.pt (J.S.A.); isabel.mafra@ff.up.pt (I.M.)

Abstract: Argan oil is a traditional product obtained from the fruits of the argan tree (*Argania spinosa* L.), which is endemic only to Morocco. It is commercialized worldwide as cosmetic and food-grade argan oil, attaining very high prices in the international market. Therefore, argan oil is very prone to adulteration with cheaper vegetable oils. The present work aims at developing novel real-time PCR approaches to detect olive and soybean oils as potential adulterants, as well as ascertain the presence of argan oil. The ITS region, *matK* and *lectin* genes were the targeted markers, allowing to detect argan, olive and soybean DNA down to 0.01 pg, 0.1 pg and 3.2 pg, respectively, with real-time PCR. Moreover, to propose practical quantitative methods, two calibrant models were developed using the normalized ΔCq method to estimate potential adulterations of argan oil with olive or soybean oils. The results allowed for the detection and quantification of olive and soybean oils within 50–1% and 25–1%, respectively, both in argan oil. Both approaches provided acceptable performance parameters and accurate determinations, as proven by their applicability to blind mixtures. Herein, new qualitative and quantitative PCR assays are proposed for the first time as reliable and high-throughput tools to authenticate and valorize argan oil.

Keywords: argan oil; authenticity; adulterant detection; real-time PCR; quantification; *Olea europaea*; *Glycine max*

Citation: Amaral, J.S.; Raja, F.Z.; Costa, J.; Grazina, L.; Villa, C.; Charrouf, Z.; Mafra, I. Authentication of Argan (*Argania spinosa* L.) Oil Using Novel DNA-Based Approaches: Detection of Olive and Soybean Oils as Potential Adulterants. *Foods* **2022**, *11*, 2498. https://doi.org/10.3390/foods11162498

Academic Editor: Maria Castro-Puyana

Received: 12 July 2022
Accepted: 16 August 2022
Published: 18 August 2022

Publisher's Note: MDPI stays neutral with regard to jurisdictional claims in published maps and institutional affiliations.

1. Introduction

Argan (*Argania spinosa* L.) is a slow-growing tree endemic only in Morocco. To protect the unique argan forest in southwestern Morocco, in 1998, the UNESCO declared the Arganeraie (the argan tree and its ecological system) as a biosphere reserve. Later, in 2014, the "practices and know-how concerning the argan tree" were inscribed in the UNESCO's Representative List of the Intangible Cultural Heritage of Humanity. The most emblematic use of this tree regards the production of argan oil, a cold-pressed non-refined vegetable oil. Argan oil is traditionally obtained from the fruits of the argan tree using a laborious multistep process that includes fruit picking, fruit peeling, nut cracking, kernel roasting, kernel grinding, dough malaxing and oil collection [1–3]. Despite the current use of modern mechanical presses for oil extraction to allow for a higher oil yield, the process still results in a very low production (approximately 4 L of oil per 100 kg of dried argan fruit), requiring laborious work corresponding to 20 person-hours [4]. Argan oil is produced in different grades, namely, for food and cosmetic purposes [5]. Edible argan oil, registered as a product with the Protected Geographical Indication (PGI) since 2011, is obtained from lightly roasted kernels conferring a hazelnut flavor to the oil, while argan oil used for cosmetics is obtained from raw kernels. In recent decades, numerous studies have shown the nutritional and dermo-cosmetic benefits of this oil [2,6,7], which have been known for centuries and transmitted among generations of Berbere women. Due to its properties and successes as an ingredient in cosmetic products, currently, argan oil is considered one of the most prized oils in the world, with a growing worldwide demand [6]. As a premium product, argan oil is highly prone to adulteration by partial or even total substitution with other vegetable oils. Therefore, different methodologies have been proposed for argan oil authentication, mostly relying on chemical markers analyzed with chromatographic approaches [8]. Hilali et al. [9] proposed the use of campesterol, a sterol present in argan oil in very low amounts (<0.4%), as an adulteration marker allowing for the detection of 2% additions of campesterol-rich vegetable oils and 5% of olive, apricot and hazelnut oils, which are naturally low in campesterol. The use of tocopherols was not so successful, as it allowed for the detection of adulterant oils only above 5% [10]. Ourrach et al. [11] suggested the combined use of 3,5-stigmastadiene, chlorophyllic pigments and hydrocarbon fractions to detect up to 5% additions of refined olive and sunflower oils and virgin olive oil. The same level of detection was achieved based on the triacylglycerol (TAG) profile determined with liquid chromatography coupled to evaporative light-scattering [12] and photodiode-array [13] detection. In addition, spectroscopic approaches [14–17] and, more recently, selected-ion flow-tube mass spectrometry (SIFT-MS) spectra [18] have also advanced as fast screening tools for argan oil authentication. Despite their rapidity and minimal sample preparation requirements, these methods involve the use of chemometrics to predict the level of oil adulteration, which demands very large numbers of samples to construct proper databases towards robust mathematical models.

Considering that several factors, such as edaphoclimatic conditions, development stage, plant part and age, among others, are known to affect the plant's chemical composition [19], DNA molecules have emerged as alternative and unambiguous markers for plant species identification in vegetable oils, which are independent from those factors. Particularly, DNA-based techniques have been successfully used for detecting genetically modified organisms (GMO) in refined oils [20–22] and for the authentication of vegetable oils, such as olive oil [23] and several refined vegetable oils [24,25]. Real-time polymerase chain reaction (PCR) combined with high-resolution melting (HRM) analysis was used by Vietina et al. [26] to detect the addition of maize and sunflower to olive oil down to a limit of 10% and by Ganapoulos et al. [27] to detect 1% of canola oil admixed with olive oil. Moreover, DNA-based methods have demonstrated their feasibility in the identification of plant species in complex matrices [28–34]. Therefore, in this work, novel approaches based on DNA markers are proposed for the first time to detect argan oil and the presence of soybean and olive oils as its potential adulterants.

2. Materials and Methods

2.1. Sampling and Reference Oil Mixtures

Fresh leaves and nuts of *Argania spinosa* L. were directly collected from trees in the region of Agadir, Morocco, while leaves of *Olea europaea* L. were collected in the region of Viseu, Portugal. Additionally, other plant species also used in the production of oil were tested in cross-reactivity studies, including walnut, sunflower, maize, almond, hazelnut, cashew nut, pistachio nut, peanut, Brazil nut, macadamia nut, pine nut, rapeseed, oat and rye (Table S1, Supplementary Materials). The fresh leaves and nuts were oven dried at 30 °C in the dark and were ground in a Grindomix GM200 laboratory mill (Retsch, Haan, Germany).

Authentic argan oil was kindly supplied by the Groupement des Coopératives Targanine (Agadir, Morocco). Commercial argan oil samples of food and cosmetic grades were acquired in Morocco (Table S2, Supplementary Materials). Samples of extra-virgin olive oil and refined soybean oil were acquired from local supermarkets in Porto, Portugal. Binary model mixtures were prepared by adding well-known quantities of olive oil to argan oil in the proportions of 50, 25, 10, 5 and 1% (*w/w*), and adding known amounts of soybean oil to argan oil in the proportions of 40, 25, 10, 5 and 1% (*w/w*). Additionally, two sets of binary mixtures containing 7.5% and 15% (*w/w*) of olive oil or soybean oil in argan oil were prepared for method validation.

2.2. DNA Extraction

Before DNA extraction, the oil mixtures were centrifuged as suggested by Costa et al. [20,21]. For that purpose, 300 g of oil was weighed into six centrifuge tubes, which were centrifuged at 18,514× g for 30 min at 4 °C. The supernatant was discarded until half of each tube was left, and the remaining oil/residue was centrifuged for another 30 min in the same conditions. The supernatant of each tube was carefully removed through pipetting and the residual pellets were collected, combined in one tube and then centrifuged for 30 min (18,514× g, 4 °C). The residual pellet was transferred to one 2 mL sterile reaction tube, centrifuged for 30 min in the same conditions and the supernatant was discarded. Afterwards, the residual pellet was submitted to DNA extraction using protocol B of the Nucleospin® Plant II (Macherey-Nagel, Düren, Germany) kit. Briefly, 300 µL of buffer PL2 pre-heated at 65 °C was added to each tube and incubated at 65 °C for 1 h with continuous mixing (1000 rpm) and occasional vortex mixing. After incubation, the procedure was followed according to the manufacturer's instructions. DNA extracts were stored at −20 °C until analysis.

2.3. DNA Quality and Purity

A UV spectrophotometer, using a Synergy HT multi-mode micro-plate reader (BioTek Instruments, Inc., Winooski, VT, USA) and the Take3 micro-volume plate accessory, was used to assess the yield and purity of DNA extracts. The nucleic acid quantification protocol for dsDNA samples in the Gen5 data analysis software version 2.01 (BioTek Instruments, Inc., Winooski, VT, USA) was used to determine the DNA content. The ratio of the absorbance at 260 and 280 nm (A260/A280) was determined as the purity parameter of the extracted DNA.

Electrophoresis in 1% agarose gel stained with 1× Gel Red (Biotium, Hayward, CA, USA) and ran in 1× SGTB buffer (GRISP, Porto, Portugal) for 20–25 min at 200 V was performed to evaluate the integrity of the DNA extracts. Agarose gels were visualized under a UV light tray Gel Doc™ EZ System (Bio-Rad Laboratories, Hercules, CA, USA) and a digital image was acquired with Image Lab software version 5.2.1 (Bio-Rad Laboratories, Hercules, CA, USA).

2.4. Oligonucleotide Primers

For the specific identification of olive and argan DNA, sequences of the chloroplastidial *matK* gene of *O. europaea* L. and the nuclear region of the internal transcribed spacer 2 (ITS2) of *A. spinosa* L. were retrieved from the NCBI database (http://www.ncbi.nlm.nih.gov/ accessed on 8 November 2021) (accession numbers AJ429335.1 and AM408056.1, respectively). Primers were designed using the Primer-BLAST software tool (http://www.ncbi.nlm.nih.gov/tools/primer-blast/, accessed on 8 November 2021) (Table 1). Primer specificity was assessed in silico using the same tool and the basic local alignment search tool BLAST (http://blast.ncbi.nlm.nih.gov/Blast.cgi, accessed on 8 November 2021). Primer properties and the absence of self-hybridization and hairpins were verified using the software OligoCalc (http://www.basic.northwestern.edu./biotools/oligocalc.html, accessed on 8 November 2021).

For soybean detection, primers targeting the lectin gene [35] were previously designed, as well as universal eukaryotic primers targeting the conserved nuclear 18S rRNA gene, used to assess the amplification capacity of the DNA extracts [36] (Table 1).

The oligonucleotide primers used in this work were synthesized by Eurofins Genomics (Ebersberg, Germany).

Table 1. Oligonucleotide primers used in this work.

Species	Target Gene	Primers	Sequence (5′-3′)	Amplicon (bp)	Reference
Argan	ITS2	ITS2A-F	CTCGTCCCGTCCCGCAAAG	117	This work (AM408056.1)
		ITS2A-R	CCACCACTCGTCGTGACGTT		
Olive	*matK*	matKO-F	GCTGTGGTTTCATCCAAGAAGGA	109	This work (AJ429335.1)
		matKO-R	GCTCCGTACCACTGAAGCGT		
Soybean	*Lectin*	LE1	CAAAGCAATGGCTACTTCAAAG	103	[35]
		LE2	TGAGTTTGCCTTGCTGGTCAGT		
Eukaryotic	18S rRNA	EG-F	TCGATGGTAGGATAGTGGCCTACT	109	[36]
		EG-R	TGCTGCCTTCCTTGGATGTGGTA		

2.5. Qualitative PCR

PCR amplifications were carried out in a total reaction volume of 25 µL, containing 2 µL of DNA extract (10 ng), 67 mM Tris-HCl (pH 8.8), 16 mM of $(NH_4)_2SO_4$, 0.1% of Tween 20, 200 µM of each dNTP, 1.0 U of SuperHot Taq DNA Polymerase (Genaxxon Bioscience GmbH, Ulm, Germany), 2.0 mM of $MgCl_2$, 200 nM (ITS2A-F/ITS2A-R, matKO-F/matKO-R) or 280 nM (LE1/LE2, EG-F/EG-R) of each primer (Table 1). The reactions were performed in a MJ Mini™ Gradient Thermal Cycler (Bio-Rad Laboratories, Hercules, CA, USA) using the following programs: (i) initial denaturation at 95 °C for 5 min; (ii) 35 cycles at 95 °C for 30 s, 62 °C (ITS2A-F/ITS2A-R, matKO-F/matKO-R) or 60 °C (LE1/LE2) or 63 °C (EG-F/EG-R) for 30 s and 72 °C for 30 s; (iii) final extension at 72 °C for 5 min. Each extract was amplified at least in duplicate assays.

Electrophoresis was carried out in a 1.5% agarose gel containing Gel Red 1× (Biotium, Hayward, CA, USA) for staining and SGTB 1× (GRiSP, Research Solutions, Porto, Portugal) was used to confirm amplicons. Agarose gels were visualized under a UV light tray Gel Doc™ EZ System (Bio-Rad Laboratories, Hercules, CA, USA) and a digital image was obtained with Image Lab software version 5.2.1 (Bio-Rad Laboratories, Hercules, CA, USA).

2.6. Real-Time PCR

Real-time PCR amplifications were carried out in 20 µL of total reaction volume, containing 2 µL of DNA (10 ng to 0.01 pg), 1× SsoFast EvaGreen Supermix (Bio-Rad Laboratories, Hercules, CA, USA), 300 nM (ITS2A-F/ITS2A-R, matKO-F/matKO-R) or 350 nM of (LE1/LE2) each primer (Table 1). A fluorometric thermal cycler CFX96 Real-time PCR Detection System (Bio-Rad Laboratories, Hercules, CA, USA) was used with the following conditions: 95 °C for 5 min, 45 cycles at 95 °C for 10 s and 65 °C for 40 s, with the collection of the fluorescence signal at the end of each cycle. The data evaluation from each real-time PCR assay was performed using the software Bio-Rad CFX Manager 3.1 (Bio-Rad Laboratories, Hercules, CA, USA). Real-time PCR assays were performed, at least, in two independent runs using $n = 4$ replicates each time.

Calibration curves were constructed using 10-fold serially diluted DNA extracts (10 ng–0.01 pg), which allowed for determining the absolute limits of detection (LOD) and quantification (LOQ). The acceptance criteria for real-time PCR assays were established according to the MIQE guidelines (minimum information for publication of quantitative real-time PCR experiments) [37] and the definition of minimum performance requirements for analytical methods of GMO testing [38]. Accordingly, the following parameters, namely, the slope between −3.6 and −3.1, the PCR efficiency within 90–110% and the correlation coefficient (R^2) $\geq$ 0.98 were established [37,38]. The sensitivity was expressed as the LOD, which was the lowest amount or concentration that could be reliably detected (the lowest amplified level for 95% of the replicates). The LOQ was the lowest amount or concentration of analyte in a sample that could be reliably quantified with an acceptable level of trueness and precision, which was determined as the lowest amplified level within the linear dynamic range of the calibration curve. The dynamic range should cover a minimum of 4 orders of magnitude and, ideally, extend to 5 or 6 log10 concentrations [37,38].

3. Results

3.1. DNA Quality and Selection of Target Region

For method development and optimization, DNA was successfully extracted from argan and olive leaves, as well as soybean flour, achieving suitable yields within 7.9–16.9 ng/µL, 5.8–8.4 ng/µL and 49.0–99.7 ng/µL, with purities (A260/A280) of 1.4–2.2, 1.7—2.0 and 1.8–2.0, respectively. The DNA yields for the oil mixtures and commercial argan oil samples were in the range of 5.2–9.3 ng/µL, with purities of 1.5–2.1. Despite the low DNA yields, all the extracts showed a suitable amplification capacity as inferred from the strong PCR fragments targeting a universal and conserved gene (18S rRNA) (Table S1 and Figure S1, Supplementary Materials).

To specifically detect argan and olive DNA, new primer sets were designed targeting the ITS and *matK* regions, respectively, because both are recognized barcode markers with high species discriminatory powers. The ITS is a robust phylogenetic marker at the species level, while *matK* has a high evolutionary rate and suitable length [39]. Moreover, both markers may provide highly sensitive methods because *matK* is a chloroplastidial gene that is present in multiple copies and the ITS is a nuclear region present in multiple ribosomes of nuclear DNA. The results of the PCR optimization for the detection of *A. spinosa* and *O. europaea* confirmed the high sensitivity of the assays, achieving 1 pg and 0.1 pg, respectively (Figure 1A,B). The specificity of the assays was firstly assessed with an in silico analysis and further confirmed experimentally using several non-target species that are commonly used in food and cosmetic oils (Table S1 and Figure S1, Supplementary Materials). The detection of soybean was carried out by targeting the lectin gene using previously designed primers (LE1/LE2) [35] to amplify a short PCR amplicon (103 bp). The choice was justified by their successful application to detect soybean DNA in refined oils [20,21]. As shown in Figure 1C, the soybean-specific PCR detection was down to 0.8 ng of DNA. The three specific PCR assays were then applied to the commercial samples of argan oil, confirming the presence of argan in all samples (Figure S2A, Supplementary Materials) and the absence of olive and soybean DNA as potential adulterants (Figure S2B,C, Supplementary Materials).

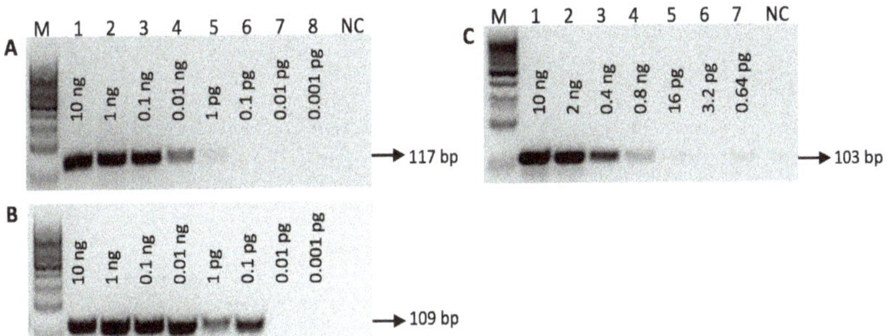

Figure 1. Agarose gel electrophoresis of PCR products targeting ITS, *matK* and *lectin* genes of *A. spinosa* (**A**), *O. europaea* (**B**) and *G. max* (**C**), respectively, using serially diluted DNA of each species. Legend: M, 100 bp molecular marker (Bioron, Ludwigshafen, Germany); lanes 1–8, serially diluted DNA; NC, negative control.

3.2. Real-Time PCR Assays

3.2.1. Absolute LOD and LOQ

After selecting the species-specific markers, real-time PCR assays using EvaGreen dye were successfully developed for each target species (Figure 2). Figure 2A,C,E show the amplification curves and respective derivative melting curves that provided the single melt peaks for each species at 89.43 ± 0.14 °C, 77.43 ± 0.10 °C and 78.79 ± 0.04 °C, supporting the specificity of the target amplification and the absence of non-specific amplicons. The calibration curves obtained for *A. spinosa*, *O. europaea* and *G. max* (Figure 2B,D,F) showed that all performance parameters, namely, the PCR efficiency (102.0 to 104.2%), slope (−3.276

to -3.226) and R^2 (0.995 to 0.998), complied with the acceptance criteria established for real-time PCR assays [37,38]. In addition, the dynamic ranges covered seven, six and four orders of magnitude, achieving limits of detection (LOD) of 0.01 pg, 0.1 pg and 3.2 pg of DNA for argan, olive and soybean, respectively (Figure 2B,D,F). Since the LOD values were within the dynamic range, the limits of quantification (LOQ) could be considered as the same values.

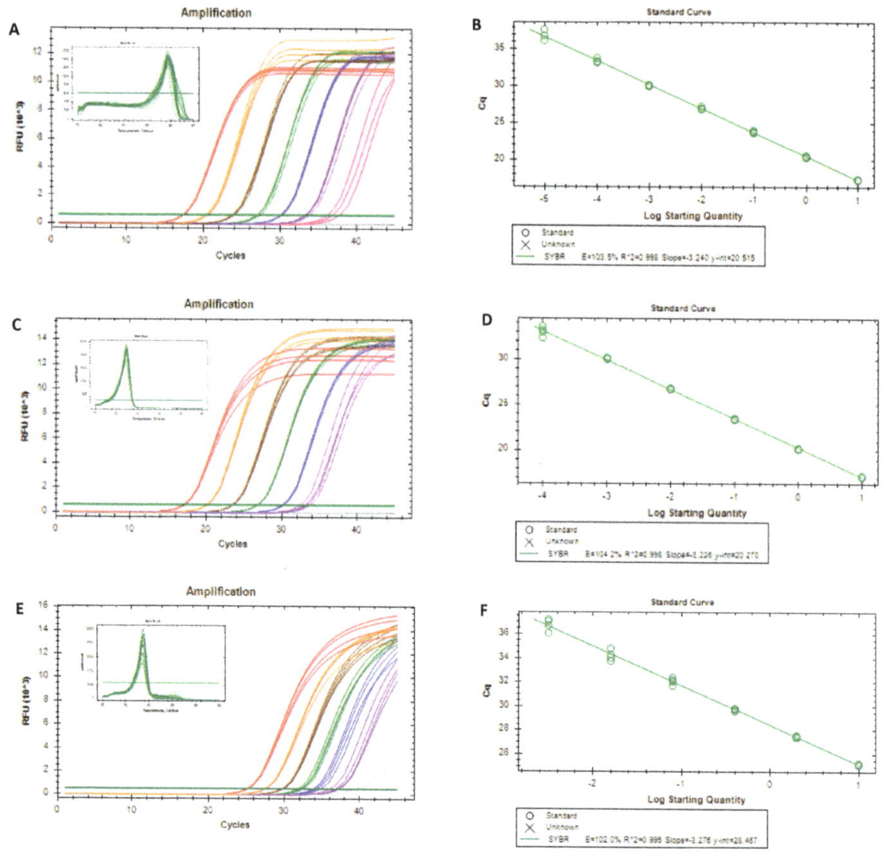

Figure 2. Real-time PCR amplification curves (with respective melting curves) (**A,C,E**) and calibration curves targeting ITS, *matK* and *lectin* genes of *A. spinosa* (**A,B**), *O. europaea* (**C,D**) and *G. max* (**E,F**), respectively, using 10-fold serially diluted DNA (10 ng to 0.01 pg) for *A. spinosa* and *O. europaea* and 4-fold serially diluted DNA (10 ng to 0.64 pg) for *G. max* (*n* = 4 replicates).

3.2.2. Construction of the Normalized Calibration Curves

To estimate the potential adulterations of argan oil with olive or soybean oils, two quantitative models were developed using the ΔCq method. This approach has been frequently applied to several complex and/or highly processed food matrices [28,29,36]. It is based on the construction of a normalized calibration curve, plotting the difference between the quantification cycle values of the target sequence and a universal reference marker (ΔCq) versus the log of the concentration of the target species. Therefore, this approach can reduce the influence of potential PCR inhibitors, DNA degradation and low DNA yields, which are critical issues when amplifying DNA from oil matrices [20,21]. For the construction of the calibration curves, two sets of binary reference mixtures of olive oil in argan oil (50, 25, 10, 5 and 1% (*w/w*)) and soybean oil in argan oil (40, 25, 10, 5 and 1% (*w/w*)) were prepared as calibrants, and the respective ΔCq values were plotted against the concentration of each target adulterant oil (Figure 3). The obtained calibration curves showed values of PCR efficiency (82.9 and 81.8%) and slopes (−3.8124 and −3.8521) slightly out of the acceptance criteria, but with acceptable correlations (R^2 > 0.98), within the ranges of 50–1% and 25–1% for olive/argan and soybean/argan oil mixtures, respectively. However, it is important to refer to that, in the case of samples where it was difficult to extract high-quality DNA, such as oils, a slope within −4.1 and −3.1 and a PCR efficiency of 75–110% were acceptable [38]. Both calibration curves provided LOD and LOQ of 1% of adulterant oil in argan oil, being able to estimate olive oil until 50% (Figure 3A) and soybean oil until 25% (Figure 3B), because above this value, the PCR efficiency and correlation values were not acceptable. However, a dynamic range of 25–1% could be considered feasible for the purpose of estimating eventual adulterations.

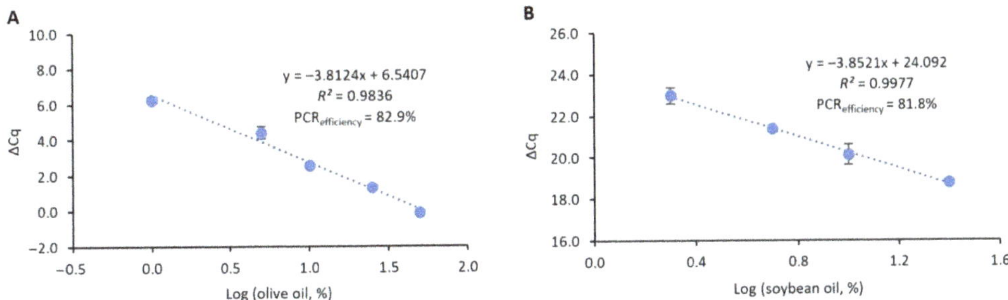

Figure 3. Normalized calibration curves obtained with real-time PCR targeting the ITS region of olive (**A**) and the lectin gene of soybean (**B**), using reference mixtures of olive oil (50, 25, 10, 5 and 1%, *w/w*) in argan oil and soybean oil in argan oil (40, 25, 10, 5 and 1%, *w/w*), respectively (*n* = 8 replicates).

3.2.3. Validation of Quantitative Real-Time PCR Systems

To assess the performance of the two quantitative normalized PCR systems, two sets of blinded samples containing 7.5% and 15% of adulterant (olive or soybean) in argan oil were used. The estimation of the respective oil contents allowed for assessing the performance of the assays regarding trueness and precision (Table 2). The coefficients of variation expressed the relative standard deviations of results and were obtained under repeatability conditions, showing acceptable values that were within 0.41% and 13.1% (≤25%) and attesting to the precision of both systems. The measured trueness was expressed as the bias, whose values ranged between −24% and 22.9%, being within ±25% of the actual values, which confirmed the closeness of agreement between the tested and the actual values of both systems [38].

Table 2. Validation results based on the application of the normalized quantitative real-time PCR approaches to blind mixtures containing olive or soybean oils in argan oil.

Samples	Adulterant Oil (%, *w/w*)		SD [b]	CV (%) [c]	Bias [d]
	Actual	Mean Predicted [a]			
	Olive oil in argan oil				
A	7.5	9.7	0.0	0.41	22.9
B	15.0	14.0	0.2	1.1	−7.5
	Soybean oil in argan oil				
C	7.5	5.9	0.8	13.1	−21.2
D	15.0	11.4	0.6	5.6	−24.0

[a] Values are the means of replicate assays (*n* = 6). [b] SD, standard deviation. [c] CV, coefficient of variation. [d] Bias = ((mean value-true value)/true value × 100).

4. Discussion

In recent years, there has been rising interest towards the authentication of foods, including vegetable oils [40] and botanicals [31], using molecular makers. DNA molecules are ubiquitously present in all cells, resistant to harsh conditions, such as food processing, and independent from plant age and tissue, climatic, geographical or agronomical factors. Therefore, DNA markers have been considered unequivocal identifiers for the traceability and authentication of food, with several advantages over proteins that are less resistant to processing and chemical markers, which might vary with edaphoclimatic conditions. In relation to their application to vegetable oils, several advances have been actualized, mainly regarding olive oil authentication [23,26,27,40,41] and the detection of GMO in soybean oil [20,21,40]. However, none of the reports addressed the authentication of argan oil, and most of them lacked any quantitative analyses, being mainly confined to the quantification of GMO in soybean oil [20,21].

In the present work, the suitability of using DNA markers was exploited for the authentication of argan oil for the first time. For that purpose, unequivocal markers were identified for the detection of argan oil and two potential adulterants, namely, olive and soybean oils. The target markers were the ITS, *matk* and *lectin* genes, providing the detection of argan, olive and soybean DNA down to 0.01 pg, 0.1 pg and 3.2 pg, respectively, with real-time PCR. The three species-specific assays provided calibration curves that complied with the acceptance criteria concerning PCR efficiency, slope and R^2 values (Figure 2). Additionally, to propose two practical quantitative methods to estimate the potential adulterations of argan oil with olive or soybean oils, two calibrant models were developed using the ΔCq method (Figure 3). The feasibility of this approach was already demonstrated in processed food matrices, such as species authentication in meat products [42], herbal products [29] and spices [36], and the detection and quantification of potentially allergenic ingredients, namely, soybean [28], lupine [43] and milk [44]. To our knowledge, the application of a normalized ΔCq approach to authenticate vegetable oil was herein described for the first time. The two calibrant models allowed for detecting and quantifying olive oil in the range of 50–1% (Figure 3A) in argan oil and soybean oil within 25–1% in argan oil (Figure 3B). Both approaches provided acceptable performance parameters, with proven applicability to blind mixtures and precise and accurate quantitative analyses. In summary, two novel real-time PCR approaches were proposed as specific, sensitive and high-throughput tools to authenticate and valorize argan oil.

Supplementary Materials: The following supporting information can be downloaded at: https://www.mdpi.com/article/10.3390/foods11162498/s1, Table S1: results of PCR amplification targeting the ITS2, *matK* and *lectin* genes of argan, olive and soybean, respectively, and a universal eukaryotic DNA region of several relevant plant species for cross-reactivity testing. Table S2: Commercial argan oil samples tested PCR amplification targeting the ITS2, *matK* and *lectin* genes of argan, olive and soybean, respectively. Figure S1: Agarose gel electrophoresis of PCR products targeting ITS, *matK* and *lectin* genes of *A. spinosa* (A), *O. europaea* (B) and *G. max* (C), respectively, for cross-reactivity assessment. Figure S2: Agarose gel electrophoresis of PCR products targeting ITS, *matK* and *lectin* genes of *A. spinosa* (A), *O. europaea* (B) and *G. max* (C), respectively, in commercial samples of argan food or cosmetic oils.

Author Contributions: Conceptualization, J.S.A. and I.M.; methodology, F.Z.R., L.G. and C.V.; validation, F.Z.R., L.G. and C.V.; formal analysis, I.M., J.S.A. and J.C.; investigation, I.M., J.S.A. and J.C.; resources, I.M. and J.C.; writing—original draft preparation, J.S.A. and I.M.; writing—review and editing, J.S.A., J.C., Z.C. and I.M.; supervision, J.S.A., Z.C. and I.M.; project administration, J.S.A., Z.C. and I.M.; funding acquisition, J.S.A., Z.C. and I.M. All authors have read and agreed to the published version of the manuscript.

Funding: This work was supported by the FCT (Fundação para a Ciência e Tecnologia) through projects FCT/CNRST (Portugal/Morocco) (FCT/6460/6/6/2017/S) and the strategic funding of UIDB/50006/2020 I UIDP/50006/2020. This work was also funded by the European Union (EU) through the European Regional Development Fund (FEDER funds through NORTE-01-0145-FEDER-000052) and the project SYSTEMIC (Knowledge Hub on Food and Nutrition Security, ERA-Net Cofund ERA-HDHL no. 696300). J. Costa and I. Mafra thank the FCT for funding through the Individual Call to Scientific Employment Stimulus (2021.03583.CEECIND/CP1662/CT0012 and 2021.03670.CEECIND/CP1662/CT0011, respectively). L. Grazina is grateful to the FCT for the grant (SFRH/BD/132462/2017) financed by POPH-QREN (subsidized by FSE and MCTES).

Institutional Review Board Statement: Not applicable.

Informed Consent Statement: Not applicable.

Data Availability Statement: Data is contained within the article or supplementary material.

Acknowledgments: The authors are grateful to the Groupement des Coopératives Targanine for supplying the argan oil sample. J.S. Amaral is grateful to the FCT for financial support through national funds FCT/MCTES (PIDDAC) to CIMO (UIDB/00690/2020 e UIDP/00690/2020) and SusTEC (LA/P/0007/2020).

Conflicts of Interest: The authors declare no conflict of interest.

References

1. Krajnc, B.; Bontempo, L.; Araus, J.L.; Giovanetti, M.; Alegria, C.; Lauteri, M.; Augusti, A.; Atti, N.; Smeti, S.; Taous, F.; et al. Selective methods to investigate authenticity and geographical origin of Mediterranean food products. *Food Rev. Int.* **2021**, *37*, 656–682. [CrossRef]
2. Charrouf, Z.; Guillaume, D. Argan oil: Occurrence, composition and impact on human health. *Eur. J. Lipid Sci. Technol.* **2008**, *110*, 632–636. [CrossRef]
3. Charrouf, Z.; Guillaume, D. Argan oil, the 35-years-of-research product. *Eur. J. Lipid Sci. Technol.* **2014**, *116*, 1316–1321. [CrossRef]
4. Charrouf, Z.; Guillaume, D.; Driouich, A. The argan tree, an asset for Morocco. *Biofutur* **2002**, *220*, 54–57.
5. Gharby, S.; Charrouf, Z. Argan Oil: Chemical composition, extraction process, and quality control. *Front. Nutr.* **2022**, *8*, 804587. [CrossRef] [PubMed]
6. Guillaume, D.; Charrouf, Z. Argan oil and other argan products: Use in dermocosmetology. *Eur. J. Lipid Sci. Technol.* **2011**, *113*, 403–408. [CrossRef]
7. El Abbassi, A.; Khalid, N.; Zbakh, H.; Ahmad, A. Physicochemical characteristics, nutritional properties, and health benefits of argan oil: A review. *Crit. Rev. Food Sci. Nutr.* **2014**, *54*, 1401–1414. [CrossRef]
8. Mohammed, F.; Guillaume, D.; Warland, J.; Abdulwali, N. Analytical methods to detect adulteration of argan oil: A critical review. *Microchem. J.* **2021**, *168*, 106501. [CrossRef]
9. Hilali, M.; Charrouf, Z.; Soulhi, A.E.A.; Hachimi, L.; Guillaume, D. Detection of argan oil adulteration using quantitative campesterol GC-analysis. *J. Am. Oil Chem. Soc.* **2007**, *84*, 761–764. [CrossRef]
10. Celik, S.E.; Asfoor, A.; Senol, O.; Apak, R. Screening method for argan oil adulteration with vegetable oils: An online HPLC assay with postcolumn detection utilizing chemometric multidata analysis. *J. Agric. Food Chem.* **2019**, *67*, 8279–8289. [CrossRef] [PubMed]
11. Ourrach, I.; Rada, M.; Perez-Camino, M.C.; Benaissa, M.; Guinda, A. Detection of argan oil adulterated with vegetable oils: New markers. *Grasas Aceites* **2012**, *63*, 355–364. [CrossRef]
12. Salghi, R.; Armbruster, W.; Schwack, W. Detection of argan oil adulteration with vegetable oils by high-performance liquid chromatography-evaporative light scattering detection. *Food Chem.* **2014**, *153*, 387–392. [CrossRef] [PubMed]
13. Pagliuca, G.; Bozzi, C.; Gallo, F.R.; Multari, G.; Palazzino, G.; Porra, R.; Panusa, A. Triacylglycerol "hand-shape profile" of Argan oil. Rapid and simple UHPLC-PDA-ESI-TOF/MS and HPTLC methods to detect counterfeit argan oil and argan-oil-based products. *J. Pharm. Biomed. Anal.* **2018**, *150*, 121–131. [CrossRef] [PubMed]
14. El Orche, A.; Elhamdaoui, O.; Cheikh, A.; Zoukeni, B.; El Karbane, M.; Mbarki, M.; Bouatia, M. Comparative study of three fingerprint analytical approaches based on spectroscopic sensors and chemometrics for the detection and quantification of argan oil adulteration. *J. Sci. Food Agric.* **2022**, *102*, 95–104. [CrossRef] [PubMed]
15. Oussama, A.; Elabadi, F.; Devos, O. Analysis of Argan oil adulteration using infrared spectroscopy. *Spectrosc. Lett.* **2012**, *45*, 458–463. [CrossRef]
16. Gunning, Y.; Jackson, A.J.; Colmer, J.; Taous, F.; Philo, M.; Brignall, R.M.; El Ghali, T.; Defernez, M.; Kemsley, E.K. High-throughput screening of argan oil composition and authenticity using benchtop ^{1}H NMR. *Magn. Reson. Chem.* **2020**, *58*, 1177–1186. [CrossRef] [PubMed]
17. Farres, S.; Srata, L.; Fethi, F.; Kadaoui, A. Argan oil authentication using visible/near infrared spectroscopy combined to chemometrics tools. *Vib. Spectrosc.* **2019**, *102*, 79–84. [CrossRef]

18. Kharbach, M.; Yu, H.W.; Kamal, R.; Marmouzi, I.; Alaoui, K.; Vercammen, J.; Bouklouze, A.; Vander Heyden, Y. Authentication of extra virgin argan oil by selected-ion flow-tube mass-spectrometry fingerprinting and chemometrics. *Food Chem.* **2022**, *383*, 132565. [CrossRef]

19. El Adib, S.; Aissi, O.; Charrouf, Z.; Ben Jeddi, F.; Messaoud, C. *Argania spinosa* var. *mutica* and var. *apiculata*: Variation of fatty-acid composition, phenolic content, and antioxidant and alpha-amylase-inhibitory activities among varieties, organs, and development stages. *Chem. Biodivers.* **2015**, *12*, 1322–1338. [CrossRef] [PubMed]

20. Costa, J.; Mafra, I.; Amaral, J.S.; Oliveira, M.B.P.P. Detection of genetically modified soybean DNA in refined vegetable oils. *Eur. Food Res. Technol.* **2010**, *230*, 915–923. [CrossRef]

21. Costa, J.; Mafra, I.; Amaral, J.S.; Oliveira, M.B.P.P. Monitoring genetically modified soybean along the industrial soybean oil extraction and refining processes by polymerase chain reaction techniques. *Food Res. Int.* **2010**, *43*, 301–306. [CrossRef]

22. Duan, Y.Z.; Pi, Y.; Li, C.W.; Jiang, K.J. An optimized procedure for detection of genetically modified DNA in refined vegetable oils. *Food Sci. Biotechnol.* **2021**, *30*, 129–135. [CrossRef] [PubMed]

23. Batrinou, A.; Strati, I.F.; Houhoula, D.; Tsaknis, J.; Sinanoglou, V.J. Authentication of olive oil based on DNA analysis. *Grasas Aceites* **2020**, *71*, e366. [CrossRef]

24. Li, Y.Y.; Wu, Y.J.; Han, J.X.; Wang, B.; Ge, Y.Q.; Chen, Y. Species-Specific Identification of Seven Vegetable Oils Based on Suspension Bead Array. *J. Agric. Food Chem.* **2012**, *60*, 2362–2367. [CrossRef] [PubMed]

25. Su, T.F.; Wei, P.F.; Wu, L.L.; Guo, Y.W.; Zhao, W.Q.; Zhang, Y.; Chi, Z.Y.; Qiu, L.Y. Development of nucleic acid isolation by non-silica-based nanoparticles and real-time PCR kit for edible vegetable oil traceability. *Food Chem.* **2019**, *300*, 125205. [CrossRef]

26. Vietina, M.; Agrimonti, C.; Marmiroli, N. Detection of plant oil DNA using high resolution melting (HRM) post PCR analysis: A tool for disclosure of olive oil adulteration. *Food Chem.* **2013**, *141*, 3820–3826. [CrossRef] [PubMed]

27. Ganopoulos, I.; Bazakos, C.; Madesis, P.; Kalaitzis, P.; Tsaftaris, A. Barcode DNA high-resolution melting (Bar-HRM) analysis as a novel close-tubed and accurate tool for olive oil forensic use. *J. Sci. Food Agric.* **2013**, *93*, 2281–2286. [CrossRef]

28. Costa, J.; Amaral, J.S.; Grazina, L.; Oliveira, M.B.P.P.; Mafra, I. Matrix-normalised real-time PCR approach to quantify soybean as a potential food allergen as affected by thermal processing. *Food Chem.* **2017**, *221*, 1843–1850. [CrossRef]

29. Grazina, L.; Amaral, J.S.; Costa, J.; Mafra, I. Authentication of *Ginkgo biloba* herbal products by a novel quantitative real-time PCR approach. *Foods* **2020**, *9*, 1233. [CrossRef]

30. Grazina, L.; Amaral, J.S.; Costa, J.; Mafra, I. Towards authentication of Korean ginseng-containing foods: Differentiation of five Panax species by a novel diagnostic tool. *LWT-Food Sci. Technol.* **2021**, *151*, 112211. [CrossRef]

31. Grazina, L.; Amaral, J.S.; Mafra, I. Botanical origin authentication of dietary supplements by DNA-based approaches. *Compr. Rev. Food Sci. Food Saf.* **2020**, *19*, 1080–1109. [CrossRef] [PubMed]

32. Grazina, L.; Batista, A.; Amaral, J.S.; Costa, J.; Mafra, I. Botanical authentication of globe artichoke-containing foods: Differentiation of Cynara scolymus by a novel HRM approach. *Food Chem.* **2022**, *366*, 130621. [CrossRef] [PubMed]

33. Soares, S.; Grazina, L.; Costa, J.; Amaral, J.S.; Oliveira, M.B.P.P.; Mafra, I. Botanical authentication of lavender (*Lavandula* spp.) honey by a novel DNA-barcoding approach coupled to high resolution melting analysis. *Food Control* **2018**, *86*, 367–373. [CrossRef]

34. Costa, J.; Campos, B.; Amaral, J.S.; Nunes, M.E.; Oliveira, M.B.P.P.; Mafra, I. HRM analysis targeting ITS1 and matK loci as potential DNA mini-barcodes for the authentication of *Hypericum perforatum* and *Hypericum androsaemum* in herbal infusions. *Food Control* **2016**, *61*, 105–114. [CrossRef]

35. Mafra, I.; Silva, S.A.; Moreira, E.J.M.O.; Ferreira da Silva, C.S.; Oliveira, M.B.P.P. Comparative study of DNA extraction methods for soybean derived food products. *Food Control* **2008**, *19*, 1183–1190. [CrossRef]

36. Villa, C.; Costa, J.; Oliveira, M.B.P.P.; Mafra, I. Novel quantitative real-time PCR approach to determine safflower (*Carthamus tinctorius*) adulteration in saffron (*Crocus sativus*). *Food Chem.* **2017**, *229*, 680–687. [CrossRef]

37. Bustin, S.A.; Benes, V.; Garson, J.A.; Hellemans, J.; Huggett, J.; Kubista, M.; Mueller, R.; Nolan, T.; Pfaffl, M.W.; Shipley, G.L.; et al. The MIQE guidelines: Minimum information for publication of quantitative real-time PCR experiments. *Clin. Chem.* **2009**, *55*, 611–622. [CrossRef]

38. ENGL. Definition of Minimum Performance Requirements for Analytical Methods of GMO Testing. European Network of GMO Laboratories, Joint Research Center, EURL. 2015. Available online: http://gmo-crl.jrc.ec.europa.eu/doc/MPR%20Report%20 Application%2020_10_2015.pdf (accessed on 10 July 2022).

39. Li, X.; Yang, Y.; Henry, R.J.; Maurizio, R.; Yitao, W.; Shilin, C. Plant DNA barcoding: From gene to genome. *Biol. Rev.* **2015**, *90*, 157–166. [CrossRef]

40. Costa, J.; Mafra, I.; Oliveira, M.B.P.P. Advances in vegetable oil authentication by DNA-based markers. *Trends Food Sci. Technol.* **2012**, *26*, 43–55. [CrossRef]

41. Muzzalupo, I.; Pisani, F.; Greco, F.; Chiappetta, A. direct DNA amplification from virgin olive oil for traceability and authenticity. *Eur. Food Res. Technol.* **2015**, *241*, 151–155. [CrossRef]

42. Amaral, J.S.; Santos, G.; Oliveira, M.B.P.P.; Mafra, I. Quantitative detection of pork meat by EvaGreen real-time PCR to assess the authenticity of processed meat products. *Food Control* **2017**, *72*, 53–61. [CrossRef]

43. Villa, C.; Costa, J.; Gondar, C.; Oliveira, M.B.P.P.; Mafra, I. Effect of food matrix and thermal processing on the performance of a normalised quantitative real-time PCR approach for lupine (*Lupinus albus*) detection as a potential allergenic food. *Food Chem.* **2018**, *262*, 251–259. [CrossRef] [PubMed]

44. Villa, C.; Costa, J.; Mafra, I. Detection and quantification of milk ingredients as hidden allergens in meat products by a novel specific real-time PCR method. *Biomolecules* **2019**, *9*, 804. [CrossRef] [PubMed]

MDPI

St. Alban-Anlage 66

4052 Basel

Switzerland

Tel. +41 61 683 77 34

Fax +41 61 302 89 18

www.mdpi.com

Foods Editorial Office

E-mail: foods@mdpi.com

www.mdpi.com/journal/foods